BIOLOGICAL ASPECTS OF THERMAL POLLUTION

Biological
Aspects of

THERMAL
POLLUTION

PROCEEDINGS OF
THE NATIONAL SYMPOSIUM
ON THERMAL POLLUTION,
SPONSORED BY THE FEDERAL WATER POLLUTION
CONTROL ADMINISTRATION AND VANDERBILT UNIVERSITY
PORTLAND, OREGON, JUNE 3–5, 1968

Edited by

Peter A. Krenkel

and

Frank L. Parker

VANDERBILT UNIVERSITY PRESS • 1969

COPYRIGHT © 1969 BY VANDERBILT UNIVERSITY PRESS

Printed in the United States of America
Standard Book Number 8265–1144–9
Library of Congress Catalogue Card Number 75–92265

This symposium and the National Symposium on Engineering Aspects of Thermal Pollution which followed it on August 14–16, 1968, in Nashville, Tennessee, were supported in part by funds from the Federal Water Pollution Control Administration, U.S. Department of the Interior.

PREFACE

THE papers contained herein are the result of the first of two symposia on thermal pollution which were co-sponsored by Vanderbilt University and the Federal Water Pollution Control Administration, United States Department of the Interior. This volume, entitled *Biological Aspects of Thermal Pollution,* includes the formal papers and discussions presented at Portland, Oregon, on June 3–5, 1968. Another volume, entitled *Engineering Aspects of Thermal Pollution,* contains the proceedings of the second symposium, which was held at Vanderbilt University in Nashville, Tennessee, on August 14–16, 1968. These symposia were made possible by a grant from the Federal Water Pollution Control Administration.

It is gratifying to note the distinguished participants who gave their time to assist in establishing the magnitude of the thermal pollution problem.

The symposia were planned to satisfy the need for exchange and promulgation of information concerning thermal-pollution problems confronting industry, consulting engineers, and regulatory agencies. The primary objectives of the conferences were to bring together all those concerned with thermal-pollution problems, to encourage an exchange of knowledge and experience, and to stimulate research.

The interdisciplinary approach to this relatively recently recognized pollution problem is essential in order to bring to bear the many talents required to solve this multi-faceted problem. The symposia were designed to encourage a maximum of discussion between participants and to elucidate areas of knowledge and types of research required.

Co-Chairmen of the Portland symposium were James R. Boydston of the Pacific Northwest Water Laboratory, Federal Water Pollution Control Administration, and Frank L. Parker, professor of Environmental and Water Resources Engineering, of Vanderbilt University. The invaluable assistance and cooperation of FWPCA personnel is also worthy of note. Other people contributing to the success of this symposium included William A. Cawley,

J. Frances Allen, James Agee, Joel Fisher, Bruce Tichenor, Frank Rainwater, Linda Bray, Priscilla Smith, Elizabeth Fletcher, and Maryann Moore.

There is no doubt that the discharge of heated water to our environment will continue to increase at an alarming rate. The best estimates show that power generation in the United States in the year 2020 will be almost twenty times that generated in 1965. The major part of this new generating capacity will be of nuclear origin, thus adding greater and more concentrated heat loads to our receiving waters.

Thus, it is obvious that the effects of this waste heat on our environment must be clearly elucidated, and corrective measures must be introduced. It is hoped that the papers and discussions contained herein have established a baseline from which future meaningful research will result and lead to adequate and reasonable control of thermal pollution.

Vanderbilt University   PETER A. KRENKEL
May 1969   FRANK L. PARKER

THE AUTHORS

JOHN S. ALABASTER
>Principal Scientific Officer, Ministry of Agriculture, Fisheries, and Food, London, England

J. FRANCES ALLEN
>Chief, Water Quality Requirements Branch, Division of Research, Federal Water Pollution Control Administration, Washington, D.C.

DAVID S. BLACK
>U.S. Under Secretary of the Interior, Washington, D.C.

JOHN CAIRNS, JR.
>Research Professor, Department of Biology, Virginia Polytechnic Institute, Blacksburg, Virginia

PETER DOUDOROFF
>Professor of Fisheries and Wildlife, Oregon State University, Corvallis, Oregon

JEFFERSON J. GONOR
>Yaquina Biological Laboratory, Marine Service Center, Oregon State University, Newport, Oregon

JOEL W. HEDGPETH
>Professor of Oceanography, Yaquina Biological Laboratory, Marine Science Center, Oregon State University, Newport, Oregon

PETER A. KRENKEL
>Chairman, Department of Environmental and Water Resources Engineering, Vanderbilt University, Nashville, Tennessee

TOM McCALL
>Governor of Oregon

DONALD I. MOUNT
>Director, National Water Quality Laboratory, Federal Water Pollution Control Administration, Duluth, Minnesota

ROY E. NAKATANI
>Director, Ecology Section, Pacific Northwest Laboratories, Battelle Memorial Institute, Richland, Washington

WHEELER J. NORTH
>   Associate Professor of Environmental Health Engineering, California
>   Institute of Technology, Pasadena, California

GERALD T. ORLOB
>   President, Water Resources Engineers, Inc., Walnut Creek, California

FRANK L. PARKER
>   Professor of Environmental and Water Resources Engineering, Vander-
>   bilt University, Nashville, Tennessee

RUTH PATRICK
>   Curator, Department of Limnology, Academy of Natural Sciences of
>   Philadelphia, Philadelphia, Pennsylvania

EDWARD C. RANEY
>   Professor of Zoology, Cornell University, Ithaca, New York

GEORGE R. SNYDER
>   Program Leader, Prediction of Environment of the Columbia River
>   Basin, Bureau of Commercial Fisheries, Seattle, Washington

J. B. STRICKLAND
>   Research Oceanographer, Scripps Institute of Oceanography, La Jolla,
>   California

DONALD P. de SYLVA
>   Professor of Marine Biology, Institute of Marine Sciences, University
>   of Miami, Miami, Florida

LARRY A. WHITFORD
>   Professor of Botany, North Carolina State University, Raleigh, North
>   Carolina

CHARLES B. WURTZ
>   Chairman, Department of Biology, LaSalle University, Philadelphia,
>   Pennsylvania

THE SYMPOSIUM COMMITTEE
NATIONAL SYMPOSIUM ON THERMAL POLLUTION

JAMES L. AGEE
>   Director, Pacific Northwest Laboratory, Federal Water Pollution Con-
>   trol Administration, Corvallis, Oregon

JAMES R. BOYDSTON
>   Director, Treatment and Control Research, Pacific Northwest Water
>   Laboratory, Federal Water Pollution Control Administration, Corvallis,
>   Oregon

WILLIAM A. CAWLEY
>   Acting Director, Pollution Control, Technology Branch, Division of
>   Research, Federal Water Pollution Control Administration, U.S. De-
>   partment of the Interior, Washington, D.C.

PETER A. KRENKEL
>   Chairman, Department of Environmental and Water Resources En-
>   gineering, Vanderbilt University, Nashville, Tennessee

FRANK L. PARKER
>   Professor of Environmental and Water Resources Engineering, Van-
>   derbilt University, Nashville, Tennessee

Co-Chairmen of the Committee:
>   JAMES R. BOYDSTON
>   FRANK L. PARKER

CHAIRMEN OF THE MEETINGS

WILLIAM A. CAWLEY

> Acting Director, Pollution Control, Technology Branch, Division of Research, Federal Water Pollution Control Administration, U.S. Department of the Interior, Washington, D.C.

FRED A. LIMPERT

> Head, Hydrology Section, Bonneville Power Administration, Portland, Oregon

DONALD R. JOHNSON

> Regional Director, Bureau of Commercial Fisheries, Seattle, Washington

J. A. ROY HAMILTON

> Pacific Power and Light Company, Portland, Oregon

FRANK L. PARKER

> Professor of Environmental and Water Resources Engineering, Vanderbilt University, Nashville, Tennessee

CONTENTS

xiii

LIST OF ILLUSTRATIONS

LIST OF TABLES

BIOLOGICAL ASPECTS OF THERMAL POLLUTION

Chapter 1   David S. Black

KEYNOTE ADDRESS

WHENEVER scientists get together, there is bound to be some electricity in the air. My pun is intentional and, I think, quite appropriate for your meetings because I hope the discussions set off some sparks, but, at the same time, create some light as well as heat.

Ironically, the problem before us is to produce more light for America with less heat.

Since industry began its phenomenal growth, the power industry has provided much of the impetus for this development and has grown apace with the increasing demands for commercial and residential power supply. The availability of large amounts of low-cost power has been in many cases the determining factor in choosing a particular site for a new manufacturing plant and consequently helped set the pattern of economic growth for many areas of the nation. The industrial development of this great Pacific Northwest is in a major way directly attributable to the power available from the massive hydroelectric system on the Columbia and its tributaries and the vast transmission network operated by the Bonneville Power Administration.

No one can deny the past contributions of the utilities to our growth. No one can doubt the need for the tremendous quantities of new generation required to meet the demands of the future.

Obviously, these growing needs must and will be met. How they will be met is another matter. It is no longer simply a technological or economic question. To an increasing extent, it is also a social question, for an entirely new factor has now entered the social consciousness of America; and that factor is the new awareness of man's total environment. There is no ducking the fact that America is faced with massive environmental problems, ranging from urban decay to the clutter and misuse of open space to the contamination of the very resources upon which life depends—the air we breathe and the water we drink. We have, right now, the technological capability of poisoning that environment from the ionosphere to the depths of the oceans.

It is hardly surprising that an operation as vast as the power industry should be caught up in this mounting wave of public concern over what the pressures of growth are doing to our shrinking landscape. In recent years, even a sophisticated new term largely associated with the power industry has found its way into the environmental lexicon. When power plants were relatively small, no one gave much thought to the problem of waste heat, much less to anything as improbable as "thermal pollution." Today, it is a subject of widespread attention—not just by the scientific community or the electrical engineers—but by the popular press, as well, despite the fact that even the experts know all too little what it actually means.

Many in the power industry react to the talk of "thermal pollution" very defensively, with grave warnings of brown-outs and power failures. On the other extreme, the emotions of some conservationists are similarly aroused by the merest suggestion that any body of water might safely absorb any heat whatever from any man-made source.

Part of the blame for the sometimes unreasonable passion with which the subject is treated rests with the term itself: "to pollute," in a dictionary sense, means to befoul, to dirty, to taint, to contaminate. These are words that elicit a strongly emotional and negative reaction. Under no circumstances can the befouling, dirtying, tainting, or contaminating of our water be defended. It needs little demonstration to establish that the discharge of raw sewage in our nation's waterways constitutes pollution in the truest sense of the term. Heat, however, does not befoul the water nor does it cause the water to become dirty. To claim that it taints or contaminates the water is assuming the answer to the real question before the house.

The ecological balance which nature has been able to maintain in the nation's waters, in spite of natural causes which frequently result in large variations in water temperature on a seasonal and even daily basis, would undoubtedly be upset by substantial and prolonged change of water temperature and our environment would be altered for the worse. I will later refer to some instances, however, in which it is claimed that a changed thermal environment would serve man advantageously. But for the moment, the significant fact is that it is not the heat which is the pollutant but rather the effect of the heat upon the ecology of the river basin or water system.

The real issue revolves around the determination of the circumstances under which this change in ecology may occur. So much for definition of terms, and I hope you will forgive this somewhat patronizing little lecture by a layman to a technical symposium.

In the days gone by, whatever effects warmed-up water had on fish and plant life were not considered important because the volume of warmed water was relatively small and the effects were not apparent. Today, the situation is entirely different. More and more power plants line our rivers, streams, and lakes, drawing on them for ever increasing amounts of cooling water.

Not only have the volumes of water used increased tremendously, but the discharge points have moved closer together. On top of that, the same stream flows we had fifty years ago must now serve almost twice as many people, and industry on a scale we could scarcely have dreamed of before the Second World War.

With these new burdens on our natural resources, the electric utility industry must now generate more power for our increasing needs and it must do this under many more restrictions. Restrictions, I mean, in the sense of land available for new plants or expansion of existing facilities, air available for exhausting products of combustion and water available for disposing of waste heat.

No longer can a site for a power plant be chosen simply because it is close to an industrial complex or because it might be cheaper to build there. In this complex age, the utility executive can't put his cane down just anywhere and say, "Build here."

Fortunately, utilities are used to taking long looks into the future to keep up with the nation's electrical appetite. But I suggest that there are new elements to be considered in this long-range forward view.

The American people now expect—indeed, demand—that all industry consider the environmental and esthetic consequences of their actions.

In this, the first session of a national symposium, you are considering one aspect of a complicated problem: the biological effects of heated water added to streams. You have some tough problems to consider and the solutions will determine the point at which we must halt the input of heat into our waters.

This solution cannot be prejudged or anticipated. Very small temperature changes might well be shown to have far-reaching effects.

I am told, for example, that an insect nymph in an artifically warmed stream might emerge for its mating flight too early in the spring and be immobilized by the cold air.

I am told that a fish might hatch too early in the spring to find its natural food organisms, because the food chain depends ultimately on the plants and these in turn upon day length, as well as temperature. Fish, generally, depend on temperature changes in specific amounts to act as a signal for migration and spawning. The entire life cycle of fish may be upset by highly unnatural changes in the temperature cycle.

Trout eggs will not hatch if incubated in too-warm water and salmon do not spawn if the temperature is too high. The sensitivity of all aquatic life to toxic substances is heightened at increased temperatures, and toxic effects of chemical substances are increased. Carp, for instance, are reported to be twice as susceptible to carbon dioxide in warm water as in water near the freezing point.

Our experts point out that the oxygen consumption by aquatic vertebrates doubles for every ten degrees' rise in stream temperature. But as those temperatures rise, the water can hold less oxygen in solution. Thus, while supplies of dissolved oxygen steadily dwindle with increased temperatures, the demand for oxygen increases. Eventually, all aquatic life would die.

I mention these things to demonstrate that I have some superficial acquaintance with the difficult tasks you face. I believe the problem confronting all of us in meeting the growing demands for more water for cooling and more concern for the aquatic environment are so intertwined that it is going to take a many-fronted assault to protect all the nation's interests.

We must *not* approach our task with the predetermined conclusion that the addition of heat, if properly controlled, to our important waterways will necessarily produce all the dire consequences which have been predicted.

The problem we face with waste heat is, in fact, many problems. From my viewpoint, we must take five basic approaches.

First, better management of waste heat can make it less harmful to the aquatic environment. One way is to increase turbulence in the water to provide aeration and cooling. Another is to introduce the heated effluent into deep portions of the receiving water and allow natural convection to promote mixing. Or it may be better to construct partial dams in the original watercourse to promote stratification, so as to permit the withdrawal of the coolest water available for cooling purposes.

The utility could try to schedule plant shutdowns for normal maintenance during the months when the climate and other water uses combine to make additions of heat most hazardous. Water releases from reservoirs should be timed to the extent that operating flexibility permits, to reduce the temperature rise resulting from the disposal of the heat. The heated effluent can perhaps be sprayed over the top of the stream or discharged from a number of outlets to disperse its full effects.

There are many alternatives, including the choice of site when building a new plant, but the magnitude of waste heat anticipated in the future is so great that stream management alone will not protect water quality. The heat itself must be controlled.

That brings us to our second approach to this problem: improving the efficiency of our thermal electric plants.

It has been said that the efficiency of a power-generating plant is about on the order of efficiency of the Franklin stove. Be that as it may, no one can deny that when 60 percent of the heat input to a coal-fired plant is wasted, we have something to shoot for in the way of improvement.

You are all familiar with the dramatic shift to nuclear power started during the past few years. While 95 percent of the thermally generated electricity is still produced by fossil fuel, the proportion is expected to decline to 65 percent by 1980.

With the advent of the nuclear power age, many persons believed the pollution problems were lessened.

But as happens so many times in our close-knit environment, the nuclear power plants did help to alleviate air pollution but added to problems in the water. Nuclear plants must dispose of more heat through their cooling water, thereby creating an even larger burden on the receiving waters.

Although we can't wait for the ultimate system whereby no warm water is created in generating power, scientists are working on the nuclear breeder reactors which could well be a short-run improvement. These reactors will produce steam at temperatures approximating those of coal-fired plants, so that amount of waste heat to be dissipated by cooling systems will be reduced accordingly.

In other words, we could keep the air-pollution advantages of the nuclear plant and seek ways to improve on its water-pollution record.

Our third approach to the problem is to dispose of the excess heat by the construction of cooling towers, cooling ponds, or spray ponds. Here, however, we create a whole new environmental problem industry must take into consideration in modern-day operations. That is the esthetic problem. How will the power plant fit into the landscape?

Cooling towers are enormous structures. Modern hyperbolic towers may rise thirty stories high and be more than a city block in diameter. And costs are another factor which tend to limit the use of towers. Wet towers may add $5 to $10 per kilowatt to the cost of plant construction. So-called dry towers, which waste almost no water and discharge no heat whatever into surrounding bodies of water, can add as much as $20 or more per kilowatt of capacity to the construction cost.

I think our next approach to this problem is the one that is the most promising, not only for the power-generating community, but for all American industry: that is, finding ways of making constructive use of the huge amounts of heat that are now going to waste in power generation. In the case of the electric utility, the productive use of the waste heat would help to make up for the relative thermal inefficiency of the steam electric generating plant, as well as reduce the pollution problem.

Many uses of heat are currently being studied by industry and by the government. For example, here in the Pacific Northwest, tests are being conducted to see if the heated water can be used for irrigation. Naturally, there have to be safeguards, because waters that are too hot or too cold may affect seedling emergence, plant growth rate, time of maturity, and crop yields. But with proper caution—management, again—it may be possible to extend the growing season and thus make constructive use of the waste heat.

Warm-water cultivation of oysters is being attempted in the East and in the state of Washington. What a treat for oyster lovers, if the experiments make it possible for oysters to spawn continuously for ten months a year and to reach their maturity in two and one-half years.

I cite these examples, not as the only answers, but to show that sometimes unwanted products can be turned into tools to heal or help some other area of our society.

What if we could warm some of the beaches along our northern shores! Just think of the added recreation for our growing population. And just think of the boost to the economy some of these areas would receive if their swimming seasons could be extended.

Utilities have been selling low-pressure steam for years to provide heat for buildings. Wider use of heat for this purpose would obviously be beneficial all the way around.

You, as scientists, can help point the way to new ways of using this excess heat by providing more basic information about the effects of waste heat in our water resources. These more complete data should make it easier to find safe ways of disposing of waste heat or turning it to constructive use.

Finally, our fifth approach to this problem is to develop new methods of power generation which are more efficient and result in less heating of the water or pollution of the atmosphere.

Fuel cells and thermal electrical systems which do not require the use of the steam cycle for power generation may be the answers for tomorrow.

Such far-out concepts as electrogasdynamics, magnetohydrodynamics, and thermionic power generation, if they can be developed economically, could help us to reach a pollution-free future.

Joining with industry and other government agencies, the Department of the Interior's Federal Water Pollution Control Administration has assigned its major research emphasis in this field to its Pacific Northwest Water Laboratory at Corvallis, Oregon. They are studying the effects of thermal power generation on water use and quality with specific reference to a study of the effects of temperature change on the Columbia River. It is estimated that the Columbia River study will require two years. When the results are available, we hope to have sufficient scientific information regarding the tolerance of the Columbia River to temperature modification so that the differences of opinions existing today can be put to rest on the basis of developed technological information.

In the interim, we have approved temperature standards based upon the best information available to us which we believe will preserve the Columbia River for present and future uses. But we recognize that the results of the pending study could require a re-appraisal of the standards so as to move them in either direction.

I have tried to suggest some of the ways we can approach the problem of excess heat. It is going to require taking into account manifold economic and social interests to be protected over the decades ahead.

I have previously noted that the problems associated with the discharge of waste heat into the nation's waterways are not new. However, the magnitude of these discharges which can be foreseen in the immediate future makes the problems acute. It will require constructive and creative thinking—with the objectivity of the scientific atmosphere—to preserve a productive and satisfying environment for America.

Chapter 2 Peter A. Krenkel and Frank L. Parker

ENGINEERING ASPECTS, SOURCES, AND MAGNITUDE OF THERMAL POLLUTION

THE purpose of this paper is to present the engineering problems associated with the discharge of heated waters to our environment and to discuss reasonable methods for preventing these heated waters from adversely affecting receiving waters. Obviously, the biologist must recommend temperature levels that are optimum for aquatic ecosystems, and the engineer must examine these recommendations, stressing their economic implications. The final decision, of course, must be made by the body politic, based on sound engineering analysis.

The vastly accelerating development of our water resources and the corresponding increase in water utilization is bringing to light phenomena requiring immediate attention by water-quality engineers. The influx of heat to our waters, whether by the discharge of artificially heated waters or by solar radiation, is causing increasing alarm to conservationists, biologists, and engineers. Environmental changes such as stratified flow, the effects of temperature on the biota, and the effects of temperature on the physical and chemical properties of our receiving waters are becoming quite critical.

The many benefits accrued by altering the natural flow of our waterways are well known. However, the deleterious effects caused by impoundments and associated power production have only recently been subjected to study.

It is mandatory that answers to the following questions be found, if adequate and equitable water-quality control is to be maintained.

1. Are the effects of stratified flow beneficial or deleterious? I.e., is it advantageous to concentrate the heat in the surface waters or should heated water be completely mixed with the receiving water?
2. What is the desirable minimum dissolved-oxygen concentration, both

spatially and temporally, in both layers, under stratified flow conditions?

3. If the waste discharge appears as an interflow (a form of stratified flow), does this constitute pollution, since it adds to the hypolimnion deficit, or has the waste been satisfactorily assimilated, since it has disappeared from an obvious location? Are not the effects of such a discharge only delayed in time or transferred to another location?

4. What is the responsibility of the power companies with respect to pollutional effects caused by the changing of the flow regime and/or the discharge of heated effluents?

5. Is the agency or company responsible for impounding a water legally liable for subsequent water-quality problems, such as the appearance of iron and manganese or the disappearance of dissolved oxygen?

6. What are the effects of temperature changes on the waste-assimilative capacity of a receiving water?

7. What temperature levels are harmful, including sublethal effects, to aquatic organisms?

8. What is the maximum rate of change of temperature that will not have deleterious effects on aquatic life?

9. What methods are available for cooling water and what are their costs?

With these questions in mind, the problems of heated water can be elucidated and it is hoped that some insight into temperature change and its effect on the environment can be gained.

STRATIFICATION IN IMPOUNDED WATERS

When discussing the effects of heat on water-quality management, it is necessary to review briefly the phenomenon of stratification in an impounded water. While the stratification process is well known, the resulting changes in water quality are not, and they are becoming increasingly serious because of the complex water-resource systems developed as a result of expanding water conservation requirements.

At the end of a winter season, the impounded water is of a fairly uniform quality and has a relatively low temperature. At the onset of higher atmospheric temperatures, the surface water and the incoming water temperatures are raised and this lighter water tends to "float" on the colder and denser water already in the lake.

Three definite strata may be formed: the surface stratum or epilimnion, the lower stratum or hypolimnion, and a transition zone called the thermocline, where the maximum rate of change of temperature with depth occurs. In the Southeast, the thermocline persists from about April to November and is approximately 10 to 20 feet in thickness. In a deep reservoir, the epilimnion may be approximately 30 to 50 feet in thickness and the hypolimnion will usually extend to the reservoir bottom.

These conditions may exist until autumn, when the lake begins to lose heat more quickly than it is absorbed. As the water becomes cooler and more dense, the thermocline sinks, unstable conditions are promulgated, and mixing or overturn of the reservoir occurs. In climates where the water temperature goes below 4° C., two turnovers may occur per year.

Many impounded waters circulate completely but some circulate only partially, these lakes being called meromictic by limnologists. This stable lower layer can be caused by either an accumulation of dissolved or suspended solids in the water and may render this lower portion of the lake unsuitable for a water supply.

A typical reservoir profile as described above is shown on Figure 1 as taken from the work of Kittrell (1959).

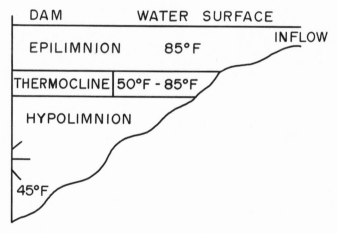

Fig. 1——.Typical summer reservoir profile
(After Kittrell)

DISCHARGES OF DIFFERING DENSITY

Density currents, density flows, under-flows, overflows, and interflows are all synonymous with stratified flow. As defined by the National Bureau of Standards (1938), "A density current is the movement, without loss of identity by turbulent mixing at the bounding surfaces, of a stream of fluid under, through, or over a body of fluid, with which it is miscible and the density of which varies from that of the current, the density difference being a function of the differences in temperature, salt content, and/or silt content of the two fluids."

The various forms of a density current are as depicted in Figure 2, the overflow, the interflow, and the underflow. Each of these currents may affect water quality in an adverse manner. Probably the most important form of a density current to the thermal-pollution problem is the overflow because of its occurrence in the discharge of cooling waters.

Stratified flow has been observed when $\Delta\rho/\rho \geqq 0.005$, where ρ is the fluid density and $\Delta\rho$ is the density differential between the fluid layers (Amer. Soc. Civ. Engrs., 1963). It is interesting to note that this condition is satisfied with water temperatures of $31°$ C. and $32.5°$ C.

In order to illustrate these various forms of density currents, the following examples, each with a possible deleterious effect on water quality, are presented.

The Underflow

Underflows can be caused by the discharge of colder, more dense water from an upstream stratified reservoir or by water containing excessive suspended or dissolved solids. Underflows caused by highly turbid waters were first noted on Lake Mead, while the underflow caused by the upstream release of hypolimnetic water frequently occurs in TVA reservoirs. Fish kills have been reported to have occurred downstream from Fort Loudon Dam, purportedly caused by the low dissolved-oxygen content resulting from the release of cold hypolimnion water (Jones, 1964).

Figure 3 shows a schematic diagram depicting the flow regime caused by an underflow as presented by Elder (1964). Two different flow regimes are noted: the lower strata, bounded by the channel bottom and the interface, with the velocity profile approximating that of distorted pipe flow; and the upper layer, bounded by the interface and the atmosphere and behaving like free surface flow.

One immediate problem that may occur is caused by the possibility of a flow reversal in the upper layers of the reservoir. Under these conditions, should the water-pollution control plant discharge be located downstream or upstream from the water-treatment plant intake, or what should be done to prevent pollution of upstream facilities?

The Interflow

The interflow results from the discharge of a fluid of an intermediate density into a stratified flow regime.

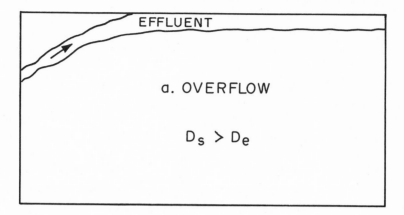

a. OVERFLOW

$$D_s > D_e$$

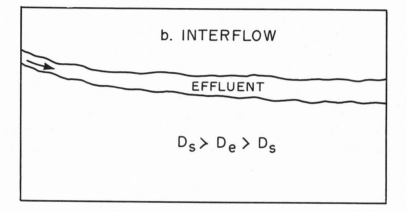

b. INTERFLOW

EFFLUENT

$$D_s > D_e > D_s$$

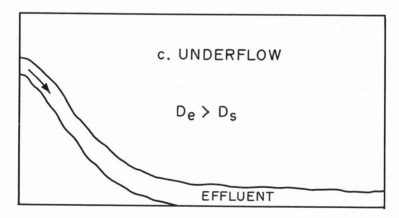

c. UNDERFLOW

$$D_e > D_s$$

EFFLUENT

Fig. 2—.Forms of stratified flow

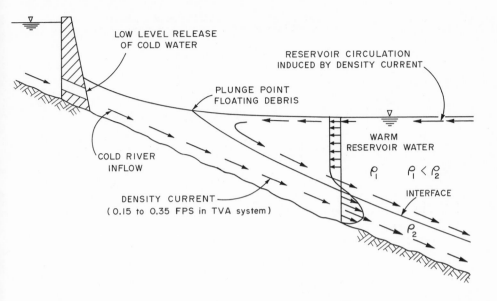

Fig. 3—.Density of current due to low-level release of cold water
(After Elder)

In order to learn more about this phenomenon, Vanderbilt University, in cooperation with TVA, continuously introduced Rhodamine B dye into the Nantahala River for a 24-hour period. The Nantahala is a stream tributary to Fontana Reservoir and had a temperature of 59° F. on the day of the injection.

Figures 4 and 5 show the history of the labeled Nantahala River water as it became a part of Fontana Lake. Note the entrance of the water, its attempt to seek a suitable level, and then its ultimate fate as an interflow at the appropriate density level. The 4° F. rise in temperature was attributed to mixing and surface heating in the shallow, upper reaches of the reservoir.

If this inflow had been heavily polluted with organic material (which it wasn't), would the reservoir have satisfactorily assimilated the waste? Since this hypothetical waste would contribute to the hypolimnion deficit, would not the pollutional effects only have been relocated and delayed? The sampling techniques for this situation also pose some interesting questions.

Another example of an interflow is the process of selective withdrawal as shown in Figure 6. Under unstratified conditions, we would withdraw water from all levels of the reservoir; however, under stratified flow con-

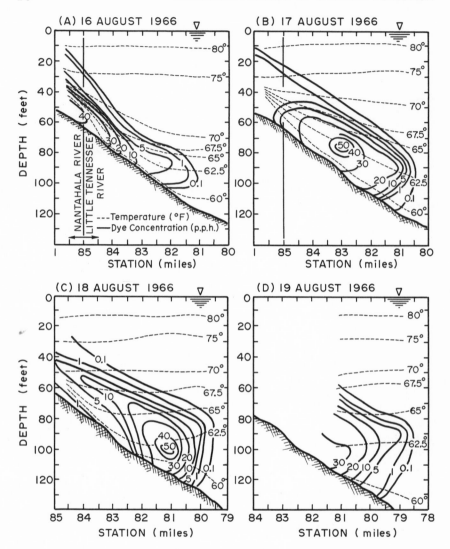

Fig. 4——.Movement of dye cloud through Fontana Reservoir

ditions, we could withdraw water from a pre-selected layer, depending on the level of the water intake, and choose desirable temperatures and other water-quality characteristics.

The Overflow

The overflow is caused by the discharge of lighter-density water and is

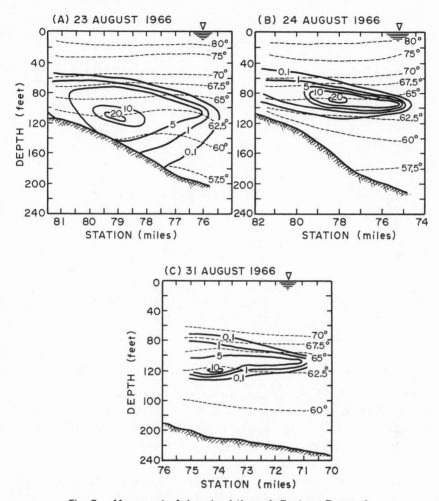

Fig. 5——.Movement of dye cloud through Fontana Reservoir

most commonly observed when cooling waters are emptied into a receiving water at its surface.

While the heat is obviously dissipated more rapidly with this means of discharge than if the cooling water were completely mixed with the receiving water, the undesirable effects of the resulting stratified flow may be totally unacceptable from the water-quality standpoint.

Further discussion of the overflow will be presented in a subsequent section.

The stability of stratified flow is quite amazing, inasmuch as it may

(A) Withdrawal from full depth in a homogeneous fluid.

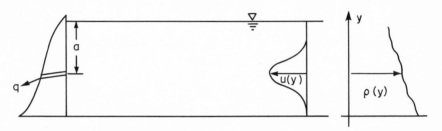

(B) Selective withdrawal in a stratified field.

Fig. 6—.The process of selective withdrawal
(After Elder)

exist for miles upstream and downstream from a power plant during all parts of the year. Mathematically, it may be demonstrated that work must be expended in accomplishing vertical displacement of a water particle. Physically, one can consider pulling a particle from the lighter upper layer into the heavier lower layer, releasing it, and having the particle return to the upper layer by the force of buoyancy. Likewise, a particle vertically displaced from the lower layer into the upper layer and released will sink back to the lower layer because of the force of gravity.

EFFECTS OF STRATIFIED FLOW ON WATER QUALITY

Dissolved Oxygen

Water containing organic material entering a stratified reservoir will deplete the oxygen resources of the reservoir, due to biological respiration. In the epilimnion, mixing by wind currents, photosynthesis, and sedimentation may render the epilimnetic water satisfactory. However, in the hypolimnion, oxygen removed by biological action is not replaced; algae cannot function; essentially, no vertical mixing takes place; and the products of sedimentation from the epilimnion may add additional organic load. The net result is a depletion of the dissolved-oxygen resources, thus making septic conditions possible.

If this hypolimnetic water with its low oxygen content is discharged from the dam, fish life may not be supported for several miles, a lower waste-assimilation capacity may exist and, in effect, the dam may be considered equivalent to a large BOD contribution.

Iron and Manganese

If there is a low oxidation-reduction potential at the mud-water interface, conditions amenable to the dissolution of iron and manganese into the hypolimnetic waters will occur. The mechanism is not clear; however, if the oxides of these metals are present in the bottom muds, troublesome concentrations may appear in the water under reduced environmental conditions.

It is interesting to note that 50 mg/l of iron and 15 mg/l of manganese have been found in the lower layers of the South Holston Reservoir. Since these concentrations do not significantly change with time, the lake can be classified as meromictic, due to chemical concentrations and corresponding density increases.

Temperature

The temperature differential in stratified lakes can be quite significant. Since a major water use in the United States is for cooling purposes, it is obvious that this cooler water is highly desirable for condenser-cooling for steam-electrical generation purposes. In fact, the use of this colder water by construction of an underwater dam at the TVA Kingston Steam Plant was reported to have saved TVA $155,000 in operating costs for the year 1956 alone, which was one-third the cost of the dam.

This then raises questions regarding the desirability of artificially overturning a reservoir. While the stratification may produce a poorer quality of water, it would appear to be less expensive to treat this water than to lose the advantage of the colder cooling water available to the steam-electrical generating plants. Obviously, what is needed is a thorough systems analysis of these alternatives.

The question posed is, "Do we want high dissolved oxygen and temperature, or low dissolved oxygen and temperature?"

Figures 7 and 8 demonstrate the conditions existing in Allatoona Reservoir during completely mixed conditions and under a stratified flow regime. Note that in the summer, the temperature varies from over 27° C. at the

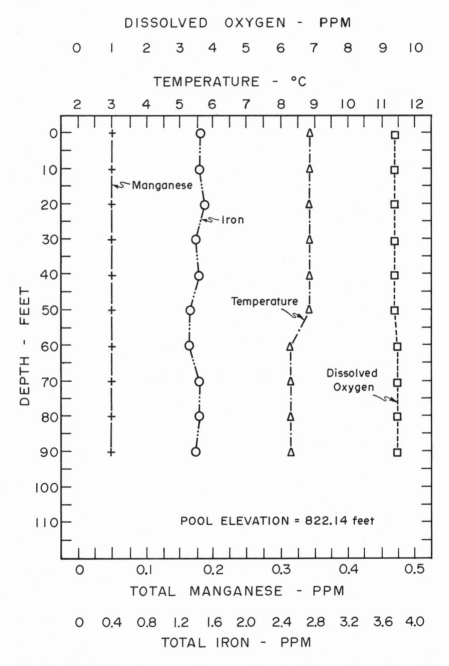

Fig. 7—.Distribution of water-quality parameters in Allatoona Reservoir on 15 January 1969

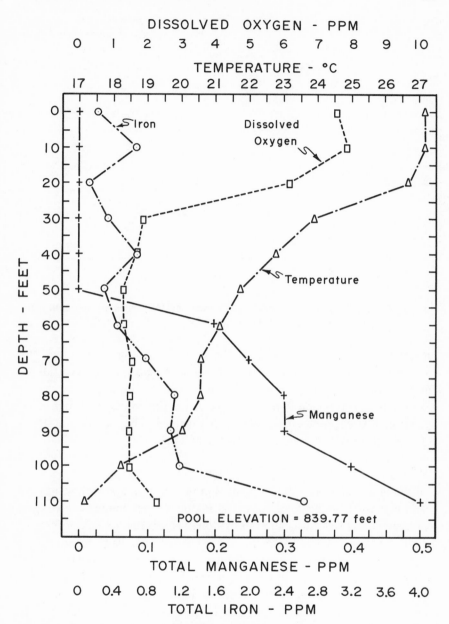

Fig. 8—.Distribution of water-quality parameters in Allatoona Reservoir on 28
August 1964

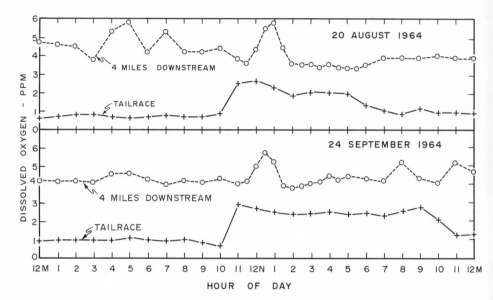

Fig. 9——.Variation of dissolved oxygen at tailrace and four miles downstream from
Allatoona Dam on 20 August and 24 September 1964

surface to approximately 17° C. at the reservoir bottom. Difficulties encountered from the low oxygen and high iron and manganese concentrations
are obvious, inasmuch as the water intake is from the lower levels.

Note also the variation in dissolved oxygen at the tailrace and four miles
downstream from the dam, as shown in Figure 9. The variation at 1000
hours is because of the peaking power operation.

Overflow Versus Completely Mixed Heated Discharges

If the discharge from a power plant is in the form of an overflow, mixing
between the upper and lower layers is inhibited, thus minimizing oxygen
replacement and self-purification in the lower layer. Due to lack of mixing,
organic wastes discharged into the lower layer do not have access to the
oxygen in that portion of the stream flowing in the upper layer. Thus, there
is less dissolved oxygen, less dilution water, and a more concentrated
organic load in the lower layer leading to an acceleration of the dissolved
oxygen depletion. The net result may be a considerable reduction in the
waste-assimilative capacity of the receiving water.

If the heated discharge is completely mixed with the receiving water,
some of the above-mentioned effects are eliminated; however, the rise in
temperature still causes a decrease in the ability of water to hold dissolved

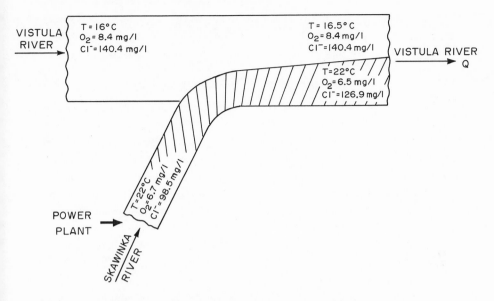

Fig. 10——.Power-plant discharge into Vistula River

oxygen, an increase in the metabolic activity of organisms, an increased rate of Biochemical Oxygen Demand exertion, and a possible reduction in waste-assimilative capacity.

The discussion as to whether the heated water should be discharged as a layer or completely mixed with the receiving water must be based on conditions existing in the receiving water, and the amount of heat to be dissipated. Obviously, the heat will be dissipated more quickly if it is concentrated in a lesser volume of water. The effects of the resulting stratified flow on the biota are not so obvious, however. While it is true that fish will have passage through the cool lower layers of water, the fish cannot utilize this route if the water is devoid of oxygen.

Another possible route for heated-water effluents is to "hug" the shoreline and not gain entry into the major portion of the river flow for many miles downstream. Such a situation was noted in Poland, as illustrated in Figure 10 (Krenkel, 1967). This separation of flow was reported to have existed for over ten kilometers downstream from the confluence of the two rivers. Note also the division of water-quality characteristics indicating the two flow entities. These conditions delight the biologists, inasmuch as the fish have a portion of the river unaffected by the heated-water discharge in which to pursue their natural activities.

The question of the effects of layered flow in the ecological system must be decided eventually by the biologist, inasmuch as the biological changes induced by the stratified-flow conditions are relatively unknown.

HEAT ADDITIONS FROM POWER PLANTS

Thus far, we have examined the causes and effects of stratified flow, emphasizing heat additions resulting from natural causes. While the contribution of heat from solar radiation is important, the primary interest of the regulatory agencies is with heat added to an environment from industrial processes.

Though the most significant contribution to the thermal pollution is from the steam-electrical generating industry, it should be noted that many industrial wet-process industries do contribute waste heat to our environment. It is more difficult to quantitize these effects; however, it has been estimated that over 70 percent of the process water withdrawn for industrial uses is for cooling purposes.

In terms of the total thermal-pollution problem, it is estimated that the contribution of industry to thermal pollution is in the order of 20 to 30 percent of the total heat rejected to our receiving waters. It should also be noted that these contributions are usually less concentrated heat loads and therefore do not cause the problems resulting from a single, large-point source of heat such as that emanating from a steam-electrical generating plant. Even sewage contributes some temperature rise to the receiving water, although its effect on the receiving water is negligible.

The heat absorbed by cooling water in some industrial processes is shown in Table 1, which was taken from McKelvey and Brooke (1959).

The really significant heat loads result from the discharge of condenser-cooling water from the ever increasing number of steam-electrical generating plants. Electrical power generation in the United States has doubled every ten years since 1945, and all indications are that the rate of increase will be even greater during the next few decades. Furthermore, the problems associated with these concentrated heat loads are compounded by the increasing size of individual power plants and the greater quantities of heat discharged by equivalent size when nuclear-power reactors are used.

It has been estimated that the 95 percent of the centrally generated electricity that is produced by fossil fuel today will be 65 percent nuclear by the year 1980 (Bregman, 1968). As will be shown subsequently, large nuclear-power plants currently require approximately 50 percent more cooling water for a given temperature rise than do fossil-fuel plants of equal

TABLE 1. Heat Absorbed by Cooling for Various Processes

Item	Heat load absorbed by cooling water	BTUs per
Alcohol	20,000	Gal
Aluminum	31,000	lb
Beer	91,000	bbl
Butadiene	31,000	lb
Cement	150,000	ton
Refined Oil	150,000	bbl
Soap	97,000	ton
Sugar	200,000	ton
Sulphuric Acid	650,000	ton
Cooling power equipment		
Air compressors		
Single stage	380	BHP-h
Single stage with aftercooler	2540	BHP-h
Two-stage with intercooler	1530	BHP-h
Two-stage with intercooler and aftercooler	2550	BHP-h
Diesel engine jacket water and lube oil (incl. dual fuel)		
Four-cycle, supercharged	2600	BHP-h
Four-cycle, non-supercharged	3000	BHP-h
Two-cycle, crank-case compr.	2000	BHP-h
Two-cycle, pump scaveng (large)	2300	BHP-h
Two-cycle, pump scaveng (high speed)	2100	BHP-h
Natural-gas engines		
Four-cycle (250 pse compr.)	4500	BHP-h
Two-cycle (250 pse compr.)	3000	BHP-h
Refrigeration		
Compression	250	min-ton
Adsorption	500	min-ton
Stream jet refrig. condenser (100 psi. dry steam supply, 2″ Hg cond.)	1100	lb of steam
Steam-turbine condenser	1000	lb of steam

size. It is expected that improved technology will reduce this added requirement to 25 percent by the year 1980. Development of nuclear breeder reactors will reduce the amount of heat dissipated to approach that of fossil-fueled plants; however, these are not expected to be available until the 1980s (Water Resources Council, 1968).

The Atomic Energy Commission (1968) stated that as of June 1968, 13 nuclear central-station electric-power reactors were in operation, 21 were being built, and 40 were being planned. The increase in size of the newer

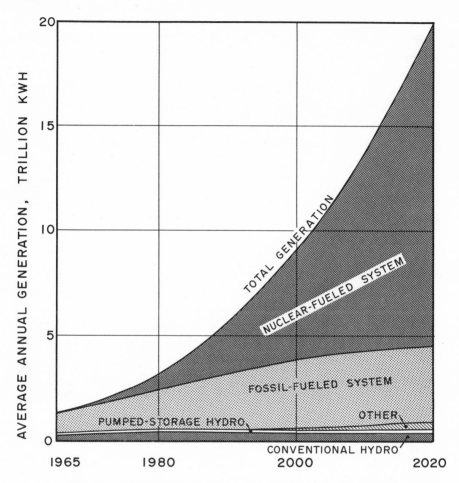

Fig. 11—.Projected electric generation by types of prime mover

power stations is indicated by noting that the nuclear-power plants planned for operation by 1973 average 624 megawatts per unit, while those retired between 1962 and 1965 averaged 22 megawatts per unit (Bregman, 1968).

Figure 11 shows the anticipated growth of the power industry until the year 2020. Even a cursory examination of this data demonstrates the need for control of the resulting cooling waters. It is also interesting to note the relatively small percentage of power production that is and will be contributed by sources other than steam generation.

In 1964, hydroelectric plants provided 19 percent of the capacity and energy produced and indications are that this will continue to decline as shown on Figure 11 (Federal Power Commission, 1964). There are very few undeveloped hydroelectric power sites remaining, although potential sites for pumped-storage installations still exist in most areas of the United States (Water Resources Council, 1968). The optimal use of pumped storage is for peaking power and reserve capacity, since the large steam-electric plants being constructed operate best at high plant factors which are complemented by the low plant factors of the pumped-storage plants.

COOLING-WATER REQUIREMENTS

It has been estimated that cooling-water requirements will increase from 50 trillion gallons per year in 1968 to 100 trillion gallons per year by 1980, and that this will be approximately one-fifth of the total runoff in the co-terminous United States. Actually, the major portion of this use will be of a non-consumptive nature, this not precluding that portion of water for other beneficial utilization. Examination of Table 2, which is the Water

TABLE 2. Average Condenser Water Requirement and Consumptive Use for Fossil-Fueled, Steam-Electric Power-Plants, 1965–2020 gallons per KWH

	Condenser Requirements	Once-through	Cooling Ponds	Cooling Towers
1965	40	0.3	0.4	0.5
1980	35	0.2	0.3	0.4
2000	30	0.15	0.25	0.35
2020	25	0.1	0.2	0.3

Resource Council estimate of cooling-water requirements and consumptive use based on a temperature rise of $15°$ F., demonstrates the consumptive use in fossil-fueled steam-electric power plants (1968). The predicted decrease in unit water requirements is based on improved technology.

The degree of thermal pollution depends on the thermal efficiency, which is determined by the amount of heat rejected to the cooling water. Thermodynamically, heat should be added at the highest possible temperature and rejected at the lowest possible temperature if the greatest amount of work is to be gained and the highest thermal efficiency realized.

Current generally accepted maximum operating conditions for conventional thermal stations are $1000°$ F. and 3500 psi, with a corresponding heat rate of 8700 BTU per kwhr, 3413 BTUs resulting in power production and 5287 BTUs being wasted. Plants have been designed for 1250° F. and 5000 psi; however, metallurgical problems have held operating conditions to lower levels.

Nuclear plants operate at temperatures from 500° to 600° F. and pressures up to 1000 psi, resulting in a heat rate of approximately 10,500 BTU/kwhr. Thus, for nuclear plants, 3413 BTUs are used for useful power production and 7087 BTUs are wasted.

Table 3 shows a typical energy balance for a steam-electric power plant producing 1 kwhr net electrical output (Cootner, 1965). Note that one British Thermal Unit (BTU) will raise the temperature of one pound of water by one degree Fahrenheit.

TABLE 3. Energy Balance: Steam-Electric Power Plant

Assumed over-all efficiency	40%
Assumed generator efficiency	$97\frac{1}{2}\%$
Heat equivalent of 1 kwhr	3,413 BTU
Fuel energy required, 3,413/0.40	8,533 BTU
Heat losses from boiler furnace, at 10% of fuel use	853 BTU
Energy in steam delivered to turbine 8,533–853	7,680 BTU
Heat loss from electric generator at $2\frac{1}{2}\%$ of generator input	87 BTU
Electric generator output	3,413 BTU
Energy required for generator equals energy output from turbine	3,500 BTU
Energy remaining in steam leaving turbine removed in condenser 7,680–3,500	4,180 BTU
Total cooling water required, 10° rise $\dfrac{4,180 + 87}{10 \times 8.33}$	51 gal.
Total cooling water required, 15° rise $\dfrac{4,180 + 87}{15 \times 8.33}$	34 gal.

For comparative purposes, a nuclear-power plant requires approximately 10,500 BTUs to produce 1 kwhr of electricity. Thus, 7087 BTUs are wasted into the cooling water or 2907 more BTUs than the fossil-fuel plant described in Table 3. It should be noted, however, that the plant efficiencies assumed are not average and the plants may not be operating at peak loads, thus somewhat modifying the comparison.

Table 4 demonstrates the increasing efficiencies that have resulted from significant improvements in the conventional steam cycle (Federal Power Commission, 1964). Future heat rate gains will be minimal, however, because super-critical pressures of 3500 to 5000 psi and temperatures of

TABLE 4. Net Heat Rates: Steam-Electric Generating Stations, 1925 to 1962

Year	Best plant[2] BTU/kwhr	Best system BTU/kwhr	U.S.[1] average BTU/kwhr
1925	15,000	NA[3]	25,000
1930	12,900	NA	19,800
1935	12,300	NA	17,850
1936	10,954	NA	17,800
1937	10,779	NA	17,850
1938	10,788	NA	17,450
1939	10,770	NA	16,700
1940	10,729	NA	16,400
1941	10,606	NA	16,550
1942	10,596	NA	16,100
1943	11,021	NA	16,000
1944	10,689	NA	15,850
1945	10,345	NA	15,800
1946	10,608	12,715	15,700
1947	10,600	NA	15,600
1948	10,588	NA	15,738
1949	10,437	NA	15,033
1950	9,378	11,876	14,030
1951	9,379	11,676	13,641
1952	9,303	11,665	13,361
1953	9,329	11,185	12,889
1954	9,113	10,660	12,180
1955	9,151	10,270	11,699
1956	9,106	9,780	11,456
1957	9,118	9,705	11,365
1958	9,130	9,760	11,085
1959	9,011	9,620	10,970
1960	8,975	9,590	10,760
1961	8,760	9,363	10,650
1962	8,588	9,390	10,558
1963			10,482
1964			10,462

1. Exclusive of Alaska and Hawaii.
2. Plants in service full year.
3. NA–Not available.

1100° to 1200° F. are reflected in the data for 1961 and 1962. Present-day central station average thermal efficiency is said to be 33 percent and the present best performance, 42 percent.

Fluctuating Discharges of Heated Water

Most steam-powered electrical generating plants are operated at varying

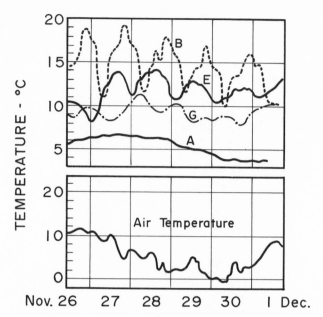

Fig. 12—.Variation in temperature of River Lea
(After Gameson)

load factors, and consequently the heated discharges demonstrate wide variation with time. Thus, the biota is not only subjected to increased or decreased temperature, but also to a sudden or "shock" temperature change.

This type of power plant operation is exemplified by Figure 12, which was taken from an investigation by Gameson (1959). At Station B, a "short distance below the outfall," the diurnal variation is as great as 8° C. Even at Station E, which was about two miles downstream, the effects are significant, being approximately one-half those at Station E. Obviously, the biota would have a most difficult time adjusting to these temperature changes with time. It is important to note that "shock" loading may be more harmful to fish than continual exposure, as observed by Cairns (1955).

It is also interesting to note that recent recommendations from the state of New York limit the rate of temperature change to 2° F. per hour for fresh water and 1° F. per hour for saline water. The state of Tennessee has proposed a 3° F.-per-hour maximum change.

Effects of Heated Discharges on Waste Assimilation

Senator Muskie has stated that: "It is the opinion of the Senate Subcommittee on air and water pollution that excessive heat is as much a pollutant as municipal wastes or industrial discharges (*Science,* 1967)." The reasons for this statement have been alluded to in a previous section. It is instructive to note the following case histories demonstrating the validity of Senator Muskie's statement.

A classic example of the change in river waste-assimilation capacity caused by both temperature and impoundment was presented by Krenkel *et al.* (1965). A paper mill of the Georgia Kraft Company, which had previously satisfactorily discharged its waste effluent into the free-flowing Coosa River, must now discharge into the backwater of a downstream impoundment and receive its flow as regulated by a peaking hydroelectric power plant upstream. The daily fluctuation in discharge may range from 200 to 7800 cfs. In addition, a steam-electric generating plant, with a capacity of 300,000 kilowatts, has been built adjacent to the mill, and the condenser water raises the temperature of the stream several degrees.

The effects of temperature on the stream self-purification process is demonstrated by Figure 13, which shows the variation of the rate constants k_1 (de-oxygenation) and k_2 (re-aeration) with respect to temperature. Examination of this relationship demonstrates that an increase in temperature causes a considerable increase in k_1. While k_2 also increases with increasing temperature, it is negated by the combination of a lesser dissolved-oxygen content and a greater rate of change of k_1 with temperature.

The over-all effects of the impoundment on the rate of oxygen recovery is demonstrated by the lower curve, which depicts the re-aeration rate constant under existing, impounded conditions. Note that, while k_1 at a given temperature is unchanged, the value of k_2 at any temperature is significantly reduced.

In order to illustrate the effect of temperature on the waste-assimilative capacity of the Coosa River, observed data were used to obtain Figure 14, which depicts the oxygen balance in the Coosa River prior to the previously mentioned water resources developments. Note that, under these conditions, the Coosa River easily assimilated 28,000 pounds per day of BOD at the existing river temperature of less than 25° C. Even under the free-flowing conditions, however, a temperature of 30° C. would cause the dissolved-oxygen level to fall below 4.0 mg/1, the minimum required to satisfy the existing stream standards.

Figure 15 is presented to show that, because of the combined effects of the water resources developments and the observed temperature increases, the Coosa River can no longer satisfactorily assimilate the 28,000-pound BOD load, as under the previously existing free-flowing conditions.

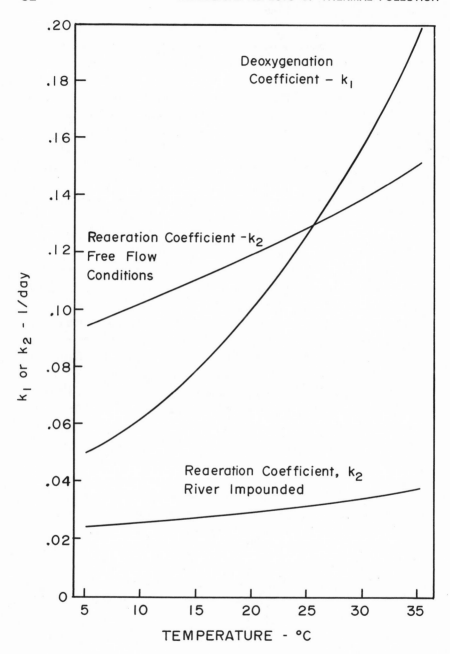

Fig. 13—.Variation of K_1 and K_2 with temperature

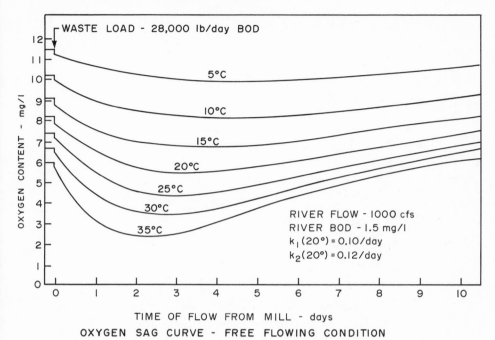

OXYGEN SAG CURVE - FREE FLOWING CONDITION

Fig. 14—.Oxygen-sag curves for Coosa River, free-flow condition

The quantitative effect of temperature on the waste-assimilative capacity of the Coosa River can be shown as on Figure 16, where the possible waste load that would not deplete the dissolved-oxygen content of the river below 4.0 mg/l is plotted versus the river-water temperature.

It may be concluded from this figure that, if the river temperature were 25° C., a 5° C. increase in temperature under free-flow conditions is equivalent to 11,000 pounds per day of BOD; and a 5° C. increase in temperature under the existing, impounded conditions is equivalent to 5200 pounds per day of BOD.

Note the significant reduction in assimilative capacity already caused by the impoundment and the observed increase in temperature, as demonstrated by comparing the curve for free-flow conditions and the curve for impounded conditions. It is obvious from these curves, which were computed from observed conditions, that the addition of heat and the impoundment of the Coosa River have had the same end result as if an equivalent amount of sewage or other organic waste material were added to the river.

Because of the previously described reduction in the waste-assimilative

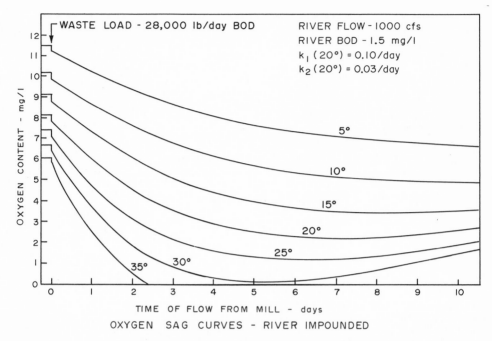

Fig. 15—.Oxygen-sag curves for Coosa River, impounded conditions

capacity of the Coosa River, the mill has reduced its waste load to the river to less than 20,000 pounds of BOD per day, which is the allowable load permitted by the Georgia Water Quality Control Board in order to maintain an adequate dissolved-oxygen content.

Bohnke (1961, 1966) states that an increase in temperature of the river water of 6° C. will cause a 6 percent reduction in self-purification capacity at a minimum oxygen content of 3 mg/l. The loss of assimilative capacity due to a specific power plant on the Lippe was said to be 10,000 population equivalents at low flow.

A recent report by Drummond (1967) estimates that the Chattahoochee River in Georgia can assimilate over three times as much organic material at 20° C. as it can at 30° C. without excessive depletion of oxygen in the water. The city of Atlanta is quite concerned because of two steam-electric generating plants, which, when operating at full load, will raise the maximum average daily river temperature from 20° C. to 34° C. at minimum flow. It was concluded that, in order to maintain satisfactory dissolved-oxygen concentrations in the river during minimum flow, either the steam-

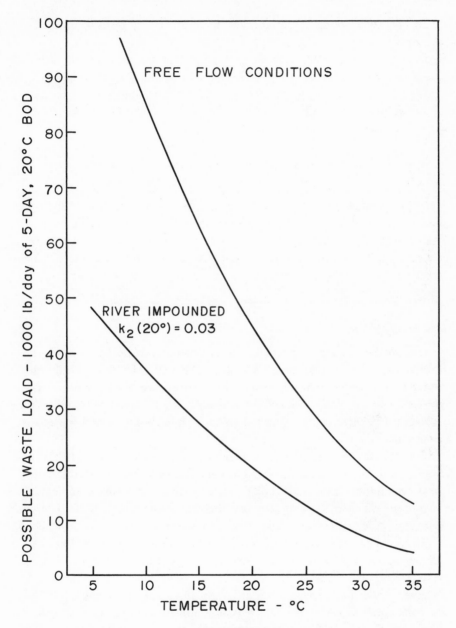

Fig. 16—.Waste-assimilative capacity of Coosa River at Georgia Kraft Mill, river flow of 940 CFS

plant operations must be curtailed or re-regulation of the river must be practiced.

It may be concluded from these studies that the addition of heated water to a receiving water can be considered equivalent to the addition of sewage or other organic waste material, since both pollutants may cause a reduction in the oxygen resources of the receiving water. This conclusion is in accord with the statement of Senator Muskie.

MECHANISMS FOR THE DISSIPATION OF HEAT

The magnitude of the thermal-pollution problem is obvious, as is the need for adequate temperature criteria that will permit the sustenance and propagation of aquatic life. After the biologists have delineated the temperature levels that must be maintained in our receiving waters, control measures to limit the amount of heat discharges must be utilized. The concepts of heat dissipation and the methodology available will now be examined.

The Energy Balance

Obviously, the simplest method of disposing of waste heat is to discharge it directly to the receiving water and then allow natural forces to bring the water back to an equilibrium temperature. This process is known as *once-through cooling,* and is shown schematically in Figure 17. In order to predict the behavior of these heated effluents, it is necessary to resort to an energy balance.

The energy- or heat-budget method was used by Schmidt (1915) to approximate ocean evaporation in 1915 and has since been applied to compute evaporation from water bodies of all sizes. The relatively recent development of more sophisticated instrumentation has allowed the energy budget to be utilized with a fair degree of reliance.

The energy budget was first tested against a water-budget control at Lake Hefner in 1950–51 (Anderson, 1954). It was concluded that satisfactory results were obtained for periods of ten days or more. A second check against a water budget was made at Lake Colorado City, Texas, in 1954–55, where inflow was extremely small and outflow was zero (Harbeck, Koberg, and Hughes, 1959). The components of the energy budget per unit surface area of a reservoir per unit time may be written as follows (World Meteorol. Org., 1966):

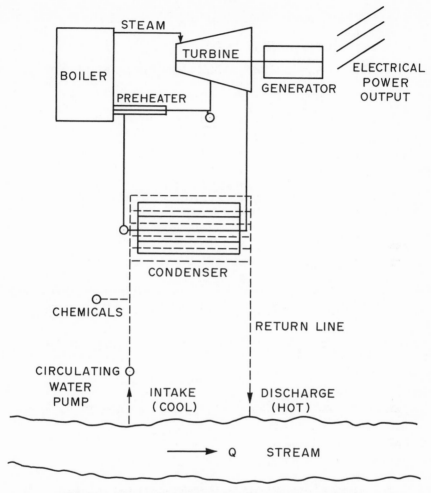

Fig. 17—.Schematic diagram of steam-electrical generation

$$Q_s - Q_r + Q_a - Q_{ar} - Q_{bs} + Q_v - Q_e - Q_h - Q_w = Q \text{ (Equation 1)}$$

where:

Q_s = *short-wave radiation incident to the water surface;*
Q_r = *reflected short-wave radiation;*
Q_a = *incoming long-wave radiation from the atmosphere;*
Q_{ar} = *reflected long-wave radiation;*
Q_{bs} = *long-wave radiation emitted by the body of water;*

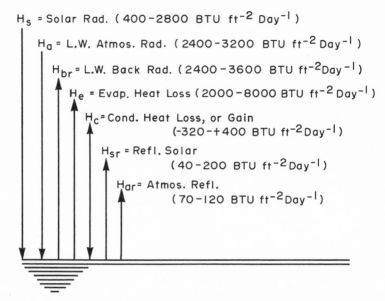

H_s = Solar Rad. (400 - 2800 BTU ft^{-2} Day^{-1})

H_a = L.W. Atmos. Rad. (2400 - 3200 BTU ft^{-2} Day^{-1})

H_{br} = L.W. Back Rad. (2400 - 3600 BTU ft^{-2} Day^{-1})

H_e = Evap. Heat Loss (2000 - 8000 BTU ft^{-2} Day^{-1})

H_c = Cond. Heat Loss, or Gain
(-320 - +400 BTU ft^{-2} Day^{-1})

H_{sr} = Refl. Solar
(40 - 200 BTU ft^{-2} Day^{-1})

H_{ar} = Atmos. Refl.
(70 - 120 BTU ft^{-2} Day^{-1})

NET RATE AT WHICH HEAT CROSSES WATER SURFACE

$$\Delta H = (H_s + H_a - H_{sr} - H_{ar}) - (H_{br} \pm H_c + H_e) \text{ BTU ft}^{-2} \text{Day}^{-1}$$

H_R

Absorbed Radiation
Independent of Temp.

Temp. Dependent Terms

$$H_{br} \sim (T_s + 460)^4$$

$$H_c \sim (T_s - T_a)$$

$$H_e \sim W (e_s - e_a)$$

Fig. 18—.Mechanisms of heat transfer across a water surface
(After Edinger and Geyer)

Q_v = *net energy brought into the body of water in inflow, including precipitation;*

Q_e = *energy utilized by evaporation;*

Q_h = *energy conducted from the body of water as sensible heat;*

Q_w = *energy carried away by the evaporated water;*

Q = *increase in energy stored in the body of water.*

Edinger and Geyer (1965) have depicted the heat-transfer terms across a water surface as shown in Figure 18, noting temperature-dependent terms and typical values obtained in English units. The addition of heated water

Heat Dissipation From Water Surface By Evaporation, Radiation, Conduction and Advection During January and June

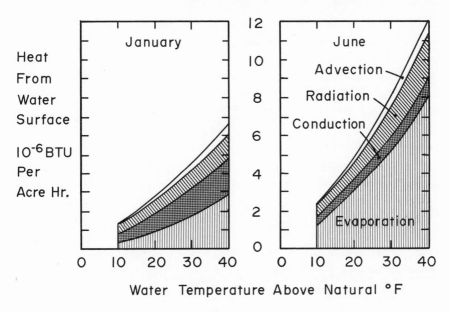

Fig. 19—.Various modes of heat dissipation from water surface
(After Kolflat)

discharges simply superimposes the heat addition upon the other dissipations and additions of energy.

Figure 19 shows the results of calculations by Bergstrom (1968) for a water surface in central Illinois. The data demonstrate the relationship of rate of heat dissipation to elevation of the water-surface temperature over natural temperature and the mechanisms by which this dissipation is achieved. It is significant to note that the rate of heat dissipation for a given rise in temperature is greater in summer than in winter and also that the heat dissipation by evaporation is much greater in summer than in winter. These calculations would appear to support the contention that heated effluents should be discharged in the most concentrated form possible (neglecting any biological or other side effects) in order to dissipate the heat most rapidly.

While the terms in the energy budget are discussed in detail by Anderson

(1954) and Edinger and Geyer, brief comments pertaining to their determination are in order.

Short-Wave Radiation, Q_s

Short-wave radiation originates directly from the sun, although the energy is depleted by absorption by ozone, scattering by dry air, absorption scattering by particulates, and absorption and scattering by water vapor. It varies with latitude, time of day, season, and cloud cover. Thus, while this quantity can be empirically calculated, it is much better to measure it using a Pyrheliometer, which will give the accuracy required for the energy budget.

Long-Wave Atmospheric Radiation, Q_a

Long-wave atmospheric radiation depends primarily on air temperature and humidity and increases as the air-moisture content increases. It may be a major input on warm cloudy days when direct solar radiation approaches zero. It is actually a function of many variables, including carbon dioxide and ozone, although it can be fairly accurately calculated by means of empirical formulation as demonstrated in the Hefner studies. It can be measured with the Gier-Dunkle Flat Plate Radiometer, although it is more convenient to calculate than measure.

Reflected Short-Wave and Long-Wave Radiation, Q_r and Q_{ar}

Solar reflectivity, (R_{sr}), is more variable than atmospheric reflectivity, (R_{ar}), inasmuch as the solar reflectivity is a function of sun altitude and cloud cover, while atmospheric reflectivity is relatively constant. The Lake Hefner studies demonstrated the atmospheric reflectivity to be approximately 0.03, while on an annual basis the solar reflectivity was 0.06. The Hefner studies used the equation:

$$R_{sr} = a\, S_a^b \qquad\qquad \text{(Equation 2)}$$

to determine solar reflectivity, where S_a is the sun altitude in degrees and a and b are constants depending on cloud cover. Note that:

$$R_{sr} = Q_r/Q_s$$
$$\text{and}$$
$$R_{ar} = Q_{ar}/Q_a.$$

Long-Wave or Back Radiation, Q_{bs}

Water sends energy back to the atmosphere in the form of long-wave radiation and radiates almost as a perfect black body. Thus, the Stefan-Boltzman fourth-power radiation law can be utilized, or:

$$Q_{bs} = 0.97\sigma \, (T_o + 273)^4 \qquad \text{(Equation 3)}$$

where:

Q_{bs} = *long-wave radiation in calories/cm²/day*

σ = *Stefan-Boltzman constant* = *1.171 x 10⁻⁷ calories/cm²/deg⁴/ day*

T_o = *Water-surface temperature in ° Centigrade*

All that is required to compute Q_{bs} is the water-surface temperature and a table giving the value of Q_{bs} for any temperature, T_o, is usually available.

Energy Utilized by Evaporation, Q_e

Each pound of water evaporated carries its latent heat of vaporization of 970 BTUs; thus Q_e is a significant term in the energy budget. The Lake Hefner study was explicitly promulgated for determining correct evaporation relationships and resulted in the following equation:

$$Q_e = 11.4W \, (e_s - e_a) = BTU/ft^2/day \qquad \text{(Equation 4)}$$

which is of the general type of evaporation formula:

$$E = (a + bW_x) \, (e_s - e_a) \qquad \text{(Equation 5)}$$

where:

a, b = *empirical coefficients*

W_x = *wind speed at some elevation, x*

e_a = *air-vapor pressure*

e_s = *saturation vapor-pressure of water determined from water-surface temperature*

E = *evaporation* = $Q_e/\rho L$

ρ = *Density of evaporated water*

L = *Latent heat of vaporization*

Many expressions have been developed for estimating the evaporation rate, the coefficients differing because of variation in the reference height for measurement of wind speed and vapor pressure, the time period over which measurements are averaged and local topography and conditions. As stated by Edinger and Geyer (1965), "It would also be expected that the coefficients would be much different for rivers and streams than for lakes and might well be dependent on water velocity and turbulence, particularly in the case of smaller rivers."

Energy conducted as sensible Heat, Q_h

Heat enters or leaves water by conduction if the air temperature is greater or less than water temperature. The rate of this conductive heat transfer is equal to the product of a heat-transfer coefficient and the temperature differential.

A single direct measurement of this quantity is not available and recourse to an indirect method is necessary. The method involves using average figures of air temperature, water-surface temperature, and humidity for the period in question and computing the ratio of Q_h to Q_e, which is known as the Bowen Ratio and expressed as:

$$R_B = \frac{Q_h}{Q_e} = \frac{0.61 \ P \ (T_o - T_a)}{1000 \ (e_o - e_a)} \qquad \text{(Equation 6)}$$

where:

P = *atmospheric pressure in millibars*
T_a = *temperature of air in ° C*
T_o = *temperature of water surface in ° C*
e_o = *saturation vapor pressure corresponding to temperature of water surface in millibars*
e_a = *vapor pressure of air at height at which T_a is measured in millibars*

Energy carried away by evaporated water, Q_w

Water being evaporated from the surface is at a higher temperature than the lake water and thus energy is being removed. While some believe this term is included in the conductive-energy term (World Meteorol. Org., 1964), it is relatively small and can be readily computed from:

$$Q_w = \rho_e \ c \ E \ (T_e - T_b) = \frac{cal}{cm^2\text{-}day} \qquad \text{(Equation 7)}$$

where:

ρ_e = *density of evaporated water, gm/cm^3*
c = *specific heat of water, cal/gm*
E = *Volume of evaporated water, $gm/cm^2/day$*
T_e = *temperature of evaporated water, ° C*
T_b = *base or reference temperature, ° C*

Advected Energy, Q_v

The net energy contained in water entering and leaving the lake may be computed from the following expression:

$$Q_v = c_{si} V_{si} \rho_{si} (T_{si} - T_b) + c_{gi} V_{gi} \rho_{gi} (T_{gi} - T_b)$$
$$- c_{so} V_{so} \rho_{so} (T_{so} - T_b) - c_{go} V_{go} \rho_{go} (T_{go} - T_b) \text{ (Equation 8)}$$
$$+ c_p V_p \rho_p (T_p - T_b) \div A$$

in which

Qv = advected energy in cal cm^{-2} day^{-1};

c = specific heat of water ($\cong$ 1 cal g^{-1} deg $^{-1}$);

V = volume of inflowing or outflowing water in cm^3 day^{-1};

ρ = density of water ($\cong$ 1 cal g^{-1} deg^{-1});

T = temperature of water in ° C;

A = average surface area of reservoir in cm^2.

The subscripts are as follows:

si = surface inflow;

gi = groundwater inflow;

so = surface outflow;

go = groundwater outflow;

p = precipitation;

b = base or reference temperature, usually taken as 0° C.

Since some of the terms in Equation 8 may not be measurable, a water budget is performed for the same period, evaporation is estimated, and the unknown terms are found by trial and error.

Increase in energy stored, Q

The change in storage in the energy-budget equation may be either positive or negative, and is found from properly averaged field measurements of temperature and the following equation:

$$Q = c\rho_1 V_1 (T_1 - T_o) - c\rho_2 V_2 (T_2 - T_o) \div At \qquad \text{(Equation 9)}$$

in which

Q = increase in energy stored in the body of water in cal cm^{-2} day^{-1};

c = specific heat of water ($\cong$ 1 cal g^{-1});

ρ_1 = density of water at T_1 ($\cong$ 1 g cm^{-3});

V_1 = volume of water in the lake at the beginning of the period in cm^3;

T_1 = average temperature of the body of water at the beginning of the period in °C;

ρ_2 = density of water at T_2 ($\cong$ 1 g cm^{-3});

V_2 = volume of water in the lake at the end of the period, in cm^3;

T_2 = average temperature of the body of water at the end of the period in °C;

T_o = *base temperature in* °*C;*

A = *average surface area in cm² during the period;*

t = *length of period in days.*

From this necessarily brief discussion of the various parameters comprising the energy-balance, it may be concluded that it is possible to predict heat dissipation using these concepts. Obviously, the reliability of the results will depend on the degree of sophistication used in the theoretical approach and the frequency and accuracy of the measurements taken.

ALTERNATIVE METHODS FOR COOLING WATER

If predictive techniques demonstrate that the heated waters must be cooled prior to their introduction into the receiving water, recourse to cooling devices must be made. A spectrum of sophistication is available for this process, the most sophisticated being the dry-cooling tower, which does not receive any cooling water, and the least sophisticated being a simple cooling pond, which may be constructed solely for the purpose of heat dissipation. These devices have been classified by McKelvey and Brooke (1959) as shown on Table 5. Relative ground areas required for the same heat load are included. Schematic diagrams of various types of cooling devices are shown in Figure 20.

TABLE 5. Devices for Cooling Water

Device	Relative Ground Area required
I. Ponds	
a) cooling	1000
b) spray	50
II. Atmospheric (natural draught)	
a) spray-filled	15
b) wood-filled	4
III. Chimney towers (natural draught)	
IV. Mechanical draught	
a) forced draught	
b) induced draught	
i) counter-flow	1.5
ii) cross-flow	1–2
V. Dry cooling	

When examining the various types of cooling devices available, it should be kept in mind that the rate of heat transfer is dependent on:

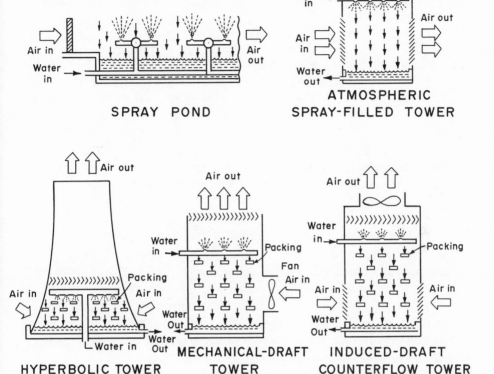

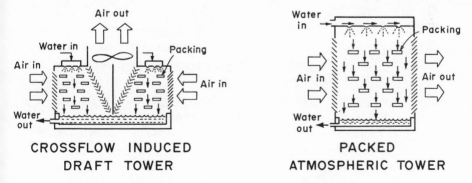

Fig. 20——.Various types of cooling devices

(1) area of water surface in contact with air;
(2) relative velocity of air and water during contact;
(3) time of contact between air and water;

(4) difference between wet-bulb temperature of air and inlet temperature of water.

Cooling Pond

The cooling pond is the simplest and most economical method of water-cooling (assuming land is inexpensive and available); however, it is also the most inefficient. It may be constructed simply by erecting an earth dike 6 to 8 feet high and may operate for extended periods with no makeup water.

Its main disadvantages are the low heat-transfer rate and the large areas required. For a still pond, the heat dissipated averages 3.5 BTU/hr/ft^2 surface/degree temperature difference between pond surface and air.

Spray Pond

Spray ponds may handle as much as 120,000 gpm of water and their low head requirements result in lower pumping costs than for cooling towers. Water is sprayed into the atmosphere some 6 to 8 feet above the pond and the water is cooled as it mixes the air and a portion evaporates.

Performance is limited, however, by the relatively short contact time of air and water-spray. Also, water loss is high and impurities may easily enter the system. Properly designed spray ponds may produce over-all cooling efficiencies up to 60 percent (Marks, 1963).

Atmospheric Towers

An atmospheric tower implies that air movement through the tower is only dependent on atmospheric conditions. A spray-filled tower depends solely on spray nozzles for increasing the air-water interface, while the packed tower sprays the water over filling or packing. The packed tower is no longer common, however.

Atmospheric spray towers are of the simplest design and may cool up to 1.5 gpm of water per square foot of active horizontal area with the wind blowing at 5 MPH (Marks, 1963). Their advantages include no mechanical parts, low maintenance costs, no subjectivity to recirculation of used air, and long, trouble-free life.

Disadvantages include high initial cost (approximately identical to a mechanical-draft tower), high pumping head, location in an unobstructed area, required great length because of rather narrow construction, high wind losses, and nozzle clogging. This design is well suited for small operations, however.

Hyperbolic Towers

The hyperbolic tower operates similar to a huge chimney: the heavier outside air enters at the tower base, displaces the lighter, saturated air in the tower, and forces it out the top. The initial cost is higher, but it is balanced against savings in power, longer life, and less maintenance. Their operation is counter-current, they can cope with large water loads, and they require a relatively small area.

These towers will probably become common in the United States as the cooling-tower requirements expand.

Mechanical-Draft Towers

As implied, the mechanical-draft tower utilizes fans to move the air through the tower. Thus, no dependence is placed on natural draft, or wind velocity. The arrangement of the fans dictates the method in which the air is moved through the system, each arrangement having certain advantages and disadvantages.

As elucidated by McKelvey and Brooke (1959), the advantages of a mechanical-draft tower are:

1. There is close control of cold-water temperature.
2. It can be maintained in a small ground area.
3. There is the generally low pumping head.
4. The location of the tower is not restricted.
5. There is more packing per unit-volume of tower.
6. A closer approach and longer cooling range are possible.
7. Capital cost is less than for a natural-draft chimney.

Likewise, the disadvantages are claimed to be:

1. A considerable expenditure of horsepower is required to operate the fans.
2. Tower is subject to mechanical failure.
3. Tower is subject to recirculation of the hot, humid exhaust-air vapors into the air intakes.
4. Maintenance costs are high.
5. Operating costs are high.
6. Performance will vary with wind intensity: unlike the atmospheric tower, the performance decreases with increase in wind strength, until a certain critical velocity is reached, after which the performance improves, due to a fall-off in recirculation.

7. Exhaust-heat loading and climatic conditions can be very prejudicial to the economic use of a mechanical-draft tower

The Dry-Cooling Tower

The dry-cooling tower is not an evaporative cooling device; instead, it cools fluids by forcing or inducing atmospheric air across a coiled cross-section. They eliminate water problems, such as availability, chemical treatment, water pollution, and spray nuisance; and there is no upper limit to which air can be heated.

However, the dry-cooling tower is much less economical than an evaporative cooling device, the specific heat of air is only one fourth that of water, and maintenance costs are high.

Thus, the cost of dry-cooling towers is presently thought to be prohibitive for most installations, even though many conservationists believe these are the only answer to the thermal-pollution problem.

Problems with Cooling Towers

In addition to the cost of installing and operating evaporative types of cooling towers, several other aspects should be considered.

The consumptive loss of water amounts to approximately 2 to 3 percent of the circulated water, including blow-down and drift loss. It should be noted that this is considerably less than the figure of 20 percent which has been stated in the literature as observed by Parker (1968).

Cooling towers may create fogging problems and may cause excessive ice formation; if located near highways, this can lead to hazardous traffic conditions. Also, towers located in the colder climates are subject to freezing. Another factor to consider is the large bearing requirement of the hyperbolic draft tower.

The Paradise Cooling Towers

An excellent example of the use of hyperbolic towers promulgated by water-quality requirements is at the Paradise Steam Plant in Kentucky. This is the first TVA plant that will provide year-round closed-circuit cooling towers. The plant capacity is 2558 megawatts and it is designed to operate with full or partial river cooling during periods of high flow and cool water.

Three cooling towers cool condensing water 27.5° F. to a temperature of 73° F. at the average air temperature of 57.3° F. dry-bulb and 52.2° F. wet-bulb. The towers are counter-flow with film-type cooling and each con-

Fig. 21——.Paradise Steam Plant and cooling towers.
(Photograph Courtesy TVA)

tains more than 500,000 cubic feet of asbestos cement sheet-type fill material.

Each tower has a cool-water basin of three million gallons and only makeup water is required during full operation. The towers are 320 feet in diameter and 437 feet high and are presently the largest in the world.

Without these impressive structures, operation of the Paradise Steam Plant would have been inimical to the biota in the adjoining Green River. Figure 21 is an aerial photograph showing the Paradise Steam Plant and the hyperbolic cooling towers.

Cooling-Tower Costs

If it is found necessary to install cooling towers to protect our aquatic life, some consideration of costs must be made. The fact that our waters must be protected against adverse degradation is irrefutable; however, treatment for treatment's sake is not justifiable. Thus, it remains for the biologist to determine what temperature levels will be inimical to aquatic life and for the engineer to examine these levels from the economic standpoint.

It is interesting to note that, in 1965, 116 out of 514 central power plants with a capacity of 100 megawatts or greater possessed cooling facilities. Furthermore, only 18 out of 347 possessed cooling facilities east of the Mississippi River (U.S., Interior, 1968).

Kolflatt (1968) states that "ball park" estimates of the additional investment for cooling towers are $7/kw for induced-draft wet type, $11/kw for natural-draft wet type, $25/kw for natural-draft dry type and $27/kw for induced-draft dry type. The Federal Water Pollution Control Administration (U.S., Interior, 1968) has estimated that the total investment costs for complete cooling of condenser waters in the United States to 1973 will cost $1800 million.

SUMMARY AND CONCLUSION

The factors involved in the heating of our waters have been elucidated, and resulting hydraulic phenomena have been discussed. Adverse effects on water quality caused by both natural and artificial addition of heat to receiving waters have been demonstrated.

The addition of heat by condenser-cooling water has been illustrated and a phenomenal growth in these discharges has been shown to be inevitable.

Various means of dissipating the excessive heat have been examined, and it may be concluded that, while it is possible to reduce the temperature of cooling water to its original temperature, it cannot be done without excessive additional costs.

Thus, it is the task of the biologist to recommend safe temperature levels and the task of the engineer to examine the basis for these levels for feasibility prior to their submission to the public.

REFERENCES

Anderson, E. R. 1954. "Water Loss Investigations: Lake Hefner Studies." Technical Report. U.S. Geol. Survey Prof. Paper 269.

Bergstrom, R. N. 1968. "Hydrothermal Effects of Power Stations." Paper read at ASCE Water Resources Conf., May 1968, at Chattanooga, Tennessee.

Bohnke, N. 1966. "New Method of Calculation for Ascertaining the Oxygen Conditions in Waterways and the Influence of the Forces of Natural Purification." In *Third International Conference on Water Pollution Research*. New York:Pergamon Press.

Bohnke, N. 1961. "Effect of Organic Wastewater and Cooling Water on Self-

Purification of Waters." Paper read at 22nd Purdue Ind. Waste Conf., 1961.

Bregman, J. I. 1968. "Thermal-Pollution Control—Need for Action." Paper read at Cooling Tower Institute Symposia, January, 1968, at New Orleans.

Cairns, John. 1955. "The Effects of Increased Temperature Upon Aquatic Organisms." Paper read at 10th Purdue Ind. Wastes Conf., 1955.

Cootner, P. H., and G. O. Löf. 1965. "Water Demand for Steam Electric Generation." *Resources for the Future, Inc.* Baltimore:Johns Hopkins Press.

Drummond, C. E., Jr. 1967. "Water Pollution Control for the Chattahoochee River." Paper read at Georgia Water Pollution Control Assoc. Conference, 1967, Atlanta, Georgia.

Durfee, C. D. 1968. "Paradise Cooling Towers and Condensing Water Facilities." Paper read at *American Soc. of Civ. Engr., Water Resources Conf.,* May 1968, Chattanooga, Tenn.

Edinger, J. E., and J. C. Geyer. 1965. "Heat Exchange in the Environment." Paper read at *Cooling Water Studies for the Edison Electric Institute,* June 1965, Johns Hopkins Univ.

Elder, R. A. 1964. "The Causes and Persistance of Density Currents." In *3rd. Annual San. and Water Resources Engr. Conf., at Nashville, Tennessee, 1964.*

Elder, R. A., and G. V. Dougherty. 1956. "Thermal Density Underflow Diversion Works for Kingston Steam Plant." In *Am. Soc. of Civ. Engr. Conf. at Knoxville, Tennessee, June 1956.*

Federal Power Commission. 1964. "National Power Survey." Washington: U.S. Gov't. Printing Office.

Gameson, A. L. H., J. W. Gibles, and M. J. Barrett. 1959. "A Preliminary Temperature Survey of Heated River." *Water and Water Engineering.* (Jan.)

Harbeck, G. E., G. E. Hoberg, and G. H. Hughes. 1959. "The Effect of the Addition of Heat from a Power Plant on the Thermal Structure and Evaporation of Lake Colorado City, Texas." *U.S. Geol. Surv.,* Prof. Paper 272–B.

Jones, S. Leary. 1964. Personal communication.

Kittrell, I. W. 1959. "Effects of Impoundments on Dissolved Oxygen Resources." *Sewage and Industrial Waste Journal.* (Sept).

Kolflat, T. 1968. "Thermal Discharges." *Ind. Water Engr.* (March).

Krenkel, P. A. 1967. "Report to World Health Organization on Poland 0026 Project." Copenhagen, Denmark.

Marks, R. H. 1963. "Cooling Towers." *Power* (March).

McKelvey, K. K., and M. Brooke. 1959. *The Industrial Cooling Tower.* Elsevier Press.

Parker, F. L. 1968. Statement in Hearings before the U.S. Senate Subcommittee on Air and Water Pollution, Thermal Pollution. Part 2. Washington: U.S. Gov't. Printing Office.

Schmidt, W. 1915. *Annalen der Hydrographic and Maritimen Meteorologic.* pp. 111–124 and 169–178.

Science. 1967. "Thermal Polluion: Senator Muskie Tells AEC to Cool it." 158 (Nov.).

Task Committee on Sedimentation, Amer. Soc. of Civ. Engr., Hydraulics Div. 1963. "Sedimentation Transportation Mechanics: Density Currents." *Progress Report, Proc. Amer. Soc. of Civ. Engr. Jour.* (Sept.).

U.S. Atomic Energy Commission. 1968. "Nuclear Reactors Built, Being Built or Planned." U.S.A.E.C., TID–8200, U.S. Atomic Energy Commission. Washington: U.S. Gov't. Printing Office.

U.S. Bureau of Standards. 1938. "Report on Investigation of Density Currents." Multilith Report. (May).

U.S. Department of the Interior. 1968. "The Cost of Clean Water." Detailed Analysis. Federal Water Pollution Control Administration, Washington, D.C. 2 (Jan.). Washington: U.S. Gov't Printing Office.

Water Resources Council. 1968. "The Nation's Water Resources." Washington: U.S. Gov't. Printing Office.

World Meteorological Organization. 1966. "Measurement and Estimation of Evaporation and Evapotranspiration." Tech. Note No. 83. Geneva.

DISCUSSION/ G. T. Orlob

DRS. KRENKEL and Parker, drawing from their considerable experience in water-quality management problems, have presented a most comprehensive review of the present "state of the art" in control of thermal pollution. They have highlighted the need to direct increased effort in research, investigation, and control in three major areas, each of importance to the aquatic biologist. These areas are:

Control of thermal waste sources

Dissipation of heat within the aquatic environment

Management of the environment itself

The latter two areas are intimately related, one concerned with the methodology of admitting a thermal load into a receiving water and the other concerned with its fate under the many influences, natural and created by man, which control heat flow and balance.

This discussion will be directed primarily to the problem of manipulation by man of thermal energy in an aquatic environment. Specific attention will be directed to one of the most complex water-storage and conveyance systems, the reservoir and the stream below.

STREAM TEMPERATURE CONTROL

The current trend in design of outlet works for dams and powerhouses is to provide flexibility to meet specific targets for downstream temperature control. This can most readily be accomplished by building into a proposed project fixed facilities which will permit releases to be made from any desired level in the impoundment. Outlets with intakes at multiple locations, skimmer walls and skirts over intakes, and intakes with elevation controls are examples of such facilities. Usually, these are provided at additional cost to the project. Costs of such facilities are generally appreciable relative to total project costs; hence, well-documented justification for such expenditures is often mandatory. Perhaps more important is the need to in-

53

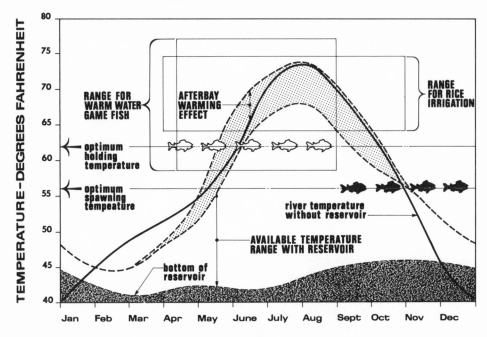

Fig. 1——.Temperature-control objectives for the Oroville Project
(After California Department of Water Resources)

sure that the investment, once made, is truly recovered in operation of the constructed facilities.

It is not a simple matter, in cases where multiple water uses must be served, to design and operate facilities to yield the maximum beneficial return. Often, multiple-use objectives, from the temperature standpoint, are incompatible with one another, and compromises must be made. Consider the case of the Oroville Project on the Feather River, for example.

Figure 1 illustrates the ranges of downstream temperature which might be achieved in operation of the Oroville outlet facilities and by re-regulation in the Thermolito afterbay. Also noted are optimum holding and spawning temperatures for migrating salmon and the desirable temperature ranges for warm-water fishes and for rice irrigation. Clearly, manipulation of release-water temperature, while meeting power and water supply demands of the California Water Project, is no slide-rule exercise. When one considers the vagaries of the impoundment behavior itself, compounded by climatological, meteorological, and operational variations, it is not surprising that the problem might be considered somewhat intractable. Let us

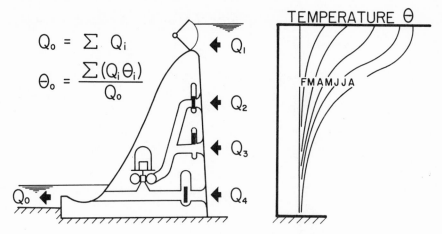

Fig. 2—.Typical multiple-outlet facility

examine briefly what the operational problem entails and how we might deal with certain of its complexities in a more optimistic vein.

Figure 2 shows a more or less typical dam section with some built-in flexibility for controlling the temperature of downstream releases. Water in the stream just below the dam will have a temperature determined by the proportion of flows delivered from each elevation. Certain restraints may be imposed on operational flexibility. In this particular case, these are determined by:

1. Excess flows over the spillway, subject only to nominal control within the flood pool.
2. Power demand flows through the penstocks, which may vary with desired output and water surface elevation (head).
3. Firm flow releases needed to meet downstream water uses.
4. Thermal structure of the reservoir, which may vary with the season and the operating schedule itself.

Let us examine in a little more detail the effects of operation on thermal structure and vice versa. First of all, it will be recognized that the supply of energy in the impoundment is uniquely determined by a balance, including the energy brought into the system by tributary flows, that supplied or extracted at the air-water interflow, and that withdrawn with outflows. The distribution of energy internally is governed by the density structure of the impoundment, its internal mixing processes, the penetration of short-wave solar radiation, and the location of withdrawals.

At the right of Figure 2, a typical history of thermal energy distribution

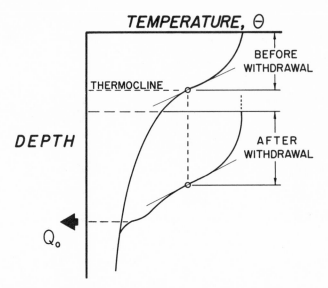

Fig. 3—.Effect of withdrawal on shape of temperature profile

is illustrated for a warming period from February to August. Assuming that no net change in volume occurs, the net energy flux to the reservoir in this period results in a transformation of an isothermal water mass into a density- (temperature) stratified system. Subsequently, the cooling of the reservoir will cause a collapse of this structure, beginning at the surface and progressing downward with isothermal convective mixing until the total mass is of uniform temperature from top to bottom. The cycle, which will be illustrated subsequently for Fontana Reservoir, is more or less typical of impoundments in northern temperate zones.

The effect of withdrawal from the reservoir on its thermal structure is illustrated in Figure 3. As the temperature profile is dropped, it becomes distorted by a change in the volumetric properties of the reservoir. The epilimnion expands in depth, and relatively little cold water is left in the impoundment. During this operation, warmer water is dropped to a position opposite the lower outlets of the structure depicted in Figure 2, and the flexibility to meet temperature-control targets is considerably reduced, if not entirely eliminated. The original profile could be restored by a corresponding influx of water with the identical thermal properties as that withdrawn. This assumes, of course, that this water could be admitted without influencing internal mixing of the water mass. Actually, all of these phe-

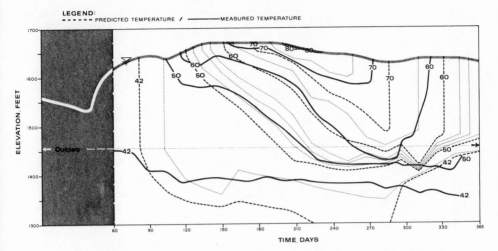

Fig. 4——.Simulation of thermal energy distribution, Fontana Reservoir, 1966

nomena, natural or operationally-induced, proceed simultaneously within the reservoir and give the thermal profile its characteristic shape. Predicting the shape of this profile, i.e., the distribution of thermal energy in time and space, is one of the more formidable problems facing us at this point in our experience with this class of pollution problems.

THE PREDICTION PROBLEM

One can gain considerable insight concerning some reservoir phenomena by observation of actual performance. However, unless we can predict future occurrences we cannot hope to cope with operational problems such as those exemplified by Oroville. What seems to be needed is a *model of the prototype,* reliable enough to permit evaluation of alternatives in operation or construction of physical facilities, *before the fact.* The model should be capable of representing the meteorological, hydrological, and climatological influences to which the reservoir may be subjected. It should also depict the internal behavior of the water mass to the extent that it may govern the distribution of thermal energy within the impoundment.

A mathematical model, recently developed for the California Department of Fish and Game (1967; Orlob and Selna, 1968), provides some of this needed capability, especially for impoundments which are likely to become well-stratified due to thermal effects. The model, which may be used to simulate reservoir behavior on a computer, was designed

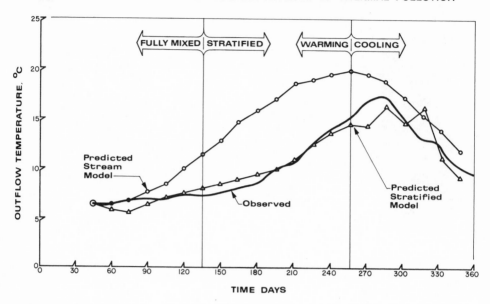

Fig. 5—.Comparison of observed and predicted outflow temperature, Fontana Reservoir, 1966

to accommodate heat transport by advection along horizontal or vertical axes and by "diffusional" mixing along the vertical axis. Any number of inflows or withdrawals may be accommodated. Convective mixing associated with either diurnal or seasonal cooling is also represented. Basic input information to the model, aside from the geometric properties of the prototype, consists of inflow rates and temperatures, reservoir release schedules (rates), and selected meteorologic and climatic parameters.

The model has been verified by simulation of Fontana Reservoir in the TVA system for a 315-day period in 1966. Figure 4 illustrates the results of the simulation run compared with the observed condition of the reservoir during the same period. Figure 5 presents a comparison of predicted and observed outflow temperatures. Over the 315-day period, the average absolute deviation of predicted values from observed values was about 1° Centigrade with about two-thirds of all values falling within the limits of ± 1.5° C. During the six-month period, 15 March through 15 September, these statistical parameters were 0.56° C. and ± 0.71° C., respectively.

OPERATION SIMULATION

The really challenging problem for the designer or operator of tempera-

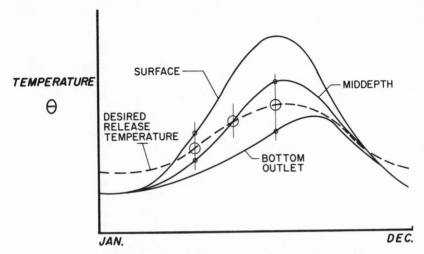

Fig. 6—.Annual variations in outlet temperature

ture-control facilities is to determine, *in advance,* the "near-optimum" configuration, or operating rule.

Given a particular outlet configuration and a fixed operating rule, it is possible, with a model such as that described above, to predict a pattern of thermal-energy distribution. It would appear, at first glance, that the achievement of a particular quality goal or "target" temperature, such as depicted by the dashed line in Figure 6, would simply be a matter of proportioning the release flows between the various outlet elevations. However, *the temperature which results in the outflow is itself dependent on* the flow withdrawn. This means that the process of achieving the best possible performance of the facility—the "near-optimal" performance—is akin to a trial-and-error technique. Actually, it is not quite that uncertain, since there are some reasonable bounds that can be placed on the problem to reduce uncertainty.

Figure 7 illustrates the general problem of predicting what the most likely performance will be over a short time span into the future. Given the temperature control target and the present state of the system at time, T, our problem is to estimate a schedule of operation so that we may follow the dotted "target" curve with the smallest possible deviation. Departures from the target are functions of the uncertainty of future climatic (and hydrologic) conditions *and* operating conditions. These factors have a variable influence on temperature predictions depending on the length of the projection interval ΔT and the "present thermal condition" of the reservoir.

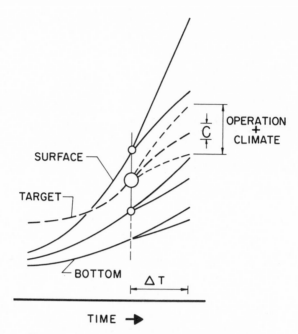

Fig. 7—.Uncertainty of temperature prediction

It will be noted that changes in temperatures for water deep in the reservoir are less likely than for water close to the surface. Also, the time of year when the projection is to be made has an effect on results. This can best be appreciated by considering the stability of the stratified water mass in summer as contrasted with the uncertain condition of the reservoir as it approaches isothermal conditions in late fall.

A practical working technique for temperature prediction might utilize the simulation capability of a model to remove a significant part of the projection uncertainty associated with operation. Using the model, the effects of operational control on the reservoir could be simulated. Based on the results of a series of carefully designed tests, a preferred alternative could be identified, one which would come closest to meeting the target. Of course, the uncertainties of future climatic conditions would remain, but since probabilities for such events may be estimated (the shorter the term, the better) the probable reliability of meeting the target could also be estimated.

In design, the simulation technique would allow one to examine critically the capability of a particular outlet configuration to meet release temperature targets. Alternatives could thus be compared for their relative effectiveness under identical environmental conditions.

In practice, for operational purposes, one might choose to perform simulations at regular intervals, say once or twice weekly during a critical period. Once "debugged" for a particular prototype, the model technique could be utilized repetitively at nominal costs, providing a quick and objective means of insuring that water temperature control objectives are met.

FUTURE DEVELOPMENT

Many promising possibilities exist for extending the capability of existing techniques to cope with complex problems of temperature control in the aquatic environment.

The simulation methodology described above may be particularly useful in dealing with thermal-pollution problems of estuarial areas where the loads of thermal power facilities may be concentrated in the future. The basic model structures for some types of estuarial systems have already been developed and should be modified to cope with thermal effects. A particularly formidable problem as yet unresolved for estuarial environments is the coupling of the hydrodynamic behavior of the system with the densimetric effects induced by thermal loads.

Characterizing thermal effects in coastal waters where the hydrodynamic behavior of the water mass is less predictable than in estuaries is probably the most formidable of all problems of this type. Yet, once simulation of weakly stratified systems, such as those of certain reservoirs and estuaries, is accomplished, the ocean problem will be susceptible of solution.

Given the capability to predict thermal effects in each of the several water environments—the stream, the reservoir, the estuary, and the ocean—the biologist should be bolstered in his confidence to evaluate the consequences of thermal pollution on systems ecology.

REFERENCES

California Department of Fish and Game. 1967. "Prediction of Thermal Energy Distribution in Streams and Reservoirs." Report to the California Department of Fish and Game. (June) WRE, Inc.

Orlob, G. T., and L. G. Selna. 1968. "Mathematical Simulation of Thermal Stratification in Deep Impoundments." Paper read at ASCE Specialty Conf., Portland, Oregon, January.

DISCUSSION FROM THE FLOOR

Krystyna Mrozinska: I am very interested in the concentration of salt and suspended solids in the sea water, because I carry on some investigation of saline waters in Poland. We have saline waters from coal mines that are not like any water as you have in Pittsburgh.

William A. Cawley: I take it that you're concerned with the differences in density that might result from differences in salinity?

Mrozinska: Yes, but—heated water using marine waters.

Gerald T. Orlob: This has not yet been accomplished. The studies of marine water heating, although there have been some specific investigations, have not been approached using this capability. That is, the mathematical approach, so far as I know, has not been fully developed for marine situations.

John Foerster: Your mathematical approach appears very reasonable if you have a reservoir. If you have a river, such as the Connecticut River, which has no reservoir to add water downstream, then what type of mathematical approach would you use?

Orlob: Depending on the particular circumstance, this problem may be either simpler or more complex than the stratified-reservoir problem. As Dr. Krenkel has pointed out, when you have a power wave and a large thermal emission, you can get stratification in a stream. This is a rather complex problem, not yet completely solved. Most of the streams with which we're concerned in the western United States, on the other hand, are more likely to be well mixed vertically, and there the problem is reduced to a rather simple one. A variety of mathematical solutions have been developed for this problem. In fact, the solution technique I have referred to here is applicable also to a case where there is full mixing. The curve I showed for Fontana, fully mixed, is the product of such a simulation. It's simply a matter of treating the segments separately under the various environmental conditions that might obtain as you move downstream. I think that problem is fairly well in hand, or at least can be approached without great difficulty.

Philip Douglas: We've heard and seen many figures on the future anticipated need for power. I would like to hear further refinement as to just how many kilowatts of power we will have to produce, and what our upper limits might be. If you can break it down, as the water people have done—originally, I think they used to say we required some fifty gallons of water per person per day on a kilowatt-per-person basis—it would help. What kind of power requirement is there for the American citizen on this sort of basis?

Peter A Krenkel: The latest projected figures (1968) are in the formal paper and taken from the United States Water Resources Council Report. Obviously, we are going to have a tremendous increase in power production in the ensuing years. Presently, we talk in terms of doubling the electrical capacity or generation every ten years. I think this is probably as good a rule as any. The rule-of-thumb figure used for water requirements in terms of power production is approximately one-half gallon per minute per kilowatt. Obviously, these figures depend on the process used and the corresponding efficiency.

Walter Glooschenko: I have a question on cooling towers. One of the arguments used by Florida Power and Light Company, now starting a reactor facility on Biscayne Bay, is that cooling towers can cause salt spray to go into areas and damage agricultural crops. Do you feel that this is a valid argument?

Krenkel: I believe that we must balance the benefits that we're going to obtain from cooling the water with any kind of cost that may occur. I think this is definitely a case where this ought to be considered. Obviously, the drift from these towers can be considerable, and I expect that since they are utilizing sea water, this could be a problem.

Again, it is a matter of economics. How much is it worth to us economically to return our rivers to the kind of condition that we all want? This is really the answer.

Orlob: I'd like to comment that this is one of those classic examples where a solution to water pollution is a problem for air pollution. There is certainly an interface that one has to consider very carefully in many instances where you're solving a water-quality problem. You end up with an air-quality problem.

Foerster: There were several rather good points that Dr. Krenkel made, some of these in the form of questions that need to be answered. One dealt with the desirability of discharging heated effluents either at the bottom or possibly at the surface, with the thought that a surface discharge might have some advantageous heat-dissipation characteristics and preserve a cooler area at the bottom. I think many of us, of course, will recognize that this is on a case-by-case basis.

The main point I'd like to make is that here, again, there could be a heavy interaction between the point and manner of discharge and what is happening to the river in terms of flow regulation upstream. Dr. Krenkel also showed a somewhat typical power-demand curve of what happens in flows of rivers from one part of the day to another. And a point here I'd like to bring out is that, if one did, in fact, couple a surface discharge with an area which was subject to tremendous flow differences —low flow at one time of day, high flow at another—so that the level of the water in the downstream area was fluctuating over a matter of several feet, the end result of a heated discharge on the surface might have much more far-reaching disadvantageous consequences than if you had a stabilized type of condition.

In other words, you would be subjecting a good deal of the littoral zone here, perhaps, to high temperatures mainly because of the daily fluctuation. The point I want to make is in terms, again, of the rather complex interaction and highly regulated streams of the kinds of biological problems which one can get into.

Krenkel: The situation that you have described actually occurred in the river that was depicted on the screen. The water level, however, did not change too much, I believe—probably only a foot or so, which would have an effect on the littoral zone. But the effect of the power wave really was not too pronounced in terms of subsequent mixing of the stratified layer.

The real problem here is whether or not we want to discharge this heat as a layer. Obviously, in the case of stratified flow, the heat is going to be dissipated at a more rapid rate as opposed to complete mixing of the waters. In this case, the unsteady-state flow condition did not appear to affect the stabilized layer; therefore, we really cannot answer the question posed.

The answer as to the desirability of stratified flow or complete mixing is one that should be obtained from the biologists in terms of the effect of the stratification or non-stratification on the biota.

W. C. Mason: Dr. Krenkel talked about industrial heat loads from the aluminum, beer, and cement industries. Will these and other industries be written up in the proceedings, and would you care to comment on the amount of this heat that actually goes out as waste heat load into the receiving waters?

Krenkel: The figures that I quoted were given as heat rejected per unit of production and are in the formal paper. One of the problems that we have with this particular aspect is that industry is not so free with this kind of information. Add the fact that many of these processes in industry intermingle cooling water usage with other processes within an industry flow-sheet itself, and we find it very difficult to put a finger on the absolute quantities utilized for cooling water.

"The Cost of Clean Water Guides," published by FWPCA, contains some estimates of the total discharges from industry and the cost of cooling these process waters.

Again, consider that these sources usually are not concentrated point sources as we are talking about with the power industry itself. When you add a lot of these together in a complex, for example, as in the Monongahela River, then we may have a problem. But individually, these are not as much of a problem as the high loads out of large power plants.

Carlos Fetterolf: The meteorological effects of some of the various methods of disposing of heat are of great concern to people in Michigan who already live in a snow-belt immediately east of Lake Michigan. We're planning two nuclear reactors, large ones—at least, large ones for us, on our shores—and we wondered where we're going to discuss these meteorological effects, either here or at Vanderbilt?

Krenkel: The Vanderbilt Conference, which is the follow-up to this one, will be oriented towards the engineering and economic aspects of thermal pollution and this will be discussed.

Fetterolf: As a biologist, I have not turned my attention to understanding the differences in the efficiency between a fossil-fuel plant and a nuclear plant. I think we're talking about something like a thirty-percent difference. Would you go into this a little?

Krenkel: The efficiency of a plant—we were talking about thermodynamic efficiency— is limited by the temperatures; that is, we want a high temperature in the boiler, and we want a low outlet temperature. Thermodynamically, this limitation is imposed by the Carnot cycle or the Rankine cycle.

For practical reasons—and we're talking in terms of the reason for the differences in efficiency in the coal-generated plants and the nuclear plants—the lower efficiency of nuclear plants is because of safety considerations. That is, we can have higher temperatures and pressures in the coal-generating plants because we're not so concerned with the implications of a nuclear accident.

The over-all limitations on efficiency are limited by thermodynamics, and we have reached almost as much efficiency as we can have, in terms of practicality, at the present time. But I believe, as mentioned previously, that as we begin to place a price on the discharge of heated water, this will perhaps give an added economic incentive to further increase the efficiency.

E. Jack Weathersbee: Dr. Krenkel raised the question, do we want high DO and high temperature, or low DO and low temperature? And I think the answer is that we don't want either of these. We actually want a combination of high dissolved oxygen and low temperature, and I'm wondering if one or both would like to comment on ways and means of achieving this?

Orlob: There have been some interesting approaches, particularly in connection with partial destratification of reservoirs, which seek to answer this problem. One scheme, for example, involves the aeration of the euphotic zone, including the epilimnion down to just below the thermocline, and recycling of that in order to avoid the disadvantages of bringing up the entire bottom of the lake.

I think that probably the answers lie not only in control of emissions, which would diminish organic loads to the system, but also in carefully planned operational control of a reservoir itself. If one can fix the targets for dissolved oxygen and temperature and other characteristics of the water downstream, and has control of releases from the impoundment, then there should be a reasonable likelihood of meeting objectives. This may, of course, involve compromises between temperature and dissolved oxygen targets, since these are not necessarily compatible, one with the other.

Another alternative involves supplying oxygen to the system in such a way that it would be most beneficial. Hence, the idea of local aeration to avoid completely disrupting the thermal structure and to provide local mixing is attractive. The epilimnion, generally speaking, is pretty well-mixed, anyway. The idea would be to bring oxygen-deficient water from below the thermocline up to the surface, aerate it, and allow it to circulate within the epilimnion and to preserve the thermal structure in the balance of the reservoir. This might, however, result in somewhat colder water that might be desirable for other purposes.

One of the bigger burdens on oxygen resources, it appears, is imposed just below the thermocline, and results, in part, from activity within the epilimnion. Eutrophication activity in the epilimnion produces organic matter which settles through that zone and into the region where bio-degradation is most active. This contributes to an oxygen deficit in that region which could probably be avoided by this local kind of circulation.

Krenkel: The Corps of Engineers this year is commencing a most significant study on artificial aeration of reservoirs. Some of the data that was used in the formal paper was from Allatoona Reservoir, one of the largest reservoirs in the East. The Corps of Engineers is installing six under-water diffusers in this reservoir at a depth of about 100 feet to add com-pressed air in an attempt to overturn the reservoir.

The real question to be answered here is whether or not we want to mix a reservoir. I definitely think that an economic study needs to be made in terms of the benefit of having the higher dissolved oxygen, lower manganese and iron content downstream, as opposed to having higher temperature releases and other problems associated with reservoir mix-ing.

Orlob: Just as a matter of information—perhaps for your reference—in February, in Los Angeles, the California Department of Water Resources held a symposium on destratification at which many of these topics were discussed. Proceedings edited by the DWR will be released soon, I under-stand. If you're interested in the discussions, which dealt rather thor-oughly with water-quality problems associated with alternatives for destratification, you might inquire of the Department.

James R. Adams: To return to thermal power plants instead of reservoirs, most of the studies that have been made at existing power plants on dis-solved oxygen levels have indicated that the dissolved oxygen levels in the outfall of the plant are higher than the levels in the intake system. This is caused by turbulence in the condenser cooling-water system.

There are a number of references on this. I think Dr. Alabaster from England can comment—or might want to comment on this, if he's here in the audience today.

John Alabaster: Have you any information on the aeration of the cooling water that takes place in cooling towers? This is being studied in England by the Central Electricity Generating Board, and is important where the water used for cooling purposes contains a high proportion of sewage effluent.

Krenkel: I have seen nothing in the American literature with respect to aeration per se inside the cooling tower as we have discussed here. There is much in the literature from England, as has been indicated, that shows an increase in oxygen in the water passing through these condensers. In the once-through cooling operation, the reason for this can be shown rather readily, which one would expect from the nature of the process itself.

Alabaster: I was thinking of closed (indirect) cooling systems used not alone but in conjunction with direct cooling, as is fairly common in the United Kingdom; the effluent discharged consists partly of water that has been aerated in a cooling tower.

Ruth Patrick: I would like to address myself to two statements. One, in regard to the use of sewage effluents as cooling water and its subsequent passage through cooling towers. These effluents, as you know, contain fairly high concentrations of minerals from the mineralized sewage. Inevitably, increased slime growths would develop on the cooling towers. Therefore, you would have to use more chlorination, pentachlorophenates, or other chemicals to control these slimes.

These substances are extremely toxic to aquatic life. Therefore, it seems to me that if you wanted to use sewage effluents, i.e., mineralized sewage effluents, for cooling water, then you should try to have a more or less closed system, i.e., make sure that the residuals from these kinds of treatments do not get into the open river.

The other statement that I wanted to make is that I think when we're considering a stratified lake or reservoir, or when we are considering a river, we should remember that in freshwater systems, the greater majority of life lives in the photosynthetic zone or a short distance below it. It is this zone that we are really concerned about keeping in a biotic condition. It should have optimum oxygen concentration and temperature ranges. We are not necessarily concerned with the deeper parts of a lake unless these are also well within the photosynthetic zone or a short distance below it.

I might also comment about the fact that an unoxidized microzone in the bottom of a lake allows the release into the water of chemicals that may be beneficial or may be toxic. If you try to oxygenate the whole reservoir so that the unoxidized microzone is destroyed, you may keep toxic chemicals from becoming available, but by the same factor, you may eliminate the availability of such things as minute amounts of molybdenum and other trace metals that are very important in the growth of aquatic life.

Orlob: I'd like to make one minor comment relative to the question of utilizing other waste effluents as coolants. As a first consideration, it would probably be most attractive to utilize such wastes as supplemental supplies where water is scarce. Of course, it is not just a question of utilizing waste effluents as makeup water. I don't think we can really solve this problem without the sliming problem at the same time. And as you point out, that could be of serious concern in certain instances. Perhaps the kinds of disinfectants used might deserve some careful consideration.

Robert D. Smith: One of the methods by which a power plant rids the intake and discharge systems of marine growth is either with a chlorination treatment or a heat process which kills the organisms. Could someone suggest yet another means of doing this, where either the chlorination method or the heat treatment is unacceptable?

Loren D. Jensen: I think that some workers have confused chlorine toxicity with thermal effects. My point would be that there are ways of controlling condenser fouling that don't encompass chlorine; one that is being used more commonly is the Amertop process, which consists of small, abrasive little rubber balls that are passed through the condensers with the cooling water.

We have had some experience with this technique and it's been very favorable in a freshwater station. I would agree with Dr. Patrick that, if you use sewage, I think you're just adding to your problem of condenser fouling and perhaps aggravating an already troublesome situation.

Foerster: I would like to pose a question to the people who run the power facilities. When we talk about dissolved oxygen contents in waters, just exactly what organisms will be using these "optimum" dissolved oxygen contents produced as a result of turbulent mixing in these condensers if the organisms die from the thermally induced effects?

Adams: The few studies that have been made on actually passing organisms through condenser cooling water systems (and there aren't very many) have indicated that organisms aren't killed by the condenser temperature change. Dr. Mihursky has been doing some more sophisticated work on this on the Patuxent in Maryland and I don't know how far his experiments have progressed. Perhaps somebody else can add to that.

Patrick: I'm familiar with the studies that Dr. Mihursky has been making. We have also been making some studies on the effect of passing organisms through condensers. I think whether organisms are damaged is often related to the size of the organism. I think that organisms such as small shrimps and some of the amphipods are hurt. On the other hand, some microscopic organisms such as rotifers and protozoa and organisms of this size-range, together with the algae, do not seem to be hurt to any great extent by mechanical damage in passing organisms through condensers.

Now, when you come to damage due to heat effects, this is a different thing, and I have no definitive answer on this. I think also, from my paper which will be given tomorrow, that you will see that the algae vary a great deal in their ability to withstand shock. I believe Dr. Cairns has been obtaining some information on this with protozoa.

Charles B. Wurtz: Since we brought up this question of invertebrate organisms going through condenser tubes, I would like to refer to the work that Markowski did in England in two or three plants there. Dr. Alabaster probably knows the plants themselves, which I don't recall.

Markowski had a list of 35 freshwater species and something like 62 marine species that went through a two-pass condenser with a fourteen-degree rise seven seconds up to a header box and about the same coming down, and had survival. There was some mechanical injury, as Dr. Patrick has mentioned, on shrimp. He had things like young lymnaea snails go through; many shrimp went through without any injury due to the effect alone. I have no idea what the chlorination pattern was in the setup.

Alabaster: Chlorination is generally carried out for short periods at intervals of 3 to 12 hours.

Foerster: In studies we're doing on the Connecticut River, sponsored by FWPCA, recent samplings done in April, at which time the atomic power generation facility was discharging water into the river at approximately 23 degrees F. above ambient river temperature, I always noted that there was a considerable change, not only in the flora, but in the rotifer organism population. It seemed to me more or less as being a more characteristic summer flora existing in this effluent. There was a definite change exhibited by the flora and the fauna in plankton samples at that time.

I can see the points of view on the condensers, but what I'm interested in is the actual happenings in the river or the estuary or the lake itself when this hot water is pumped out into it.

Robert L. Cory: I've been working on this toxic project that's been mentioned here, and I think the problem in this temperature business is a seasonal one. We found very little effect, and in my opinion, the organisms can be passed through the plant and can withstand high rises. The chlorination does cause death, but in 1964 we had a high temperature year—high weather temperatures and the heat temperatures out of the canal were over 100 degrees. I've been studying the epifauna, and we experienced a complete wipe-out of these animals in the effluent canal.

Also, the other studies on the microfauna indicated death. Mihursky now is doing laboratory studies on the organisms and it seems that, between 95 degrees and 100 degrees, everything starts to die. I think this is the problem here. These organisms knew their upper limit naturally; and then, when you get over that 95-degree limit, they all start to go, some of them faster than others.

Charles Hodde: In Dr. Krenkel's talk, he dismissed the use of cooling ponds on the basis that they take a lot of space, of which we have a lot out here. We also have the situation in the West, where ponds are created, where water does not return back to the original stream—at least, until it's been used for irrigation. We're contemplating spending quite a bit of money on the investigation of the use of hot-water effluents or warm-water effluents from thermal plants for irrigation.

In several questions—it would be helpful to us if we had your expression about the biological operation effect of one of these cooling ponds. That is, the size and temperature at which they may be practically operated, if you have that experience. Do you require a volume as well as a surface area of a certain character? I know lots of figuring is being done in regard to the economics of this choice in relation to a cooling tower, but is there any substantial research on the use of cooling ponds as just a part of the machinery of operating a power plant, and to what size do we have to go in order to be practical? What are the space requirements and volume requirements?

Cawley: In the FWPCA thermal-pollution program, a project is included on just this particular subject, the design, economics and optimization of cooling ponds. This project is just about under way and I estimate that we will have a report available sometime about the end of this calender year.

Krenkel: On the question about ecology of the ponds, I think this could be more properly answered by an ecologist. Obviously, the use of ponds depends primarily upon economics. If you have a lot of land, then you may very well justify the use of a pond. I did not intend to infer that ponds are not used, because they are. Rule-of-thumb design criteria are available in the literature.

Norris R. Fitch: I think the utilities generally think in terms of an acre per thousand kilowatts as a reasonable cooling-pond size. In other words, if you had a 600-megawatt unit, you'd need about a square mile of area. This applies to a fossil-fuel generating plant. A light-nuclear operating plant will require fifty percent more area.

REMARKS ON
THE EFFECTS OF
HEATED DISCHARGES ON MARINE ZOOPLANKTON

OF COURSE I think one of the main troubles with this whole game is that one has to be very clear as to what one is trying to do. This was brought home to me last week at Woods Hole when I was at a meeting of the International Association of Biological Oceanographers. This meeting was basically concerned with people interested in taking plankton samples.

The sum total of this meeting was that, for the first time in the ten years that I have been involved in this profession, I heard a group of people, most of whom were professionals in the field, honestly admit that they really didn't know what they were doing, why they were doing it, and how they were doing it. This started out to be one of the most refreshing four or five days that most of us had spent.

The trouble is, of course, I think, that, basically, one is being really expected to play God on a very limited budget. We don't really know to what extent the entire food chain and its indirections and feedbacks are affecting any particular part of it. And if a politician or a power company starts asking us questions about thermal pollution or any other sort of pollution, we have to, for political and common-sense reasons, look as if we knew what we were talking about and pontificate.

But really, the business is in such a state that we can never be certain. I think that if anybody were to tell a Congressman that he had to cure cancer or send a man to Mars on some small budget of a few million, this would be

This is an edited version of an impromptu talk made by Dr. Strickland, who was most accommodating in substituting for Dr. John D. Costlow, Jr., who was unable to attend the meeting due to illness.—P.A.K.

ridiculed. But basically speaking, this is precisely what we're being asked to do. And I say this in all humility, because some of the patrons of science of Washington are here, have been extremely generous to me and to many of my colleagues at other universities, and I think we have done the best within our capabilities to understand the principles involved in the case of the marine plankton and the food chain.

And of course, we clearly know more now than we did a decade ago. But I think the more we get to know about it, the more complicated and interreacting it becomes.

First of all, I don't know the first thing about the way in which thermal pollution of the marine environment will react in a physical sense, having never looked at the literature on this. I am in absolutely no position to summarize the work that has been done in an intelligent way.

But I do presumably know something about the way in which the non-polluted, thermally or otherwise, marine environment works, especially close in to the coast. I know perfectly well that I don't really know this, so it seems to me extremely difficult to know how anybody can really predict what can happen in a meaningful way. Before we can do almost anything, we must know a lot more about the sheer physics and hydrodynamics of the discharge process. It's becoming increasingly apparent to those of us who are interested in this near-shore area that this knowledge is very badly needed.

For some peculiar reason, universities, especially on the West Coast of America, seem to be desperately interested in the physical and geological problems around Tahiti, but very few of them have shown an enormous amount of interest in what happens half a mile off their own coasts. In fact, my own group at Scripps recently made probably the only fairly sophisticated investigation of the environment outside La Jolla for a few miles, and ended up with a large amount of data and many more problems than when we started.

But this is the area, of course, which is going to be important, and I believe that its physics is going to be what dominates the situation. One can relatively easily culture a few marine organisms, take them into the laboratory and warm the beaker up three degrees and see whether they jump out or look unhappy; however, this really isn't solving the problem of what's going to happen when you have thermal pollution.

Probably the first thing that's going to happen is that a ribbon of hot water will disappear a few miles offshore, zooplankton will look at it askance, depart the area, and nothing really bad will happen. On the other hand, it may mix in an estuary in which the temperature is raised considerably, resulting in some rather delicate feedbacks into an unbalanced equilibrium which would have quite a disastrous effect.

I don't think that we can basically do anything about this until we have had enough research to understand the environment before any pollution takes place. And here we're up against the whole problem, i.e., to what extent one wants to know the details. And this I find is the main problem. Until one knows the system well enough, one is never sure to what detail one needs to know it, resulting in a series of approximations. After five or ten years you suddenly find that you really looked at the wrong species.

Now, we have accumulated, over the last few decades, a vastly increased amount of information about basic physiological responses of some of the planktonic species, and a considerably larger amount of literature on some of the more important fishes and how they react. Insofar as the fish food is presumably at one stage or the other, the importance of knowing the status of plankton is very difficult to assess. In some fish stocks, any thermal effects may be almost directly on the fish. In some cases, one may have significant interreactions where the larval fish do not survive because of premature production or the delayed production of the right type of food.

I think all I can say here is that if I had to make some generalized statements as to what knowledge I think is the most important for us to determine, the following would result. Obviously, we do need to know more about the kinetics of the remineralizing processes which produce the mineral nutrients which supply the plankton.

In the in-shore areas in particular, where you may obtain stabilized systems, we are very dependent on recycling nitrogen much more than phosphorus, which is very rarely limiting. Thus, the whole concept of heat kinetics, of how quickly ammonia is liberated by grazing and bacterial action, is important.

On the physical oceanographic phenomenon concerned with the breaking of internal waves, we know pathetically little. The textbook example of winds puffing on the shore and material coming up is a gross oversimplification. This type of feeding into the environment of the critical nutrients, mainly nitrogen, is obviously a thing we need to know, along with the reaction of plot species. We are beginning to accumulate some evidence.

However, animals do not eat primary production or C^{14} or chlorophyl or anything else. They eat plants. And they are very concerned about what sorts of plants they eat. The whole question of the species succession of the plankton is the type of thing which may well be materially affected by pollution of various sorts.

The ignorance here is simply fantastic. After twelve years or so in this profession, I really don't know why, when you have fairly nutrient-rich hot running water, you produce diatoms. The more I think about diatoms, the more surprised I am that they really are existing. If I were a flagellate, I should have thought I could have beaten them to the draw on any occasion. But nevertheless, in temperate waters and also down a few degrees from the equator in Peru or in Panama, every time fairly nutrient nitrate-rich water comes to the surface, there the bulk of the biomass is generally tied up in relatively large diatoms.

I don't think anybody really has a clue as to why this occurs, and certainly all the talk about diatoms liking nitrate and diaflagellates not, is complete nonsense. This isn't the case. We don't really know even why they produce red tides. We've had ten, fifteen, twenty years of erudite papers talking about phosphate, B_{12}, and other factors. Yet, every time we find a paper explaining one red tide, we can go to another area where it couldn't possibly apply, and see another red tide forming.

So these types of things, specifically shown by things like diatom blue and red tides, really aren't understood. The patchiness situation in the sea is becoming progressively worse. We can see, off California, an almost ribbon-like situation as one goes from the coast with respect to flora and fauna. One perhaps might expect populations to be quite different half a mile offshore than two or three miles, but being different by quite striking degree is not immediately explicable. We know practically nothing about the population dynamics and the breeding of any of the zooplankton. And of course the fish larva can be considered to be zooplankton. This becomes a problem of survival of larvae.

In practice, we find patchy distributions of the little animals called copepods. Of course, the primary reason that we know nothing about them is the size factor. We are still basically running around with a bunch of butterfly nets trying to do a population study, which would be ludicrous if it were done with insects on land. However, large amounts of money are being put into doing this in the sea.

The first thing one has to know if one is going to study anything meaningful about thermal pollution—and this is presumably going to be in estuaries in the near-shore—is more about the physical oceanographic processes occurring very near the coast.

We need to know all aspects of the physiology of the phytoplankton and the zooplankton and this is, in terms of manpower and time, an enormous problem, which is only just beginning to be examined.

My final thought is that I don't think we're going to learn anything really worthwhile until someone faces up to the manpower situation. It is getting progressively easier and easier to mechanize and automate. I am now in the position in my group where I can automate most chemical and physical parameters in studies that are expected to demonstrate why the phytoplankton and zooplankton are present. In the same day, we have presented on the computers practically everything we want to know chemically and physically. About four weeks later, we finished counting the plankton.

The question posed is: is plankton-counting now going to be instrumented, so that we are going to produce enormous amounts of data with a mechanized inaccuracy, which nobody will recognize, and with nobody to look at it? This is a situation which cannot possibly be remedied shortly, and one that people are going to have to recognize. Federal pollution people and other agencies that are required to find practical solutions to these problems are going to have a very, very grave time. I think we do a disservice to the profession by giving either politicians or our bosses in various forms any cause for over-optimism. One has to combine this with an intelligent approach, so that one doesn't appear to be too completely pessimistic.

But on the other hand, sooner or later, these problems that we're going to hear about today and tomorrow and Wednesday are going to be upon us. Sooner or later, the statements that we now make about what will and what will not happen will be proven or disproven the hard way. And every time we are wrong, which we are going to be if we don't have enough basic data, it's going to be a very large nail in our coffins. So I suggest, at least collectively, that we try to have an insurance policy against this by very firmly telling everybody that, at the moment, although we are doing our best, the situation on the marine and probably all freshwater ecology is such that we are being asked to do almost the impossible; that we will do the best that we can do to give an intelligent answer, but that the basic problem is still the training of sufficient people of sufficient caliber. Also, enough money has still got to be put into basic research on the environment so that we can make intelligent predictions. Until that is done, no amount of money spent on ad hoc fire-fighting will really be effective.

DISCUSSION FROM THE FLOOR

Daniel F. Krawczyk: Today, with automation, we can analyze on sight. Perhaps we can develop preservatives to analyze back in the laboratory. What bothers me, when I look at a sample of seawater preserved with acid, mercury, and freezing, is that I get three different answers on the same sample. What are we going to do with the data that has been collected previously and is still being referenced?

J. B. Strickland: I think one would have to decide on what type of data. There are many types; some are biologically oriented and chemically oriented, oceanographic data. I would feel inclined to put nothing into a data center or data bank prior to about 1960. I don't mean that no one should look at data earlier than that, but if one is going to study data earlier than that, the variability in quality is such that the only safe thing for the investigator to do is to go to the original literature and have a good look at it.

The question of preservatives for plankton is a peculiar little backwater which everybody seems to overlook. I suppose it's because it's nobody's main business. The boss tells you to go out and take plankton but is not very clear how to do this or how to preserve it, once it is taken. We have a peculiar situation here with the marine phytoplankton in that we have no preservative, certainly not one single addition that can possibly be considered satisfactory.

If you're prepared to call every little round blob seen in the phytoplankton a monad and every little round blob with a waist in it a dinoflagellate, then you can get away with certain preservatives. If you are concerned with the volume of the plankton in order to assess the actual biomass, which I think is very important, then you must show that there has been no change of size of the organisms during preservation. There has been very little work done with this; however, most investigators indicate serious size changes.

In fact, if you use weak formalin, organisms that don't explode in front of your face often shrink down to half their initial volume. Clearly, we should do more research on this, although I doubt if there is such a thing as a single good preservative. The amount of money involved in taking two or three bottles of sample and preserving each with two or three different chemicals is trivial compared to the amount of money that's involved in paying people working with the microscope for days counting each sample.

David R. Schink: I'd like to suggest the possibility of radiation sterilization for oceanographic samples. Radiation sterilization has not yet been approved for foods, due to a quite proper concern for absolute safety in foods people eat. But certainly we could be bolder in applying radiation to preserving oceanographic samples than we can with food. Portable irradiators are not terribly expensive. They might be placed on shipboard for a few thousand dollars.

Strickland: Have you actually tried that? What happens to the internal enzyme systems of cells, apart from any living bacteria?

Schink: I'm not an authority on what goes on in these systems. There are some research programs going, particularly for seafoods, but I'm not familiar with the details. Enzymes seem to offer the most serious problem. Also, it has been argued that you will raise the temperature of the sample if you irradiate it. It seems to me this could be handled fairly easily by chilling as you irradiate. The specific details of what happens do require further study. Radiation preservation probably won't be perfect, but it may prove to be the best available technique.

Chapter 4   Joel W. Hedgpeth and Jefferson J. Gonor

ASPECTS OF THE POTENTIAL EFFECT OF THERMAL ALTERATION ON MARINE AND ESTUARINE BENTHOS

THERMAL alteration or "pollution" of aquatic environments is a comparatively new factor in man's heedless alteration of his environment. Yet it may, within a very few years, be one of the most significant, since our growing power technology will require more and more cooling water for power plants, and the not-so-boundless seas washing our coasts will be increasingly involved. We are thus confronted with new problems concerning the quality of our environment and the possible survival of potentially useful organisms in it. There is very little time left for biologists to play a role in the planning of control and management of this new effect of man's technology on the natural world. The engineering and construction technology develops at a rapid pace, yet the available relevant ecological information is, for the most part, an inadequate base for recommendations on protective measures or controls or for suggestions of means for utilizing the heat energy to be produced so abundantly.

In view of the recent detailed reviews by Naylor (1965) and Kinne (1963, 1964), we will make no effort to present a comprehensive summary of the literature of the effects of high or warm temperatures upon marine organisms in general. Instead, we shall present an eclectic review of a more limited literature and evaluate these examples and approaches as they relate to an understanding of the effect of both naturally and artificially induced environmental temperature regimes on the marine benthos. From this evaluation we can then recommend the types of information and studies needed to evaluate the effects of thermal effluents on the marine biota,

Original data discussed in this paper were obtained during research supported by the Office of Naval Research, Project No. NR 104–936.

and possibly to establish criteria for the control of such effluents. For the most part, our discussion will concern aspects of the thermal factor related to the marine and estuarine benthos, which properly includes marine bacteria, algae, bottom fishes, and demersal eggs of such fishes as the herring, as well as mobile, sessile, and infaunal bottom invertebrates.

As ecologists, we are in the broad sense studying the flow of energy through ecosystems, and it seems most unfortunate that our society must waste some of the energy resource available to it in the particularly useful form of heat in water in a manner which may cause further damage to other resources of value. Therefore we will also discuss some ideas concerning the possible economically beneficial use of heated marine water in the aquiculture of marine benthic animals.

NATURAL SEA TEMPERATURE CHANGES AND THEIR KNOWN EFFECT ON THE MARINE BENTHOS

It is generally conceded that "temperature is the most important single factor governing the occurrence and behavior of life" (Gunter, 1957), although Dunbar (1968, p. 1) suggests that consideration of temperature as the prime factor "has handicapped rather than helped, and that we have to examine the matter from new viewpoints and with less unthinking acceptance of the nineteenth-century assumption that what is true of nonliving things must also be true of living organisms." While Dunbar is concerned primarily with life in polar regions, his suggestion should be accepted as a salutary warning to all who are concerned with the study of the relations of organisms to temperature. It may therefore surprise some readers when we state that one of the difficulties in interpreting temperature effects is that we may not have enough information concerning the temperature regime in which many marine organisms actually live. In part, this difficulty stems from the very ease with which temperature data may be accumulated, so that adequate attention has not been paid to the manner of obtaining data, or to the possibility that comparatively small variations or changes of short duration might be significant. More accurate temperature-sensing instruments, with the capacity for obtaining readings within small spaces or distances, are opening up a new field of ecological and physiological investigation.

Before departing too deeply into a discussion of temperature and marine benthic organisms, we should caution the reader who is not a biologist that animals and plants do not compartmentalize the various physical and biotic factors in their environment the way we do in this review. Instead, they

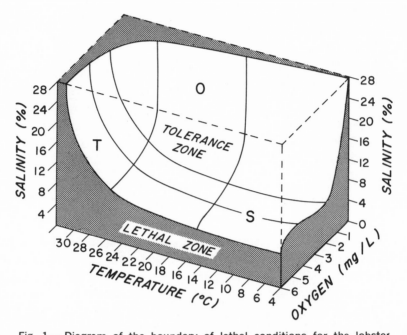

Fig. 1——.Diagram of the boundary of lethal conditions for the lobster
Homarus americanus for various combinations of temperature, salinity,
and oxygen.
(Modified from McLeese 1956)

respond to them in complex and integrated ways as a perusal of the review
by Kinne (1964) on combined effects of temperature and salinity will show.
As Figure 1, taken from McLeese's (1956) study of the interplay of temperature, salinity, and oxygen levels on survival in the North Atlantic lobster succinctly demonstrates, the nature of a response to a change in one
major environmental physical factor is strongly influenced by the state of
others at the time. We have separated out the biological effects of temperature for this discussion; the organisms in nature are not so accommodating to the biologist.

The first attempt to measure near-surface and shallow-layer temperatures in the sea appears to be that by Lorenz (1863) in the Adriatic near
Trieste. He measured temperatures at the surface by immersing only the
bulb of his thermometer, and at depths of 10, 20, and 30 fathoms, with
a specially designed apparatus in the Gulf of Quarnero during the years
1858–59. Lorenz did not believe that organisms living in the variable surface layer required such variation for their continued existence but that

they were there because they could tolerate the temperature ranges at the surface. He wrote:

. . . we are not justified in designating the species of shallow regions as species related to large temperature differences as such, since these differences are certainly not a part of the requirement for life, but indicate rather their degree of tolerance. The true relationship of species to temperature differences applies only to the extent that these temperature differences are not harmful; they are not a requirement for existence. Mobile animals are less influenced by this unfavorable factor than are the sessile seaweeds which cannot retreat to more favorable levels. This is strongly indicated by the occurrence of littoral seaweeds at their maximum abundance only in the months of January, February, and the beginning of March, when daily temperature differences are least (Lorenz, p. 11, translated).

One wonders whether Lorenz would have come to this conclusion had he been working on a coast with pronounced tidal differences, but the possibility that many shallow-water and intertidal animals may actually require the seasonal temperature differences they experience, rather than tolerate them, does not seem to have been seriously investigated since the time of Lorenz. Yet it should be obvious that, not only do organisms of the shallow seas and intertidal regions adjust to comparatively wide temperature fluctuations, they may also be, somewhat paradoxically, controlled in their geographic distribution by comparatively narrow ranges of sea temperature as expressed on the basis of means alone. There is a large literature on the relation of comparatively slight changes in sea temperatures to the abundance, success of year classes, and perhaps survival of commercially valuable fisheries stocks. Some examples of this literature are discussed in the following paragraphs.

The association of cooling or warming trends in the sea with fluctuations in populations of marine organisms has been made by a number of investigators. Blacker (1957), for example, has discussed changes in the populations of benthic organisms which indicate altered hydrographic conditions associated with the shift between Arctic and Atlantic water masses near the bottom in the Svalbard area. He was able to suggest possible indicator species useful for monitoring these changes. Southward (1960) notes that there has been a warming trend in the surface waters of the English Channel in the past 50 years. The mean annual surface temperature in the western Channel has risen about 0.5° C. and, at one place, the mean bottom temperature has risen about 0.25° C. Years of higher bottom temperatures may be associated with poor recruitment of herring stocks, possibly because schooling and spawning are inhibited by the warmer water, and the unfavorable conditions for herring may be accompanied by ecological replacement of the herring by pilchard (Southward, 1963).

The warming trend in the New England region has been noted since 1900, but it is not a steady increase and there have been cool intervals. According to Taylor *et al.* (1957), there seems to be no obvious general alteration in the composition of the fish or invertebrate fauna of the Gulf of Maine region although some northward extensions of distribution and abundance have been noted; perhaps this apparent lack of major change is associated with the oscillations of temperature or the persistence of the amplitude of annual ranges. A few of these southern species established resident populations north of their previously known limits, and Dow (1964) finds relationships between the temperature changes in the Boothbay Harbor area and the relative abundance of shrimp, lobsters, and scallop. The catch of *Pandalus borealis* fell to nothing two years after the all-time high temperatures of 1953, but lobsters increased. Dow concludes that the temperature changes operate in somewhat different ways, that spawning of shrimp is depressed by high or widely fluctuating temperatures and good catches are associated with minimum ranges or consistently low temperatures during the spawning period, whereas the increased lobster catch is associated with warm temperatures in winter and spring during the growing period.

In a recent analysis, Dow (1969) was able to correlate fluctuations in abundance of *Homarus americanus* in Maine with temperature changes since 1905, and to predict that optimum conditions should reach the southern limit of the lobster's range in the vicinity of Cape Hatteras by the mid-1970s, if the trend of falling temperatures continues in the northeastern area.

Comparable fluctuations in the northeastern Pacific have been discussed by Radovich (1961), who lists northward extensions of ranges of various fishes associated with the warming of north Pacific waters during 1957 through 1959; this warming is in the order of an increase of 2° C. in the annual means from 1956 to 1959. Ketchen (1956) has discussed the relation between temperature fluctuations and fish stocks in British Columbia waters, but emphasizes that he does not wish to imply that there is necessarily a causal relationship involved. Nevertheless, he concluded that "The relation of trends in the success of some marine fisheries to climatic trends is too suggestive to be ignored, and points to the need for more trenchant investigations of variations in the marine environment and its effects on abundance" (p. 370). The significance of the possible relationship between temperature and fishery production has been questioned by Bell and Pruter (1958) for the British Columbia area, especially with regard to heavily fished stocks.

While there may well be valid grounds for disagreement concerning the exact relationships of fluctuations in seawater temperatures and the success of certain fishery stocks, there is certainly evidence that comparatively small perturbations of the temperature regime do influence the presence or absence of many marine animals and that these little-understood, long-term natural fluctuations must be taken into consideration in any study of the possible effect of thermal alteration by industrial effluent.

Because of the inevitable shore location of outlets, heated effluent discharges into the sea have their most direct affect on the biota of inshore shallow water and intertidal areas. Fish may move in and out of such areas, but the less motile benthic fauna cannot do so. At the same time, these discharge sites are often in areas of scientific, educational, recreational, and esthetic value. In addition, the fauna affected includes many species of economic value, such as crabs, shrimp, clams, and oysters. In order to predict realistically the effects of thermal discharges in such areas, it is necessary to have certain information about the natural environmental temperature regime and the thermal responses of the organisms involved. Unfortunately, while detailed long-range engineering plans and predictions are being worked out for large oceanside power plants in response to our ever growing demands for power, nearshore and intertidal ecological studies are not being supported and carried out in a manner to provide an adequate scientific basis for the prediction and control of thermal effects on the natural system.

Our own studies of the ecology of nearshore and intertidal benthic populations have demonstrated that, despite the ease of acquisition of such information, inshore temperature data are not available in sufficient detail for predicting natural biological events and certainly they are not adequate for predicting the possible biological effects of the flow of thermal effluents into this system.

In the following pages we will describe the coastal area most familiar to us as an example of the current state of knowledge of the temperature regime of coastal waters and evaluate our information on environmental temperatures actually experienced by the coastal benthic biota in an essentially natural site. This will serve as a basis for recommendations for the type of investigation necessary to evaluate thermal disturbances in near-shore environments.

The California Current is the major oceanographic feature dominating the marine climate off Oregon and California. This is a great offshore oceanic stream about 350 miles wide, originating in the North Pacific Ocean as an eastward flow, influenced by the strong westerly wind of that latitude. It turns south along the coast, moving slowly at about half a knot, as a flow of surface water about 200 meters deep which is cooler and less saline than the ocean surface waters west of it. The current flows over colder and more saline deep oceanic water, which, off Oregon, is located rather close inshore because the continental shelf here is only 9 to 30 nautical miles wide (Byrne, 1963a, 1963b). The cool temperature of the Current is the basis for the generally cool Pacific coastal climate of both the land and the inshore surface waters derived in part from the Current.

Between the California Current and the coast is a narrow region in which occur complex water movements showing great seasonal variation, compared to the steady southward flow of the Current. The environmental water temperature of the coastal intertidal animals is dependent upon the complex and variable temperature characteristics of this inshore region.

During the summer months, even close inshore, the currents move southward along with the California Current movement offshore. The coastal temperatures of this period of the year are not those of the California Current, however. During the spring and summer, strong northwest winds from the prevailing North Pacific high-pressure system blow parallel to the coast, in the same southerly direction as the flow of the Current. This wind and the rotation of the earth cause surface water of the California Current to be transported west, offshore, and away from the coast. This water is replaced from below, close inshore, by upwelling of the deeper, colder water. The occurrence of upwelling and its low summer temperature is a major influence on the annual environmental temperatures experienced by the coastal organisms. Upwelling is most intense during the spring and summer (May to September, off Oregon), lengthening the cold-water period of the year and reducing the range of seasonal coastal water temperatures well below that of the offshore water which would otherwise approach the coast.

The largest Pacific coast upwelling area extends from just north of San Francisco Bay to central Oregon, centering on the Cape Mendocino and Cape Blanco areas, for upwelling is greatest on the southern sides of large projecting points. Here, water temperature throughout the year is lower than that of more southerly areas which are still within the California Current influence but which have little or no upwelling. Even at times of the year when the northwest wind is weak, it may still cause some upwelling, adding to the stability of the low inshore temperature.

During the fall, the northwest winds along the coast diminish and, after a transition period, are replaced by variable winds mainly from the south. A seasonally occurring current flowing from the south, the Davidson Current, appears in the inshore region between the coast and the California Current and can be detected off Oregon from October through March (Burt and Wyatt, 1964). The variable surface flow of this counter-current is increased by the southerly winter gales. The counter-current brings from the south more saline and warmer water close inshore, where it is intensively mixed with cooler water by the eddies and currents of the near-shore region. Since the Davidson Current occurs in the period just following the summer, when atmospheric warming of the surface waters is greatest, it has the effect of prolonging the warm period of the year. At shore stations, the highest surface temperatures of the year may occur during the early fall. An abrupt drop is characteristic of surface temperatures in the early winter along the Pacific Coast. Bolin and Abbott (1962) attribute this to the sinking of heavy Davidson Current water and its replacement by an onshore flow of cold offshore surface water. Insolation is also sharply reduced by cloud cover in the fall and the abrupt drop in Oregon waters may also be related to seasonal cooling at a rate faster than the Davidson Current water can bring heat into the area.

Cooling and mixing continue in the winter, and surface temperatures reach their annual minimum during this period. Organisms in the intertidal zone experience both air and sea low temperatures of a narrow range at this time of year.

Temperature and salinity observations at many shore stations on the Oregon coast have been made at weekly intervals by the Department of Oceanography of Oregon State University for several years. Average monthly maximum and minimum and monthly mean temperatures derived from these measurements for several stations are given in Figure 2.

The annual mean surface temperature at these beach stations is between 11° C. and 12.5° C., with a range of from 5 to 7 degrees. At all stations, the yearly low of 8° C. is reached in December and January, which is also a period of narrow temperature range. Increasing temperature of greater range in the early spring is followed by a drop and a narrowing of range in June, as upwelling commences. Differences in the summer tem-

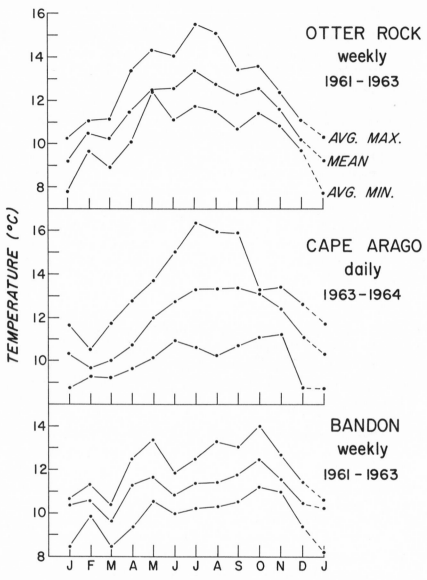

Fig. 2—.Monthly mean temperature curves for various Oregon shore localities.
(From O.S.U. Data)

peratures of these stations is largely due to local variations in the intensity of this upwelling. Increased insolation begins to warm the upwelled water faster than it is displaced, and a slow increase in summer temperature and a more rapid increase in its range is evident. The annual maximum of about

16° C. occurs in August and September at most stations, while the minimum is about 10° C. An unexplained drop in September is followed by a slightly warmer October with a mean of about 12.5° C., and a narrow range, marking the onset of the Davidson Current period. This is followed by a steep decline to the annual low range of late winter.

It is evident that the Cape Arago curve derived from daily measurements would be influenced by the thermal character of the tidal prism of Coos Bay as tidal exchange brings this water past the Cape. A heat budget representing the situation implied by this curve would differ from one representing the curves derived from weekly measurements more than we would have expected from the hydrographic differences.

In order to evaluate the effect of sampling frequency on shore-station sea-temperature curves, we have made daily afternoon measurements at a sample shore station for the last year, supplementing the single measurement this summer with additonal measurements throughout the day. As examination of the general nature of the local hydrographic conditions should indicate, we have observed very variable and quite unexpected temperatures not predictable from previously available data. In the month of June 1968, for example, we have recorded temperatures from 7.8° C. to 14.6° C., spanning the entire predicted annual range.

The type of shore-station annual surface-water temperature curves given in Figure 2, based on monthly or, at best, weekly measurements, are often used in attempts to correlate environmental temperature conditions with biological activities of the benthic biota, such as reproduction, growth, and distribution. Frequently, data for a site removed from the actual biological study is used. We feel that such an approach is no longer an adequate one and recommend that it be abandoned and that conclusions based on this type of meager data be rejected, a priori. We strongly urge that there is, instead, an urgent need for long-range, detailed recording of shore-station temperatures, both for use in fundamental ecological research and for planning future power-station outfalls. We also recommend the abandonment of naive biological comparisons to temperature curves and urge, instead, the examination of possible biological correlations to daily, seasonal, or annual environmental heat budgets as a minimal approach to the problem of temperature relations of the marine benthic biota. Prediction of the effect of thermal effluents on inshore-sea heat content from the typical scanty, short-term study is scientifically irresponsible. An implementation of our recommendations requires detailed temperature studies of the actual habitat of the animals and of the animals themselves.

The temperature conditions experienced by the intertidal biota are almost equally the combination of air- and ocean-temperature conditions. At locations of intense upwelling, such as Cape Blanco in Oregon, the air temperature may be a more important influence on gonad development and spawning than the more constant cold sea temperatures. There is sufficient information available to indicate that the influence of air temperature on basic biological processes in intertidal species is of major importance to these animals. Spawning in *Patella depressa,* an intertidal British limpet, appears to be related to air temperature rather than sea temperature (Orton and Southward, 1961; Southward, 1958). Spawning of *Arenicola marina* was found (Howie, 1959) to coincide with the first sharp fall in the air temperature to the autumnal minimum for populations living high on the beach, while a different population at a lower tidal horizon, apparently protected from low winter air temperatures by infrequent exposure, was able to mature gametes during the winter period and also spawn in the spring. Barnes (1958, 1959) considered that one of the most important factors determining the southern limits of the circumboreal barnacle *Balanus balanoides* is a temperature block to the final ripening of gametes. The final stages of gametogenesis are inhibited at temperatures above 10° C. in this species, and at the southern European limits of its distribution, air temperature is probably of more critical importance than sea temperature; air temperatures below 10° C. permit gonad maturation even when sea temperatures are above this level. The zoogeographical distribution of other barnacles may be similarly limited, affecting their competition when they co-occur (Southward and Crisp, 1954).

In an examination of the relation between environmental temperatures and growth in marine animals, Taylor (1959, 1960) used air-temperature data for his studies of two Pacific coast intertidal clams, because of the lack of sea-temperature information for the collection localities. He showed that the growth parameters of the cockle, *Clinocardium nuttalli,* were quantitatively associated to mean air temperatures from Alaska to Tillamook Bay, Oregon. Similar statistical correlation was obtained for the growth of the razor clam, *Siliqua patula.* It is likely that these correlations go beyond the general relation of sea and coastal air temperatures.

The abundant data on air temperature available from weather agencies cannot be used to construct an accurate composite annual temperature regime for intertidal species, because this data is not taken under conditions which sufficiently approximate the microclimate of the substrate surface on the exposed shore where the organisms actually exist. Coastal weather stations are usually sufficiently near the shore for the seasonal air temperatures to follow those of the adjacent open ocean, but those of the intertidal exposed at low tide do not inversely correspond to those measured for the air at the weather stations (Lewis, 1964, p. 32).

The only published information available on temperatures continuously recorded in a situation approximating the intertidal concerns a marine marsh locality in southern California. Air and water temperatures were continuously recorded by Phleger and Bradshaw (1966) for one year. The temperatures measured were not those of the substrate itself as it was covered and exposed by the tides, but rather of the air and, separately, of water in a permanent tidal stream. However, these records revealed a remarkably broad range of annual water temperature (5° C. to 33° C.) and an 8° C. range within a typical 24-hour period. Air temperature greatly influenced water temperature at low tide. A marine marsh is highly atypical of the entire intertidal zone, but these results clearly indicate that when the environmental variables of this zone are actually measured over a long period, much greater variations occur than is usually assumed. Even when only monthly maximum and minimum temperatures are measured *in situ,* as Johnson (1965) did on a sand flat in central California, disparity appears between the actual intertidal substrate temperatures and those of the surface waters of the area, the degree of which is proportional to the depth below the sand surface at which temperature is measured.

Records of temperature maxima and minima without time-course information are not satisfactory for comparison with reproductive cycles and may be misleading (Beckman and Menzies, 1960). A brief exposure to a sublethal temperature extreme is probably of less effect on reproduction of intertidal species than the cumulative effect of longer periods of exposure to temperatures occurring more frequently during the same period, but such relations have not been examined.

We have been unable to discover in the literature any attempt consistently to record temperature conditions in the rocky intertidal *in situ,* on the substrate among the organisms at both high and low tide. Lewis (1964, p. 38) also found such information lacking. Such scattered direct measurements that have been recorded are usually of single measurements and often of tide-pool water (e. g., Ambler and Chapman, 1950). The most detailed continuous record available for a rocky shore is a series of hourly measurements for an eight-hour period. On a summer day at Monterey, California, Glynn (1965) found that weather-station air temperature, the temperature on an exposed rock surface, and that in a clump of algae

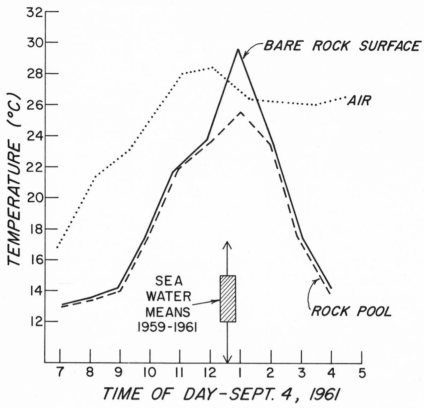

Fig. 3—.The temperature range in the Endocladia habitat at Pacific Grove on September 4, 1961, compared with the seawater means and extremes for Monterey.

(Modified from Glynn 1965, Figs. 17 and 21)

(*Endocladia*) rose and fell in a similar but not identical fashion. However, the range between the highest and lowest of the three during these eight hours was 8° C., which is three times greater than the range between surface water maximum and minimum for the entire year at that locality. (Figure 3, compiled from Glynn). Changes in the intertidal substrate temperatures occurred at a later time in the day than the air-temperature changes. Further, when these and other temperatures measured directly in the intertidal are compared to air temperatures of the nearest weather station, the intertidal temperatures often show greater extremes. On rocky surfaces, the initial temperature of the wet substrate surface is that of the sea, and lower than that of the air, in summer. After tidal exposure, the surface temperature may rise, along with air temperature, as the day progresses, but with a rate which is highly dependent upon such habitat variables as rock type and algal cover, which control wetness and evaporation. On exposed rocks where many species live, insolation in summer at low tide produces surface temperature conditions significantly above air-temperature levels. We have observed, for example, a July rock-surface temperature, adjacent to barnacles or limpets, of 33° C., while the air 3 feet above the surface measured 24° C. and, at a weather station a few miles away, 20° C. (Gonor, 1968).

The difficulty of using either sea-surface or weather-station air temperatures in studies of intertidal species is further complicated by the difference in duration of tidal exposure experienced by organisms at different tidal horizons. The important discontinuities in the degree of exposure experienced at different vertical levels is reviewed in detail by both Doty (1957) and Glynn (1965). The existence of critical levels at which abrupt changes in duration of exposure occur is thought to be the major environmental factor controlling the observed zonation of intertidal organisms whose vertical zones of occurrence correspond to the critical levels. Clearly, differences in relative durations of exposure to sea and air produce very different temperature regimes. The annual temperature curve for the level at which an intertidal organism occurs is the only meaningful one to use in correlations with reproductive cycles and other biological activities; such information is wholly lacking for rocky shores anywhere, at this time. Basic intertidal ecological studies require this information now; the use of intertidal animals as indicators of thermal pollution will add a great practical and economic need for it also.

After some preliminary observations on sea, substrate surface, and near-surface air temperatures at low tide, we have concluded that it is necessary to record these temperatures continuously, preferably as black-bulb readings. They will only serve, however, as an index of the nature of the intertidal temperature regime. They alone are not adequate to determine the true temperature relations of intertidal species at low tide. Only measurements of the rate of rise or fall and the levels reached by the internal temperatures of animals will suffice, as the first serious study of this problem (Southward, 1958) demonstrated.

In the last year, we have completed some 5,000 internal temperature measurements on intertidal invertebrates. From this, we have concluded that most ideas and statements concerning the temperatures which intertidal animals experience are in error and that a great deal of the experimental work on temperature tolerance and other heat effects on these animals is of very limited value, because the methods used in these experiments bear so little relation to the actual ecological situation.[1]

The details of our information will be published separately, but we give some examples. In the mussel *Mytilus californianus,* exposed for approximately 6 hours in early summer, the mean internal temperature in the population is commonly found to be between 19° C. and 24° C. and often between 27° C. and 31° C. at a time when the sea-surface temperature ranged between 10° C. and 12° C. at the study localities. Individual internal temperatures as high as 35° C. have been recorded without signs of lethal effects in the population. Populations of the goose-neck barnacle *Mitella (Pollicipes) polymeris* also reached the same range of mean internal temperatures, while the black turban snail, *Tegula funebralis,* has been found to have mean internal temperatures between 22° C. and 29° C. after about five hours of tidal exposure on a sunny day. Other common species of barnacles, chitons, and anemones show similar internal temperatures, well above air temperature and, even on overcast days or in shadowed places, significantly above the sea temperatures they experience on the same days.

Even the sea urchin, *Strongylocentrotus purpuratus,* which usually inhabits hemispherical burrows in rock and is thus kept cool on the bottom side, commonly shows a population mean (n = 30) internal temperature from 15° C. to 23° C. at low tide during the day in early summer in Oregon. On several successive days in May 1968, mean internal temperatures of 26° C. were recorded in beds of urchins exposed to the sun and maximums of 27° C. to 30° C. were measured in some urchins. Successive days of this type of heating at low tide for periods of three to five hours led to a heat kill at the study site and other areas along the central Oregon coast, with many urchins dying at each place.

1. The idea that thermostability of certain tissues may be an index of the environmental temperature conditions critical enough to separate closely-related species has been proposed by Ushakov (1959) and Zhirmunsky (1967). This property is determined by the temperature required to induce narcosis in epithelial tissue at stated times, or the temperature at which muscle tissue ceases to twitch under electrical stimulus. It is, essentially, another measure of the thermal death point, applied to isolated tissue. The temperature for thermostability is higher than the thermal death point for the intact organisms. (But the time may be much shorter.) The thermostability values obtained by these authors are higher for tropical animals and for intertidal ones than for temperate or deeper-water species, as might be expected. The authors suggest that thermostability values may not only be used for diagnosing species (Ushakov), but for interpreting past climatic regimes (Zhirmunsky). Be that as it may, the utility of the thermostability value for ecological analyses remains to be demonstrated, and the assumption that a physiological property is genetically stable for geological time is debatable.

Several important things can be learned from this example. Heat kills of marine animals are usually noted when they are very massive and spectacular (Brongersma-Sanders, 1957) and are often thought to be characteristic of the tropics, such as the mass heat kill of reef echinoids in Puerto Rico recently reported in an excellent study by Glynn (1968). Even in cold-temperate seas, natural heat kills may occur, and while perhaps less spectacular than the cold kills of these latitudes, such as that of 1962–63 (See *Helgol. Wiss. Meeresunters. 10,* 1964), they may be a regular occurrence. We do not have an estimate of the percent of the total *S. purpuratus* population killed this May, but hundreds of dead animals could be seen near the end of the period. Dead and damaged urchins have been noticed in early summer in past years here, but we had not noted the cause. Before this event in May, the urchin beds were very crowded; many animals were without burrows, on bare rock, and some were in burrows high on vertical rock faces. These animals were the ones showing progressive heat damage on successive days.

There are many implications of these observations for the study of effects of heated effluents on nearby littoral populations. The true natural annual temperature regime experienced by the animals must be known from long-term studies which include measurement of sea, intertidal surface, and animal internal temperatures. Natural events, such as the heat kill described above, must be noted by regular ecological monitoring and distinguished from the effects of erratic discharge of heated waters.

Information on the temperature tolerance of marine benthic animals is of importance to both general marine ecology and to assessing the biological effects of thermal effluents. Our studies on the normal increase in internal temperature upon tidal exposure indicates that such experimental work must be based on the natural situation if it is to have any ecological relevance. For intertidal forms, the rate of heating of experimental animals must correspond to naturally occurring rates of change, and the durations of exposures to test temperatures should be selected to correspond at least to the periods of shortest, average, and longest summer tidal exposure for the level of occurrence of the species. Parallel studies on animals totally immersed and in air of controlled humidity are essential for intertidal forms if ecological predictions are to be made from the work.

Of even greater ecological significance would be temperature tolerance and acclimation experiments modeled after those of Heath (1967), who simulated the natural thermal curve measured in the habitat where the animals were actually seen. These experiments with fish in the northern

Gulf of California demonstrated, among other things, that the fishes normally experience, for short periods of time, water temperatures that are lethal for longer intervals. For intertidal species, simulation of tidal curves would allow examination of the effects of tidal exposures which occur for varying periods of time at different parts of the day throughout the year.

Experiments designed along these lines would permit examination of the role the normal range of environmental temperature plays in the life of the animals. The venerable idea that marine intertidal organisms "tolerate" the normal seasonal temperature extremes of their habitat is prejudicial to development of new ideas and should be abandoned. These animals cannot be less than adequately adapted to their present environment, since they have survived selection and flourish there. It is equally likely that they require the seasonal variations in temperature, including both low and high extremes and that they utilize rather than tolerate them.

There is some evidence in support of this idea. Bartholomew (1966) found that the body temperature of the Galapagos marine iguana fell to that of the sea while feeding upon submerged algae but was maintained $10°$ C. to $15°$ C. above this on shore by a basking behavior. The large black intertidal chiton *Katharina tunicata* feeds on algae at high tide but remains motionless, tightly affixed to the rock, at low tide, when it may be exposed six or more hours during a summer day. We have measured mean internal body temperatures of this chiton in June and July $10°$ C. to $15°$ C. above the sea-surface temperature. In both of these herbivores, the gut is filled with algae when the animal is at the higher temperature. Digestive enzymatic reactions generally have a Q_{10} value between 2 and 3. In both of these otherwise dissimilar animals, having a body temperature $10°$ C. or more above sea temperature must facilitate digestion, since the reaction rate at this temperature level would be some 96% faster than at sea temperature. Short-term exposure to summer low-tide temperatures could be required by intertidal invertebrates because it would increase digestive and assimilative efficiency at this season, when the greatest increase in gonad reserve storage is taking place.

The possibility that the normal environmental temperature range is a requirement of existence has implications for prediction of the effect of thermal effluent discharges on the biology of benthic marine invertebrates. Much of the discussion of the effects of thermal effluent centers around immediate lethal effects on adult organisms which might occur in the immediate vicinity of outfalls. Such effects would have little importance from an ecological or population-biology view if they were entirely contained within restricted localities. But it is the possibility that widespread, subtle, pervasive, and permanent changes in the ecosystem might be caused by the wide dispersal of heat which makes thermal pollution of our cold-water coasts a potentially very serious problem. Rather than easily-seen local killings, events more difficult to detect, such as changes in reproductive success, growth, the slow decline and disappearance of species, the eventual disruption of and change of local food chains leading to our valuable commercial sources, and undesirable faunal shifts are the potentially more undesirable effects of thermal alteration of the inshore environment.

In one of the few studies known to us in which laboratory and environmental conditions are compared over a period of time, Sieburth (1967) examined the seasonal cycle of estuarine bacteria as related to water temperature for a period of more than two years in Narragansett Bay.

He departed from the usual bacteriological technique of culturing at a uniform single temperature:

Multiple incubation temperatures were used in an attempt to observe all segments of the populations and to cultivate all thermal types present. More than twenty-five hundred isolates of these different thermal types have been determined to taxonomic groups and correlated with the distribution data. Growth curves at 2° C. intervals of over 600 isolates were also determined in order to characterize the different thermal types. This has permitted an evaluation of the influence of water temperature on the selection of thermal and taxonomic types of bacteria and the results may, therefore, be discussed in relation to their possible relevance to ecological problems (p. 117).

Sieburth found that the temperature range of maximal growth forms a curve comparable to that of the environment, but lags by two months (Figure 4). The growth rate of the microflora, including cold-adapted forms, was suppressed during cold-water periods, and Sieburth asks whether this reduction in bacterial activities might not be related to the winter-spring diatom flowering of Narragansett Bay, and suggests, conversely, the possibility that "maximal activities during warm water periods either bind nutrients or recycle them in such a manner as to prevent sustained phytoplankton blooms" (p. 118). The possibility that accelerated bacterial activity from warm water might inhibit phytoplankton productivity in an estuary bears directly on the thermal-effluent problem, in that it could materially affect the food supply of invertebrate larvae whose development is accelerated by warming water.

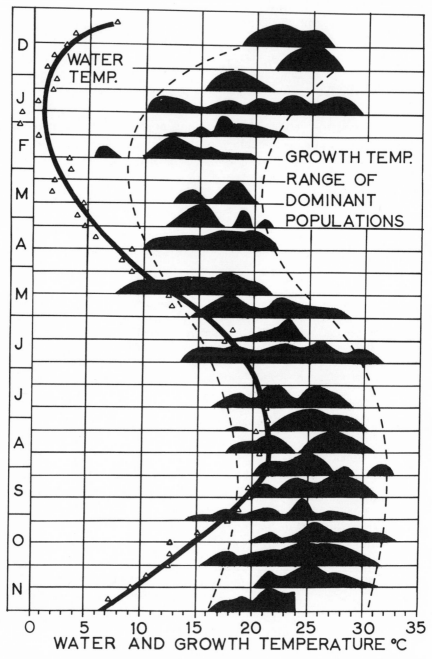

Fig. 4—.Comparison of the water-temperature curve with the range of maximal growth for the dominant microflora in Narragansett Bay, December 1960 through November 1961.
(From Sieburth 1967)

ANNUAL REPRODUCTIVE CYCLES AND TEMPERATURE

Most benthic marine invertebrates have cycles of reproductive activity, with periods of nutrient storage and maturation of gametes alternating with restricted periods of natural spawning. These breeding cycles have a constant relation to the seasons. Spawning in populations may occur over a protracted period but almost always shows definite peaks of activity. The natural factors controlling these events, both internal to the animal and external in the environment, are very imperfectly known. Local variations in spawning time of species probably related to differences in environment are known to occur, but causative environmental factors or physiological differences between populations of the species involved have seldom been clearly identified. A number of the advances which have been made in this field since Giese's (1958, 1959) reviews are mentioned in the following discussion.

The environmental factor most important in synchronizing reproductive periodicity with the seasons appears to be temperature, with light and other factors playing a variable secondary role. Seasonally-breeding marine invertebrates usually have as ecological requirements an annual minimum time period when the environmental temperature is at a level which initiates or allows gametes to be produced and also a temperature level which triggers release of gametes (spawning). Species which are seasonally reproductive, usually in cold-temperate seas, can produce gametes only if, during the year, there occurs a long enough period of time when water temperature is sufficiently elevated for gamete production to take place, and they can spawn only if the temperature reaches a critical level, usually higher than that required for gametogenesis (e.g., *Mytilus edulis,* Chipperfield, 1953; Bayne, 1965). Production of gametes and spawning are also limited by a maximum temperature sub-lethal for the species. Thus temperature regulation of reproductive cycles restricts the geographical distribution of these species to areas with optimum regimes of minimum and maximum temperature periods (Barnes, 1959; Beckman and Menzies, 1960; Fritchman, 1962). An excellent example is afforded by the work of Crisp (1957) and Patel and Crisp (1960), who demonstrated that, in British waters, summer breeding of barnacle species of warm-water origin and winter breeding of boreo-arctic species was based on reproductive temperature adaptations of opposite nature; the species had respectively warm and cold optimum-temperature requirements.

Physiological races within some cosmopolitan species show, as local adaptations, differences in these temperature requirements similar to those

found between related species (Sastry, 1963; Loosanoff and Davis, 1952).
It is also evident that the temperature requirements for successful breeding of many species will govern annual production of larvae and recruitment of settled young to the population, especially during years when the
water-temperature regime deviates significantly from the optimum.

The diversity of reproductive temperature requirements and their effects on the biology of different species is illustrated in the following examples. On the Atlantic coast of the United States, the soft-shell clam,
Mya arenaria, has both a spring and fall gamete-producing period in the
southern end of its range (Pfitzenmeyer, 1965), while, in the north, this
clam has one summer period of reproduction (Ropes and Stickney,
1965). Summer temperatures in the south appear to exceed the optimum
range required for gamete production, and the winter temperatures fall
below this range, thus producing two short southern optimum-reproductive
temperature conditions. Ropes and Stickney obtained some supporting
experimental evidence, which, in addition, indicated that there was some
genetic adaptation to the southern conditions by populations there.

In northern Europe, temperate, boreal, and arctic species occur together
in the same habitat (Kinne, 1963), but their periods of spawning differ;
the forms with a more southerly principal distribution spawn later in the
year than the northern species. Such species are apparently adapted to
southern conditions and require the warmer temperatures of summer and
spring for gamete production and release, while the northern species can
produce gametes at winter temperatures and release them at colder temperatures.

On the Pacific coast of the United States, Fritchman (1962) demonstrated that the temperature requirements for gamete production and
spawning of limpets (*Acmaea* spp.) was the principal factor limiting their
north-south geographical distribution. The temperature dependence and
response of gamete production and spawning in bivalves have been exploited in the culture of commercial species (Loosanoff and Davis, 1963;
Walne, 1964), where manipulation of breeding in seasons not normal for the
species is attained by temperature control of the laboratory stock. This
work, done primarily to provide a continuous supply of larvae for culturing experiments, remains as the earliest and best experimental evidence that
temperature directly affects the reproductive cycle of these species and not
some other temperature-dependent or related environmental factor. The
more recent studies of Bayne (1965) and Sastry (1966) have confirmed
this relation.

Laboratory investigations on many pelecypods (Loosanoff and Davis, 1963) Ropes and Stickney, 1965; Bayne, 1965; Sastry, 1966) and barnacles (Patel and Crisp, 1960) have amply demonstrated that accelerated gamete maturation and spawning can be induced by regulating water temperature to that optimum for breeding even during parts of the year when the natural populations are not breeding, provided adequate food or stored nutrient is available.

Sastry (1966) also showed that temperature affected all stages in the activation, growth, and maturation of the gonad in the scallop *Aequipecten irradians*. His experiments covered stages throughout the natural cycle, while other studies have been restricted to the non-breeding season. Sastry's studies, however, involved only starved animals, and it appears from his results that this scallop, unlike many other clams, cannot store sufficient nutrients to produce gametes and also supply metabolic needs during starvation at any temperature. Consequently, the results on the influence of temperature on early gonadal changes did not reveal what role temperature might normally play when nutrition is adequate.

Holland and Giese (1965) demonstrated that in the sea urchin, *Strongylocentrotus purpuratus,* the time course for the later events in spermatogenesis is the same throughout the seasonal period of sperm production. In this species, any factor influencing the rate of sperm production must therefore act early in spermatogenesis.

These last two reports indicate that it is therefore necessary to conduct experiments on animals with adequate nutrition throughout the entire period of the natural cycle if the role of temperature influence is to be understood.

Studies of *Mytilus edulis* (Bayne, 1965) and other clams (Loosanoff and Davis, 1963), the chiton *Katharina tunicata* (Nimitz, 1964), and the starfish *Pisaster ochraceus* (Mauzey, 1966) indicate that in these species nutrient stores accumulated by normal feeding before gamete production begins are adequate to support gonad development and complete gamete production without further feeding. In such species, where accumulation of nutrient stores is normally displaced in time from gamete maturation, it should be possible to separate experimentally the effects of annual temperature and light cycles upon gonad maturation alone from additional effects on feeding, digestion, and assimilation rates and thus more precisely define the temperature requirements of these animals at different seasons.

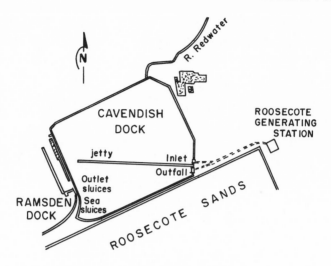

Fig. 5——.Plan of Cavendish Dock
(After Markowski 1959)

SOME STUDIES OF THE EFFECT OF THERMAL EFFLUENT ON THE BENTHOS

To date, there have been few studies specifically related to effect of change of temperature on the composition of bottom communities, although there have been extensive studies, particularly on the California coast, directed towards the analysis of possible changes in the composition of the bottom fauna as related to sewage and industrial pollution. See, for example, Hartman (1960), Filice (1959), and Reish and Winter (1954). At the same time, the potential for thermal alteration of near-shore environments along the Pacific coast is steadily increasing. These sources represent several types of outfalls, from surface or across-the-beach outlets to submerged pipes that produce a warm-water boil (Squires, 1967). So far, only surface effects have been observed and there is no information available indicating the situation in England discussed by Naylor (1965), where deep-water temperatures were slowly rising over a period of several years. Studies of the relation of cooling water from power plants to marine or estuarine benthos have been made in England by Markowski (1959, 1960, 1962) and Naylor (1965a), and in the Chesapeake Bay by Warriner and Brehmer (1966).

Since the papers by Markowski have been frequently cited, somewhat out of context, as demonstrating no ill effect, or possibly beneficial effects, of heated effluent water upon aquatic life, it should be emphasized that the principal situation investigated is a somewhat peculiar dock which is essentially a divided enclosure containing diluted sea water (Figure 5) and that Markowski himself stated that his first paper consisted of a "few sporadic observations" (1960). In the latest available paper of the series, Markowski (1962) states that "Although a 5-year period of biological investigations is not long enough, some observations, however, on changes in specific composition of the dock fauna have been noticed." It is unfortunate that Markowski was not as cautious at the outset in setting forth his conclusions, since such a statement as "No difference in fauna composition between the intake and outfall was found from the qualitative and quantitative points of view" (Markowski 1959, p. 258) could easily mislead those who might read only the summary and conclusions of his paper. Just how sporadic the data in support of this conclusion were can be plainly demonstrated by preparing a summary table (Table 1) based on the three tables of

TABLE 1. Summary of the data in paper by S. Markowski, "The Cooling Water Outfall of Marine and Estuarine Power Stations," Journal of Animal Ecology 28 (2), p. 34, November 1959.

Power Station	Stated Days of Observation	Salinity	Temperature ° F. Intake	Outfall	Diff.	Vol. of Flow 10^6 gal./ hour	No. of Species of Organisms Recorded (except protozoa)
Brighton	one	not stated	68.5	79	10.5	not stated	6
Carmarthen	one	31.6	61.9	72	10.1	12	3 (two dates given)
East Yelland	two	32.5	65	75	10.0	3	26
Plymouth	two	31.0	67	62.5	—4.5*	4.8	20
Poole	five	26.9	58	75	17	8	12
		29.0	59	76	17		
			59	75	16		
			48	not stated			
Roosecote	three, + six months	13–25**	62.5	73.8	11.3	40	17

*It is possible there is a mistake in the table published, but both Fahrenheit and Centigrade temperatures agree in being lower at the outfall.

**After 1957, salinities were all less than 12°/oo.

data provided by Markowski (1959, Tables 2, 3, 4). It will be noted from this summary that, for five of the six power stations discussed, the total time of observation was eleven days, and that the information on salinity and volume of flow is not complete. The flow data would suggest, however, that one could be equally justified in concluding that warm-water effluent has a marked effect in reducing the number of species, since the largest numbers of species were recorded at the smallest power stations. (Markowski's Table 4 simply records "Species found in the intake and outfall of stations using sea water").

In the second paper of his series, Markowski (1960) discusses observations made at Cavendish Dock, from which the Roosecote Generating Station draws water on one side of a dividing jetty and discharges it on the other (Figure 5). Studies were carried out for one calendar year, mainly of organisms on fouling plates. In all, 19 species were observed on the plates on the intake side and 26 (one of them the European eel) on the outfall side. No vegetation was observed in the intake, but there was a

prolific growth of algae in the outfall. On the other hand, denser animal populations (Coelenterata, Polyzoa) were found in the intake than in the outfall. . . . It is difficult to say whether the settlement of the organisms takes place during the operation of the station or in the periods when there is no water movement in the intake and in the outfall (p. 356).

Nevertheless, from this sort of information, the author concluded that "It is supposed that the outfall area is beneficial in some respects as the benthonic forms appear earlier than in the intake." The possibility that this earlier appearance may be related to the proximity of the outfall area to the sea and to the invasion of pelagic larvae from stock outside the docks was apparently not investigated, and it must be said that the summary statement deserves the Scotch verdict of "not proven."

This discussion of these papers is not intended so much as criticism of this work as a warning to those unfamiliar with ecological field work to read more than the conclusions of a paper. Unfortunately, Markowski's papers have been taken more seriously than is justified; they have been cited in support of industry claims of no potential effects at hearings of the California Public Utilities Commission for the proposed power plant on Bodega Head. Naylor (1965b, p. 76) rather tactfully tried to dismiss these papers as applying to "organisms in brackish regions which might be presupposed to be rather more tolerant of heating than fully marine species" and remarking that in "a number of other papers upon the effects of heated effluents in both marine and estuarine waters, several types of change have been observed." Naylor points out later (op. cit., p. 78) that his own work (Naylor, 1965a) in "an almost fully saline locality" contradicts Markowski's conclusions and demonstrates that native species of benthic invertebrates were excluded from the Queen's Dock area. How-

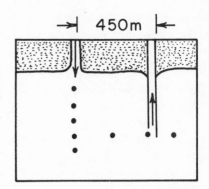

Fig. 6—.Location of sampling stations in reference to intake and discharge of power plant on York River.
(Modified from Warriner and Brehmer 1966)

ever, it seems obvious that Naylor's more detailed and careful studies are also concerned with the closed system of a dock and cannot be cited as indicative of what may happen in a more open situation, such as a large estuary or the seacoast.

The work by Warinner and Brehmer (1966) in the York River (part of the Chesapeake Bay system) is a serious attempt to assess the effects of a warm-water discharge in a more open situation. The data were obtained by bottom grabs from a pattern of stations near the intake and outfall of the power plant (Figure 6). Two $\frac{1}{20}$-square-meter grabs or samples were taken at each station each month for a year; this was considered reasonable after a pilot study involving several grabs at a station, and it was admitted that, while two grab samples do not yield all the species that may occur in a given region, they did get most of them and the remainder were species represented by very few individuals. The finest screen used was 1-mm. mesh. The information was converted to two indices. First of these is the redundancy index, in which minimum diversity is expressed as maximum redundancy ($R = 1$), and maximum diversity as low redundancy ($R = 0$), that is, all species present would be represented by an equal number of individuals when redundancy is low. The other index employed is that obtained by dividing the numbers of species minus 1 by the natural logarithm of the number of individuals in the sample ($S - 1/lnN$).

The authors found that community composition and abundance of the benthos were affected by the thermal discharge, and that the diversity index was lower at the stations nearer the effluent source during the summer months, but the diversity index was higher at these stations during the winter. According to their interpretation, "the redundancy index suggests that the greatest diversity occurs in the heated water even though the number of species was only half the number taken at the remote stations"

(p. 289). They concluded that "there is clear evidence of stress on the benthic population over a limited area during the months of high normal river temperatures."

What Warinner and Brehmer have really said is that there seems to be more effect attributable to heated effluent in the summer than in winter, a circumstance already reasonably well known. The serious question concerning the study on the York River is again that of methodology. Their station pattern does not provide them with a natural "control," a comparable source of information for conditions related to other factors than temperature near shore, since their stations form a straight line from the effluent source. What happens elsewhere along the shore at the same depth? Is the mat of detritus mentioned at the two near-shore stations opposite the outfall a general phenomenon over the entire area, or is it a result of concentration (or conversely, of scour) along the current produced by the discharge? While one must concede that each station produces that much more material and requires more man-hours of drudgery, a better-designed station grid would have made it unnecessary to raise such questions.

Warinner and Brehmer's interpretation of diversity and redundancy indices suggests that these abstractions of ecological field data may not be applied as simply as implied by Pearson et al. (1967), who find a greater diversity in the mid-part of San Francisco Bay and lower diversity at the extreme ends and declare: "Thus, the diversity index confirms quantitatively our subjective appraisal of the situation." [2] What should be empha-

2. San Francisco Bay: Estimation of pollution effects in San Francisco Bay, especially on the ecological grounds of "replacement fauna" and the depauperate areas of the upper arms of the bay are difficult to assess because the bay has been so extensively abused in the past. According to Gilbert's (1917) estimates, more than a billion cubic yards of sediment, most of it mining debris, were deposited in San Francisco Bay from 1849 to 1914, and over a period of 41 years, the average deposition over shoals in Suisun Bay was 3.3 feet and in San Pablo Bay, 2.5 feet. Unfortunately, we have no good information on the original or pre-mining benthos, except that the evidence of Indian middens on the bay shores indicates a rich molluscan fauna in prehistoric times. At this time, more than 40 percent of the 1860 surface area of San Francisco Bay has been filled for airports and subdivisions, and further extensive fillings are being planned. Such a radical change in the dimensions of the bay must have effects on the benthos which should be taken into consideration as part of the pollution problem.

Another effect of hydraulic mining in California streams was the virtual extirpation of the resident stock of Chinook salmon; stocks were so reduced by the 1930s that it was assumed that the salmon were done for, and no measures to protect them would be necessary or effective. By 1940, however, a remarkable resurgence became obvious, and extensive fish-protection measures, including controlled reser-

sized is that, at this state of our knowledge of bottom communities, the various diversity indexes have value as a means of comparing samples within a given area or set of conditions (Wilhm and Dorris, 1968). Care should be taken not to apply such an index, separated from the conditions under which it was developed, as an absolute criterion to an unknown situation, like the concentration of a pesticide. It does not follow that an index associated with unfavorable conditions in one region means the same thing in another. The idea that a diversity index may be developed, independent of sample size, as suggested by Wilhm and Dorris, should also be received with caution, since the problem of ideal sample size has yet to be resolved in benthic ecology, and selection of an inadequate sample size would limit the usefulness of data obtained for future analysis. Overeager use of diversity indexes has inspired one of the principal architects of such indexes to warn: "All points in an ecosystem have unique properties, and it is a bad beginning to assume that one is working with samples from a 'uniform universe' " (Margalef, 1968, p. 20).

In spite of a number of reviews of the problems of benthic ecology (Holme, 1964; Hopkins, 1964; Knox, 1961; Longhurst, 1959, 1964; Thorson, 1957), there is no indication that agreement will be reached on the matters of sample size, type of gear, or statistical procedures in the near future, if ever. The greatest difficulty of this branch of ecology is that it is hard work, involving the careful sorting, identifying, and processing of hundreds of specimens. Few researchers can command the manpower and funds required to undertake long-term, repeated studies of the benthos; yet, such studies as those of Henriksson (1968) and Ziegelmeier (1963) are essential to developing the basic information for evaluating the effects of

---

voir flow and hatcheries, were instituted in construction projects. At the present time, the population of Chinook salmon in the Sacramento-San Joaquin River system may be larger per unit area of stream bed than that of the Columbia River system. Average annual landings for the years 1954–1964 of Chinook salmon in the Columbia were 6,242,000 pounds; there were 7,000,000 pounds for San Francisco and northern California, combined (the combined Columbia River and adjacent coastal landings from Washington and Oregon 2343 were 10,000,000 pounds). While these catch statistics (from the Bureau of Commercial Fisheries statistics for 1965) do not support the idea that the Chinook salmon population of the Sacramento-San Joaquin system is greater than that of the entire Columbia River system, they do indicate a major recovery of the stocks in California. This seems more an indication of the remarkable vitality of the fish, perhaps genetic "acclimation" to changed river conditions, than of the beneficial ministrations of the power-generating establishments. The fish had started to come back before, not after, such installations as Shasta and Friant Dams and the Antioch generating plants were constructed.

changes induced by man. Standardization of benthic studies is difficult to attain because of the differences in sediment type, requiring different kinds of sampling apparatus, and the diverse nature of the aquatic environment, which may govern the number and frequency of samples possible. The most essential prerequisite to a benthic study (as well as to all ecological investigations) is a carefully thought-out campaign of action. Particular care should be given to avoiding a program based on an a priori statistical approach, since this could lead to an inappropriate sampling method. The procedure at a given locality must, obviously, depend on a reconnaissance to determine the type of gear and sampling pattern, always bearing in mind that sampling a control situation is as essential in this type of field ecology as controls for laboratory experiments.

PROSPECTS OF MARINE AQUICULTURE OF BENTHIC ORGANISMS IN HEATED WATERS

The immense quantities of heat to be generated has inspired proposals for its use in the cultivation of desirable marine organisms. Although the matter has been under careful study in England for some years, it has recently received some notice in this country. Martino and Marchello (1968), for example, suggest that cooling water could be piped to the sea bottom, perhaps five miles offshore, to produce a thermonutrient pump which would carry rich bottom water to the surface, as the warm water rises, to increase phytoplankton production and, consequently, the production of fish in the vicinity of the artificially-produced upwelling. As the authors admit, the scheme assumes a relatively constant nutrient supply and would require consideration of the possibility that the fish would move away, to be caught by someone else.

It would seem that a pipe five miles long, even if insulated, might result in considerable cooling of the effluent and might yield lower heat-flow values than calculated; but what is most conspicuously absent from the calculations is consideration of the habits and behavior of the organisms sought. Such a method of stimulating production requires not only the phytoplankton and the fish: it also requires that the intermediate links in the food chain linger in these nutrient-rich patches. Induction of significant upwelling would probably result in cooler surface water above the outlet than in the surrounding oceanic area. The demonstrated susceptibility of many zooplankton species to comparatively small temperature differences suggests that if such a thermal pump were situated in the sort of ocean described by LaFond and LaFond (1967), steady growth and increase of the system would be somewhat uncertain. At the present state of our knowledge, such proposals as this would seem to be a more expensive way of wasting heat.

Gaucher (in press) discusses the possibilities of what he terms "raceway aquiculture" to utilize heated effluents and, at the same time, diminish the amount of heat finally released into the sea. Assuming a production of 50 pounds per gallon per minute of flow, Gaucher estimates an annual yield of 4500 x 10^6 pounds per year of protein-rich fish and invertebrates in the U.S.A. by the year 2000. The most formidable difficulty in such a scheme as this is that of nutrients to support this production. In the pilot studies of Ansell *et al.* (cited below) flue gas from the steam-generating plant was used as a source of CO_2 and the water was fortified with domestic sewage for nutrients to culture algae for food for the clams to be cultured in the warm effluent waters. This neat system would be inconvenient to arrange for remotely situated nuclear reactors.

The difficulties of promoting a common property resource, alluded to in the thermal-pump scheme by Martino and Marchello, are discussed in economic terms by Scott (in press); Gaucher's scheme would provide full control of the system and thus, according to his assessment, be attractive to private capital.

The problems involved in the culture of a benthic organism are indicated by the work of Ansell and his colleagues (Ansell, *et al.*, 1964a, 1964b; Ansell and Lander, 1967) on experimental colonies of *Venus mercenaria* in areas subject to warm-water effluent in Poole Harbor. Although this species has been extensively studied in the United States and is the subject of an appreciable literature, there was no information on the seasonal changes of condition and little on reproductive cycles. The authors found it necessary to undertake basic biological studies to obtain information necessary to carry on experiments with culture. Selection of this species for study was prompted by the discovery of a naturally-established population in Southampton Water, and by the availability of hatchery-reared young clams from the Milford Laboratory in Connecticut. Animals in water warmed by power stations had extended growth periods and an increase in the instantaneous growth rate, compared to animals reared in cooler natural waters. An important discovery was that rearing the clams within the influence of the warm water from the power station had the effect of producing a population comparable to that naturally occurring in the southern part of the range of this species in the United States: that is, it spawned in the spring, as well as in late summer. However, hydrographic conditions in England during the spring are not favorable to survival of the larvae so produced, with the result that this increased spawning activity has no advantage to production unless the adult clams and the larvae they produce be maintained in artificially-warmed enclosures. It was also observed that reproductive success was poor during periods of zooplankton abundance, which suggests that these larvae are being consumed by other zooplankters. It is obvious from this experience that it will not be enough simply to release warm water into an open area populated by a desirable species in order to stimulate reproduction and cause increased yield. Aside from the loss of larvae to predation and unfavorable temperature conditions outside the influence of the effluent, there is the possibility that induction of spawning at an unfavorable period might impair the success of spawning at a more favorable time later in the year in an animal such as *Venus mercenaria,* which apparently has only one or two spawning periods a year.

Nevertheless, this work is an important step in the direction of utilizing the potential resources of heat that might otherwise be wasted, and its progress will be watched with interest.

Maintenance of a warm, open environment for desirable bivalves may have the fringe drawback of increasing the activity of predatory gastropods, since such oyster and mussel pests as *Urosalpinx* and *Thais* increase activity at warmer temperatures. *Urosalpinx cinerea* feeds most actively at 25° C. (Hancock, 1959) and the European *Thais lapillus* at 20° C. (Largen, 1967). This also suggests that, in a closed-culture system, vigilance against such pests cannot be relaxed, for, once established, they might flourish better than the bivalves being cultured.

This recent work indicates that the use of heated marine effluent water to increase production in natural and uncontrolled populations of economically important benthic invertebrates is limited, but that the potential for such use in closed or restricted controlled systems deserves further inquiry in this country. If natural areas having the requisite conditions for successful larval production and settling can be preserved, they could serve as sources for young animals which could then be transferred to warm-effluent areas for faster and possibly greater growth for a time. The practice of moving oyster spat from good natural production grounds to other areas more favorable for adult growth is centuries old.

But the potential for aquiculture of marine benthic invertebrates in heated marine effluents will be fully realized only by controlled hatchery production of larvae and subsequent transfer of young to rearing enclosures

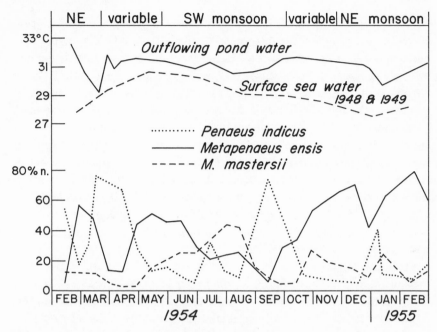

Fig. 7—.The "equatorial alternation of populations" of penaeid prawns in a Singapore pond.
(From Hall 1962, figs. 37, 39)

where water temperature and food supply can be controlled to produce the most economic yield. The technology of oyster husbandry is very nearly advanced to this stage. Such close control of larval production would be necessary if the potential use of heated effluents for growing valuable southern species of shrimp and bivalves in northern latitudes is to be exploited.

In planning aquicultural projects, the idea that maintaining a constant temperature is a desirable feature of the system should be entertained with caution. There is a growing body of information indicating that there are well-defined breeding and growth cycles in animals of both the uniformly cold and uniformly less variable tropical regions of the seas. The idea originally suggested by Semper (1881) that organisms in tropical seas reproduced continuously was based on field observations, perhaps of the sort of situation described from the Singapore Straits region by Hall (1962). In this area, there are indeed abundant larvae of the same species of shrimp at most times of the year, but these actually represent successive recruitments from different breeding populations of the same species, as indicated in Figure 7. This periodicity seems to be associated with the different monsoon conditions, which in turn may reflect differences in surface run-off and nutrient supply for these near-shore species, since the temperature range appears to be very narrow. Hall's example suggests that, if the culture of organisms from a narrowly variable temperature regime is to be attempted, the stock would have to be selected from several breeding populations to take full advantage of an artificially stable thermal environment.

Maintenance of a stable temperature for organisms from temperate regions would not produce the best results, in the light of Kinne's experiments with culturing the amphipod *Gammarus duebeni*. This animal could not be cultured at a constant temperature but required temperature fluctuations similar to those of its environment; as Kinne (1963) observed from this experiment: "A constant temperature of 20° C. and temperatures fluctuating between 15° C. and 25° C. with an average of 20° C. do not necessarily have the same biological effects."

Studies of the possible application of heated effluents on the Pacific coast to aquiculture are yet to be made, or if they have been, are not available to the public. There are some sporadic observations, but we are unaware of careful studies of gonadal cycles and the relations of larval stocks inducted out of phase with the environment to predators or ambient temperature conditions after the manner of the work done in England. There is, for that matter, no clear indication of the effect of the effluent at Morro Bay upon the stocks of Pismo clams on the nearby beach, and it is possible that much of the warm effluent rises to the surface and is dispersed rather rapidly. This action could possibly draw in colder water on the bottom, near the outfall, by induction, which might depress temperatures for nearby stocks and thus have an inhibiting effect on spawning, but this possibility does not seem to have been investigated. The most noticeable effect of heated effluent in California on benthic organisms is the congregation of round stingrays south of Alamitos Bay; their accidental "culture" has evidently been all too successful. So many of these undesirable and potentially dangerous fish have come into the warmer water of a thermal discharge from a steam-generating plant that a regular seining program to protect bathers, surfers, and lifeguards by removing the stingrays has been carried out for the past several years (John E. Fitch, personal communication).

SUMMARY AND RECOMMENDATIONS

Perhaps the best summary of our present understanding of the temperature relations of marine benthos is the admonition by Bullock (1955):

Without further work we cannot assume that the monthly average, daily average, or any given measure short of the complete curve of the temperature against time, is biologically appropriate, since we do not know how the organism weights equal departures from the average in the two directions for different lengths of time.

At least, we have progressed far enough to realize that laboratory experiments on thermal tolerances, death points, and the like, without reference to the natural conditons, including previous temperature experience and state of tide or season at which experimental material was gathered, have questionable utility in reference to what the organisms may actually do in nature. Intensive field studies, with *in situ* measurements of the environment and the organisms, combined with continuous monitoring, especially of thermal gradients, are needed. Ideally, such a study of an area that may be subjected to thermal pollution should be carried out for several years in advance, so that we would have the information necessary for making the most of what should be a field experiment rather than an experience. Our preliminary information again indicates the dubious value of such casual temperature data as that compiled from readings made once weekly or even daily, or sporadically, from selected shore stations. Too often, this data produces its own artifacts, which cannot be related to the experience of the organisms.

Field evidence indicates that some intertidal herbivores are well adjusted to the temperature ranges from seawater to rather high air temperatures for varying periods of time, and that indeed such a temperature range may be an ecological requirement, rather than an environment to be endured. How extensive this may be, and to what degree it may apply also to subtidal organisms, remains to be determined.

There is evidence indicating that some marine organisms do not flourish in a stable temperature regime but require the variation around the statistical mean. The significance of such a requirement for a scheme involving an artificially stable warm environment for aquiculture is obvious. However, such temperature requirements are not yet established for many organisms, since most experiments involving laboratory culture are of comparatively short duration.

In contrast to the observed ability of many marine organisms to withstand wide ranges of environmental temperatures at some stages of their life cycles, there is the growing body of evidence that comparatively small fluctuations in oceanic temperatures may influence the distribution and abundance of many species, especially those that occur in large populations.

A thorough knowledge of such fluctuations is necessary in order to separate population changes associated with them from changes induced by artificial thermal alteration of restricted environments.

There seems to be no prospect of agreement on standard methods of studying benthic associations, or upon minimum sample size. The varying types of environment to be studied indicate that each survey of benthic communities must be programmed according to the circumstances. While some of the various diversity and redundancy indexes proposed may have value in comparing samples at different stations and seasons in the same environment, the application of such indexes without reference to the sampling methods, or independently of a specific environment, is inadvisable. A successful benthic survey requires careful assessment of the environment and the most feasible sampling methods before actual work is begun, and a guarantee that funding, adequate for the technical support or processing the samples, will be provided.

Full knowledge of the reproductive and biochemical cycles of organisms that may be amenable to mass culture in thermally-enriched situations is necessary to ensure success, and this knowledge must be applied to the ecological circumstances of the environment into which the young from artificially-cultured organisms may be released. The experience in England indicates that it may be wasteful to induce spawning if larvae are to be subjected to predation or lack of food at inappropriate times of the natural biological cycle. Hence, it would seem that the best hope for utilization of excess thermal energy in aquiculture will be under conditions of environmental control of the entire system involved. Changing the thermal characteristics of near-shore waters may induce unanticipated and possibly undesirable changes in the flora and fauna, rather than stimulate production of potentially useful species.

REFERENCES

Ambler, M. P., and V. J. Chapman. 1950. "A Quantitative Study of Some Factors Affecting Tide Pools." *Trans. Roy. Soc. New Zealand,* 78(4):394–409.

Ansell, Alan D., and K. F. Lander. 1967. "Studies on the Hard-Shell Clam, *Venus mercenaria,* In British Waters. III. Further Observations on the Seasonal Biochemical Cycle and on Spawning." *J. Appl. Ecol.,* 4(2):425–435.

Ansell, Alan D., and F. A. Loosmore, 1963. "Preliminary Observations on the Relationship between Growth, Spawning, and Condition in the Experimental Colonies of *Venus mercenaria." Jour. du Conseil.,* 28(2):283–294.

Ansell, Alan D., K. F. Lander, J. Coughlan, and F. A. Loosmore. 1964a. "Studies on the Hard-Shell Clam, *Venus mercenaria,* in British Waters. I. Growth and Reproduction in Natural and Experimental Colonies." *J. Appl. Ecol.* 1:63–82.

Ansell, Alan D., F. A. Loosmore, and K. F. Lander. 1964b. "Studies on the

Hard-Shell Clam, *Venus mercenaria,* in British Waters. II. Seasonal Cycle in Biochemical Composition." *J. Appl. Ecol.* 1:83–95.

Barnes, H. 1959. "Temperature and Life Cycle of *Balanus balanoides* (L)." In *Marine Boring and Fouling Organisms,* edited by D. L. Ray, pp. 234–245. Seattle, Washington: University of Washington Press.

Bartholomew, George A. 1966. "A Field Study of Temperature Relations in the Galapagos Marine Iguana." *Copeia* (2):241–250.

Bayne, B. L. 1965. "Growth and Delay of Metamorphoses of the Larvae of *Mytilus edulis.*" *Ophelia* 2(1):1–47.

Beckman, C., and R. J. Menzies. 1960. "The Relationship of Temperature and the Geographical Range of the Marine Woodborer, *Limnoria tripunctata.*" *Biol. Bull.,* 118(1):9–16.

Bell, F. H., and A. T. Pruter. 1958. "Climatic Temperature Changes and Commercial Yields of Some Marine Fisheries." *Fish. Res. Bd. Canada Jour.* 15(4):625–683.

Blacker, R. W. 1957. "Benthic Animals as Indicators of Hydrographic Conditions and Climatic Change in Svalbord Waters." *Ministry of Agr. Fish and Food, Fishery Invest. Ser. II* 22(10):1–49.

Bolin, Rolf L., and D. P Abbott. 1963. "Studies on the Marine Climate and Phytoplankton of the Central Coastal Area of California, 1954–1960." *Calif. Coop. Oceanic Fish. Invest. Repts.* 9:23–45.

Brongersma-Sanders, Margaretha. 1957. "Mass Mortality in the Sea." In *Treatise on Marine Ecology and Paleoecology,* vol. 1, pp. 941–1010. Geo. Soc. Amer. Memoir 67.

Bullock, Theodore Holmes. 1955. "Compensation for Temperature in the Metabolism and Activity of Poikilotherms." *Biol. Revs.* 30:311–342.

Burt, W. V., and Bruce Wyatt. 1964. "Drift-Bottle Observations of the Davidson Current off Oregon." In *Studies on Oceanography,* pp. 156–165.

Byrne, J. V. 1963a. "Geomorphology of the Continental Terrace of the Northern Coast of Oregon." *Ore Bin* 25(12):201–209.

———. 1963b. "Geomorphology of the Oregon Continental Terrace South of Coos Bay." *Ore Bin* 25(9):149–154.

Chipperfield, P. N. J. 1953. "Observations on the Breeding and Settling of *Mytilus edulis* L. in British Waters." *J. Mar. Biol. Assoc. (U.K.)* 32(2):449–476.

Crisp, D. J. 1957. "Effect of Low Temperature on the Breeding of Marine Animals." *Nature* 179(4570):1138–1139.

Doty, Maxwell S. 1957. "Rocky Intertidal Surfaces." In *Treatise on Marine Ecology and Paleoecology,* vol. 1, pp. 535–585. Geo. Soc. Amer. Memoir 67.

Dow, Robert I. 1964. "A Comparison Among Selected Marine Species of an Association Between Seawater Temperature and Relative Abundance." *Jour. du Conseil.* 28(4):425–431.

———. 1969. "Cyclic and Geographic Trends in Seawater Temperature and Abundance of American Lobster." *Science,* 164(3883):1060–1063.

Dunbar, M. J. 1968. *Ecological Development in Polar Regions. A Study in Evolution.* Englewood Cliffs: Prentice-Hall.

Filice, Francis P. 1959. "The Effect of Wastes on the Distribution of Bottom Invertebrates in the San Francisco Bay Estuary." *Wasmann Jour. Biol.* 17(1):1–17.

Fritchman, Harry K. 1962. "A Study of the Reproductive Cycle of the California *Acmaeidae.* Part IV." *Veliger* 4(3):134–140.

Gaucher, Thomas A. "Thermal Enrichment and Marine Aquiculture." In *Conference on Marine Aquiculture.* Corvallis, Ore.: Oregon State University Press (in press).

Giese, Arthur C. 1958. "Annual Reproductive Cycles in Marine Invertebrates." *Ann. Rev. Physiol.* 21:547–576.

———. 1959. "Reproductive Cycles of Some West Coast Invertebrates." In *Photoperiodism and Related Phenomena in Plants and Animals,* edited by R. B. Winthrow, pp. 625–638. Washington: Amer. Assoc. Adv. Sci.

Gilbert, Grove Karl. 1917. "Hydraulic-Mining Debris in the Sierra Nevada." U.S. Geol. Survey, Prof. Paper 105.

Glynn, Peter W. 1965. "Community Composition, Structure, and Inter-relationships in the Marine Intertidal *Endocladia muricata—Balanus glandula* Association in Monterey Bay, California." *Beaufortia* 12(148):1–198.

———. 1968. "Mass Mortalities of Echinoids and other Reef Flat Organisms Coincident with Mid-day, Low-Water Exposures in Puerto Rico." *Marine Biol.* 1(3):226–243.

Gonor, Jefferson J. 1968. "Temperature Relations of Central Oregon Marine Intertidal Invertebrates: A Pre-publication Technical Report to the Office of Naval Research." Department of Oceanography, Oregon State University, Date Report No. 34, Ref. 68–38, 49 pp. (Mimeo.)

Gunter, Gordon. 1957. "Temperature." In *Treatise on Marine Ecology and Paleoecology,* vol. 1, pp. 159–184. Geol. Soc. Amer. Memoir 67.

Hall, D. N. F. 1962. "Observations on the Taxonomy and Biology of Some Indo-West Pacific *Penaeidae* (Crustacea, Decapoda)." Colonial Office Fishery Publ. 17.

Hancock, D. A. 1959. "The Biology and Control of the American Whelk Tingle, *Urosalpinx cinerea* (Say)." *U.K. Fishery Invest.* 22:1–66.

Hartman, Olga. 1960. "The Benthonic Fauna of Southern California in Shallow Depths and Possible Effects of Wastes on the Marine Biota." In *Waste Disposal and the Marine Environment,* pp. 57–81.

Heath, Wallace G. 1967. "Ecological Significance of Temperature Tolerance in Gulf of California Shore Fishes." *J. Ariz. Acad. Sci.* 4(3):172–178.

Henriksson, Rolf. 1968. "The Bottom Fauna in Polluted Areas of the Sound." *Oikos* 19(1):111–125.

Holland, N. D., and A. C. Giese. 1965. "An Autoradiographic Investigation of the Gonads of the Purple Sea Urchin, *Strongylocentrotus purpuratus.*" *Biol. Bull.* 128(2):241–258.

Holme, N. A. 1964. "Methods of Sampling the Benthos." *Adv. Mar. Biol.* 2:171–260.

Hopkins, Thomas L. 1964. "A Survey of Marine Bottom Samplers." *Progress in Oceanography* 2:213–256.

Howie, D. I. D. 1959. "The Spawning of *Arenicola marine* (L.) I. the Breeding Season." *J. Mar. Biol. Assoc., U.K.*:38.

Johnson, R. G. 1965. "Temperature Variation in the Infaunal Environment of a Sand Flat." *Limnol. and Oceanog.* 10(1):114–120.

Ketchen, K. S. 1956. "Climatic Trends and Fluctuations in Yield of Marine Fisheries of the Northeast Pacific." *J. Fish. Res. Bd. Canada* 13(3):357–374.

Kinne, O. 1963. "The Effects of Temperature and Salinity on Marine and Brackish Water Animals. I. Temperature." *Oceanogr. and Mar. Biol., Ann. Rev.* 1:301–340.

———. 1964. "The Effects of Temperature and Salinity on Marine and Brackish Water Animals. II. Salinity and Temperature Salinity Combinations." *Oceanogr. and Mar. Biol., Ann. Rev.* 2:281–339.

Knox, G. A. 1961. "The Study of Marine Bottom Communities." *Proc. Roy. Soc. New Zealand* 89(1):167–182.

LaFond, E. C., and K. G. LaFond. 1967. "Temperature Structure in the Upper 240 Meters of the Sea. 'The New Thrust Seaward.'" In *Trans, Third MTS Conference and Exhibit,* pp. 23–45. n.p.

Largen, M. J. 1967. "The Influence of Water Temperature upon the Life of the Dog-Whelk *Thais lapillus* (Gastropoda: Prosobranchia)." *J. Anim. Ecol.* 36(1): 207–214.

Lewis, J. R. 1964. *The Ecology of Rocky Shores.* London: English Univ. Press.

Longhurst, Alan R. 1964. "A Review of the Present Situation in Benthic Synecology." *Bull. Inst. Oceanog. Monaco* 63(1317).

Loosanoff, V. L., and H. C. Davis. 1952. "Temperature Requirements for Maturation of Gonads of Northern Oysters." *Biol. Bull.* 103(1):80–96.

———. 1963. "Rearing Bivalve Mollusks." In *Advances in Marine Biol.,* vol. 1, pp. 1–136. n.p.

Lorenz, J. R. 1863. "Physicalische Verhaltnisse und Vertheilen der Organismen im Quarnerischen Golfe." Wien: K. K. Hof- und Staatsdruckerei.

McLeese, D. W. 1956. "Effects of Temperature, Salinity, and Oxygen on the Survival of the American Lobster." *J. Fish. Res. Bd. Canada* 13(2):247–272.

Margalef, Ramón. 1968. *Perspectives in Ecological Theory.* Chicago: University of Chicago Press.

Martino, P. A., and J. M. Marchello. 1968. "Using Waste Heat for Fish Farming." *Ocean Industry* April:36–39.

Markowski, S. 1959. "The Cooling Water of Power Stations: A New Factor in the Environment of Marine and Freshwater Invertebrates." *J. Animal Ecol.* 28: 243–258.

———. 1960. "Observations on the Response of Some Benthonic Organisms to Power Station Cooling Water." *J. Animal Ecol.* 29:349–357.

———. 1962. "Faunistic and Ecological Investigations in Cavendish Dock, Barrow-in-Furness." *J. Animal Ecol.* 31:43–51.

Mauzey, K. P. 1966. "Feeding Behavior and Reproductive Cycles in *Pisaster ochraceus.*" *Biol. Bull.* 131(1):127–144.

Naylor, E. 1965a. "Biological Effects of a Heated Effluent in Docks at Swansea, S. Wales." *Proc. Zool. Soc. London* 144:253–268.

———. 1965b. "Effects of Heated Effluents upon Marine and Estuarine Organisms." *Adv. Mar. Biol.* 3:63–103.

Nimitz, M. A., and A. C. Giese. 1964. "Histochemical Changes Correlated with Reproductive Activity and Nutrition in the Chiton, *Katharina tunicata.*" *Quart. J. Micro. Sci.* 105(4):481–495.

Orton, J. H., and A. J. Southward. 1961. "Studies on the Biology of Limpets. IV. The Breeding of *Patella depressa* P. on the North Cornish Coast." *J. Mar. Biol. Ass., U. K.* 41:653–662.

Patel, B., and D. J. Crisp. 1960. "The Influence of Temperature on the Breeding and the Moulting Activities of Some' Warm-Water Species of Operculate Barnacles." *J. Mar. Biol. Assoc., U.K.* 39(3):667–679.

Pearson, E. A., P. N. Storrs, and R. E. Selleck. 1967. "Some Physical Parameters and Their Significance in Marine Waste Disposal." In *Pollution and Marine Ecology,* edited by Theodore A. Olson and Frederick J. Burgess, pp. 297–315. Interscience.

Pfitzenmeyer, H. T. 1965. "Annual Cycle of Gametogenesis in the Soft-Shell Clam, *Mya arenaria* at Solomons, Maryland." *Chesapeake Science* 6(1):52–59.

Phleger, F. B, and J. S. Bradshaw. 1966. "Sedimentary Environments in a Marine Marsh." *Science* 154(3756):1551–1553.

Radovich, John. 1961. "Relationships of Some Marine Organisms of the Northwest Pacific to Water Temperatures, Paricularly During 1957 through 1959." *Calif. Fish. Bull.* 112.

Reish, Donald J., and H. A. Winter. 1954. "The Ecology of Alamitos Bay, California, with Special Reference to Pollution." *Calif. Fish and Game* 40:105–121.

Ropes, J. W., and A. P. Stickney. 1965. "Reproductive Cycle of *Mya arenaria* in New England." *Biol. Bull.* 128(2):315–327.

Sastry, A. N. 1963. "Reproduction of the Bay Scallop *Aequipecten irradians* Lmk. Influence of Temperature on Maturation and Spawning." *Biol. Bull.* 125(1):146–153.

Sastry, A. N. 1966. "Temperature Effects in Reproduction of the Bay Scallop, *Aequipecten irradians* Lmk." *Biol. Bull* 139(1):118–134.

Scott, Anthony. "Economic Impediments to Marine Development." In *Conference on Marine Aquiculture*. Corvallis, Ore.:Oregon State University Press (in press).

Semper, Karl. 1881. *Animal Life as Affected by the Natural Conditions of Existence*. New York:D. Appleton and Co.

Sieburth, John. McN. 1967. "Seasonal Selection of Estuarine Bacteria by Water Temperature." *J. Exp. Mar. Biol. and Ecol.* 1(1):98–121.

Southward, A. J. 1958. "Notes on the Temperature Tolerances of Some Intertidal Animals in Relation to Environmental Temperatures and Geographical Distribution." *J. Mar. Biol. Assoc., U.K.* 37:49–66.

————. 1960. "On Changes of Sea Temperature in the English Channel." *J. Mar. Biol. Assoc., U.K.* 39:449–458.

————. 1963. "The Distribution of Some Plankton Animals in the English Channel and Approaches." *J. Mar. Biol. Assoc., U.K.* 43:1–29.

Squires, James L., Jr. 1967. "Surface Temperature Gradients Observed in Marine Areas Receiving Warm Water Discharges." Tech. Paper, *Bur. Sport Fish. and Wildlife* 11:8pp.

Taylor, C. C. 1959. "Temperature and Growth—the Pacific Razor Clam." *Jour. du Conseil. Internat. pour l'Explor. de la Mer* 25:93–101.

————. 1960. "Temperature, Growth, and Mortality—the Pacific Cockle." *Jour. du Conseil. Internat. pour l'Explor. de la Mer* 26(1):117–124.

Taylor, C. C., Henry B. Bigelow, and Herbert W. Graham. 1957. "Climatic Trends and the Distribution of Marine Animals in New England." *Fishery Bulletin of the Fish and Wildlife Service* 57:291–345.

Thorson, Gunnar, 1957. "Bottom Communities." In *Treatise on Marine Ecology and Paleoecology*, vol. 1, pp. 461–534. Geol. Soc. Amer. Memoir 67.

Ushakov, Boris. 1959. "Thermostability of the Tissue as One of the Daignostic Characters in Poikilothermic Animals." *Fifteenth Int. Congr. Zoology*, papers read in title, 37, 5pp.

Walne, P. R. 1964. "The Culture of Marine Bivalve Larvae." In *Physiology of the Mollusca*, vol. 1., edited by K. M. Wilbur and C. M. Yonge, pp. 197–242. New York:Academic Press.

Warinner, J. E., and M. L. Brehmer. 1966. "The Effects of Thermal Effluents on Marine Organisms." *Air and Water Poll.* 10(4):277–289.

Wilhm, Jerry L., and Troy C. Dorris. 1968. "Biological Parameters for Water Quality Criteria." *BioScience* 18(6):477–481.

Zhirmunsky, A. V. 1967. "A Comparative Study of Cellular Thermostability of Marine Invertebrates in Relation to Their Geographical Distribution and Ecology." In *The Cell and Environmental Temperature*, edited by A. S. Troshin, pp. 209–218. New York:Pergamon Press.

Ziegelmeier, Erich. 1963. "Das Makrobenthos in Ostleil der Deutschen Bucht nach qualitativen und quantitaven Bodengreiferuntersuchungen in der Zeit von 1949–1960." Veröffent. Inst. Merresf. Bremerhaven, Sond. herausg. W. Hohnk.

DISCUSSION/ Wheeler J. North

THE paper presented by Dr. Hedgpeth has certain features in common with the remarks heard from Dr. Strickland. To some extent, what I have to say will apply to both papers. Dr. Strickland and Dr. Hedgpeth have had a wealth of experience in the marine environment and both see clearly how difficult it is to predict consequences from altering even one basic parameter, such as temperature. I can only concur with their misgivings. Certainly, a majority of marine biologists would also concur. In fact, some of my comments today will add to the list of unknowns that Dr. Strickland and Dr. Hedgpeth have mentioned.

During the next several decades, our population and economy will continue their free expansion. Demands for additional power will not be ignored. However much we may object to it, when the public need conflicts seriously with conservationism, the public need is always given preference. The outcome could only be changed by extending the voting franchise to fishes, lobsters, and abalone!

In the years ahead, coastal waters will be required to receive and absorb enormous additional quantities of wastes. Thirty years hence, natural and undisturbed habitat will probably exist only within a few scattered preserves. Biology of the rest of the coast will be altered, often profoundly. The impact of man could be mitigated substantially if communication and cooperation between biologists and engineers takes place. Our past mistakes—the desolate underwater wastelands now surrounding many of our coastal cities—are mute testimony, not just to our lack of knowledge, but to our failure to apply existing knowledge.

An urgent goal in thermal-pollution research should be establishment of general ecological guidelines as an aid to designing and locating discharge systems. The following suggestions are based on my own experiences with heated discharges in southern and central California.

1. *Ecological advice should be sought early in the planning stages of a*

new installation. An important corollary is that ecological surveys should start as soon as possible, even before a site is definitely selected.

2. *Zones where substantial temperature fluctuation occurs should be minimized.* The heated discharges I have examined indicate that, where warm conditions persist, luxurious growths of heat-tolerant organisms develop. Relatively barren conditions tend to appear in the region of the transition from hot to cold, particularly when wide temperature fluctuations are frequent.

3. *Significant interference with dispersal of planktonic larvae should be avoided.* Persistent tongues of hot water extending considerable distances from shore might provide serious barriers to the longshore migration of sensitive planktonic larvae from benthic parents. Sitting in bays and inlets or behind small islands or headlands could help solve this problem.

4. *Potential needs for mitigation should be considered during planning.* If expensive construction for protection of wildlife or programs of culture or transplantation might be necessary, early recognition is desirable. Mitigation costs are an incentive to proper design if they are considered during the planning stages. After construction, such costs become powerful reasons for doing as little as possible.

In addition to applying existing knowledge to solution of our pollution problems, we must obviously seek new knowledge. In the field of thermal pollution, a fair amount of laboratory research has been conducted, evaluating tolerances of various organisms. Certainly this is useful and valuable information; but it is my belief that such information applies only to a rather minor portion of the total problem in the marine environment. The areas where temperatures exceed tolerance limits are usually relatively small. Nonetheless, effects may extend over much wider areas. In many cases, changes are not the result of physiological damage, but result from alterations in ecological relationships within communities. These "ecological problems" may well be the most significant manifestations of thermal pollution and certainly they deserve far more attention than they are receiving. We know from observations in natural communities that shifts in temperature can have profound effects.

For example, kelp beds in southern California change substantially with season. During winter, the dominant organism, giant kelp (*Macrocystis*), tends to appear much more "healthy" than during summer. In summer, the blades are frequently heavily encrusted and portions are often grazed. In winter, the blades tend to be free of encrustation and are usually intact. The shabby summer condition develops at water temperatures well below tolerance limits of kelp, as indicated by the photosynthesis *vs.* temperature curve. The principal encrusting species are present throughout the year. Presumably, their growth rates rise more rapidly than that of kelp as temperature increases, enabling them to colonize larger surfaces of the blades during summer. Increased animal encrustations make the kelp blades more attractive to fish grazers, accounting for the chewed remnants commonly seen on the plants during summer.

The "ecologically damaging" temperatures usually persist for four to six weeks, in an average summer. From 1957 to 1959, however, warm conditions lasted through most of the year, with devastating results. Some kelp beds lost 90 percent or more of the standing crop. The losses had economic repercussions, since the kelp beds are important marine resources in southern California.

Quite interestingly, kelp communities in central Baja California appear to have adjusted to the high summer temperatures that occur there. For example, at Turtle Bay, Baja California, temperatures of 25° C. to 26° C. are common in August and typically exceed 20° C. for six months of the year. In southern California kelp beds, temperatures rarely exceed 22° C. to 23° C. and are above 20° C. for about two months. Clendenning (in press) has shown that the kelp in Turtle Bay has a photosynthetic capacity per unit-blade area which is about 50 percent greater than the same species of plant in southern California. Likewise, the temperature of maximum photosynthesis for Turtle Bay kelp was 26° C. *vs.* 21° C. for southern California kelp. These differences may be critically important for survival in the Turtle Bay environment.

We have transplanted the Mexican strain of kelp to Newport Bay in southern California with some success. A single plant was obtained from cultured spores in 1965. Natural reproduction occurred in the bay, and by 1967 we had six plants. By February 1968, the small bed numbered 184 plants. To the best of our knowledge, the southern California strain of kelp has never colonized Newport Bay (except for a few plants that occasionally establish temporarily at the tip of the jetties). During summer, water temperatures in Newport Bay rise to 25° C. and higher for weeks and probably prevent establishment of the cold-water strain of kelp.

An interesting and unique group of animals has become associated with our little colony of Mexican kelp. Included are species commonly found in the bay, as well as animals that usually prefer the open sea. Naturally, the permanent members of this community are forms that tolerate the warm summer temperatures. Cold-water species appear in winter but disappear in late spring, when temperatures begin to rise. In essence, introduction of a dominant plant form that is able to cope with the ecological consequences of slightly elevated temperatures has encouraged development of a warm-water community of unusual composition. The applications to areas influenced by thermal discharges are obvious. Even crude studies such as this are useful, since they provide some insight into ecological relationships, they furnish clues for promising lines of research, and they frequently have immediate practical applications.

The nuclear power industry is just coming of age. Only a few of the thermal discharges that will eventually be built are now in operation. Improvement of techniques by research during the next few years is particularly important, since it should benefit the majority of thermal-discharge systems constructed before the year 2000.

Of the various human activities that affect marine organisms—fishing, sewage disposal, coastal construction, gross oil pollution, etc.—thermal discharges offer perhaps the best opportunity for mitigation and circumvention. Most systems discharging heated water into the open sea are still relatively crude, compared to the elaborate systems currently in use for sewage disposal. Thermal discharge systems could, if necessary, employ long outfalls and well-designed diffusers to disperse their effluents. The thermal waste product mimics, to some extent, the natural change from winter to summer conditions and the rise in temperature that occurs geographically proceeding from temperate to tropical climates. Few, if any, of man's other wastes have such a close parallel in nature. Obviously, the closer we can keep to natural conditions, the better are the chances for avoiding irreversible changes.

DISCUSSION FROM THE FLOOR

Ray T. Oglesby: If you look at the program, you'll find that we have some of the leading experts in the world, certainly here in the United States, who are capable of addressing themselves to this problem of thermal pollution. I think it's rather significant that, at least to my knowledge, these people are not working together on any particular problem. They each follow their own research interest, which is as it should be.

But I think one of the problems is that we have a hard time bringing enough people together to spell out problems relating to complex ecosystems. This relates to Dr. Strickland's plea for more manpower, but I think it also points out a possible danger. It seems quite attractive, perhaps, to define these ecosystems, or try to define them, as completely as possible, perhaps build mathematical models, deterministic models that are predictive in nature. This is very attractive and it's certainly a trend that we have in ecology today.

I would caution against putting all of our eggs in one basket, however, and studying one problem to the exclusion of all others. I think perhaps we will find, as time goes on, that our ecosystems, even the simplest of our natural ecosystems, are too complex to model with electronic computers, at least as they exist today. So we are chained to a trial-and-error method, to some extent, as Dr. North has indicated.

This trial-and-error has to be rationally directed, and it has to be based on good fundamental knowledge. But I think we will have to take some steps and find out where we're going. The worrisome thing here is that we might take steps that we can't retreat from, and this means that we back ourselves into a corner, much as we have done with the chlorinated-hydrocarbon insecticides. We've spread these around the environment and we're still spreading them. There's nothing we can do to take these back, except wait for very long periods of time.

123

This is the thing to be most concerned about in the case of thermal pollution, and I think that the biggest problem here is to get a crash program going. I hate the thought of crash programs as much as anyone, but a very concerted effort in the early stages is needed, so that we have some intelligent direction for planning. Then it might be hoped, for once, that the biologist could ask the engineer for information and not the other way around. He could say, "All right, we've got some good ecological guidelines. Now you tell us where the thermal discharges are going and what effect they're going to have on water temperature and then we'll try to tell you what effect we expect this will have on the aquatic ecosystems."

Wheeler J. North: I would say that I heartily agree with that.

Joel W. Hedgpeth: No, I'm not against sin, either.

Walter Glooschenko: I would like to make two comments, one in relation to Markowski's work. I completely agree with Dr. Hedgpeth, inasmuch as I had occasion to read these papers quite critically for a presentation for Senator Muskie's hearings in Miami, Florida. I would like to make a few comments that Dr. Hedgpeth didn't mention.

One: Markowski did a lot of his observations on plankton coming through the cooling pipes, mainly on the basis of visual observations. I don't know how you can visually tell if a little beastie is "healthy." You can perhaps tell it's living, but how long will it last afterwards? What will be the long-term effect? That was my main comment on the subject. I think that a lot of his observations were visual and most of his studies, again, were in a temperate system, not a tropical system.

Also, a comment on Dr. North's statement is in order. He asks why ecologists have not said more about some of these effects. My question here, is: Why aren't ecologists asked to make comments on many of these particular things? I can state several examples. First of all, there have been many studies in the past, many construction projects, where ecologists seemed not to get asked, quite purposely, because of some of the answers they might give.

I think we have many good examples of Army Corps of Engineers' projects where ecologists were never asked, and we can see what happens now to the environment. I believe that it is not a matter of why the ecologists didn't state their opinions, but instead, why were they not asked?

I believe industry is also much to blame. There are many particular examples, where industry says, "No harm will be done, don't get ecologists to testify for themselves." It is very interesting that ecologists who work for industry never seem to have any disagreements with a lot of the industrial opinions, at times, then pick things out of context. But it's not why aren't ecologists saying anything: why aren't they asked to say more by the agencies? Or why aren't they asked by industry to discuss things, such as site selection for power plants?

I've seen this in the state of Florida quite a bit, now, where state biologists seem to be neglected until a lot of public pressure by conservationists occurs, saying, "Let's have the state biologists testify here." And once they do testify, it's usually different from what the politicians say it is.

Hedgpeth: I am more or less substantially in agreement with these Socratic questions. I don't think I can really add to them at this point.

North: I am in agreement. I might say that I think this is changing. Perhaps we might even see ecology taught at West Point. But certainly, ecology in recent years has been taught in engineering schools, sanitary engineering and environmental health engineering, and I know from personal experience that power companies are beginning to make larger and better ecological surveys. I think this is, in part, due to a general public interest in ecology and in the problems of pollution. This interest has been growing, recently, and the power companies are becoming aware of this feeling. It may be hoped that this trend will continue and accelerate.

John Foerster: Dr. North, concerning this transplantation or increasing the range of a warm-water species of kelp: What has this done to the original community that was in the area before you began the introduction?

North: I didn't quite understand your question. You want to know if this new, rather unique community that is building up is due to exotic species, or to local species moving in?

Foerster: You indicated that there were endemic species associated with this warm-water kelp. When you transplant the kelp, do the endemics come with it?

North: No. These species may or may not occur with the kelp down in Turtle Bay, Mexico, where we get it from. Many of them don't occur in Turtle Bay. They are species that are, say, common in the open sea off Southern California, perhaps in Central California, but they aren't ordinarily found in the bays. However, once we put this open-sea plant in Newport Bay, we are beginning to find things such as kelp bass and half-moon and some starfish and Gorgonian corals, and other peculiar things that we have never seen in the Bay before. They are suddenly appearing in this kelp bed.

Foerster: Would you assess this as being beneficial when compared to the original habitat?

North: I would say yes. The original habitat as it existed twenty years ago was pretty good and was very nice bay habitat, but it's going downhill until there were only a few species there, up to a year ago. Now we're beginning to get a number of other species. I think that this is an improvement over what it was a year ago. But it may not have been an improvement over what it was twenty years ago.

Steven Carson: By way of introduction, I am a marine biologist who got out of his aquatic medium and went on into pharmacology and toxicology and rarely, if ever, looked at an aquatic beast. I'm struck however, because of the background that I stem from, with the similarities that have come up in the last few years. The discussion of biologists and engineers is a unique sort of discourse. I've lived through this in the biomedical engineering groups. When this was first getting a hold, the engineer would get up and say, "You tell me what your problems are and I'll solve them for you."

This was the beginning of a discourse that didn't go very far until the engineer learned a little biology and medicine; and, by the same token, the biologist learned a little of the engineer's medium. Just two weeks ago, I took part in a national research council session that was held on behalf of Congress where the legislators asked for guidance on orthopedic transplants, orthopedic media. The same problem there.

The metallurgist is developing his alloys, the biologist has the problem of handling them, and we're first getting to meet at the same table to solve the same problems. I'm struck by the similarity of some of the discussions that are taking place today.

The ecological systems that are poorly described in certain senses but very richly described in very limited areas are being disturbed. They are being disturbed in part by natural forces that are continuous and, in part, even by the forces of the discussants, by the last discussant. Yours is being done with good reason: you are attempting to transplant an area. But nevertheless, the ecology of this system is modified. And with it come a whole variety of changes that you hadn't counted upon. This I would comment from the last statement that you have made.

What is lacking is an apparent integrated approach that can only come with continuous discussions, similar to those that have taken place here.

The action of the Department of Interior, the water group that has sponsored this particular symposium, is perhaps to be congratulated because we find ourselves, for the first time, beginning to discuss a variety of problems, perhaps not hitting on all eight cylinders, but nevertheless, the problems are being laid out for all to see. With time will come some of the answers. Here, of course, is a basic question that has been raised, i.e., how much time do we have to get at the problem?

I would wonder, for example, in some of the discussions that have taken place; and let's not talk about marine ecology, let's talk about a freshwater system. We talk about an ecology that will be modified by thermal pollution, if I may use the term. But you're going to do more than to modify that ecology. You're going to provide a water source which may modify an ecological system. But I would like to remind you that this water must be re-used by a variety of industries that must make use of this water. Some of it will go into the human diet, because it is being used to prepare foods. Other water will be used for processing. Other portions of it will be used for pure washing. But in every instance, this water is going to be modified by each of the industrial users so as to achieve the final products which they have in mind. Therefore, you have not a *single* ecological problem because of a thermal effluent, but a *continuous* problem associated with man's need for this water.

Where the biologist and the engineer fit in together is perhaps an association at this stage of the game, in large part, to determine what some of the factors are. How far you can approach their control is something that only time will tell.

North: Thank you for your discussion on the discussion. I might say that there is one other need which I think Dr. Strickland beautifully outlined, and that is: you can't play God on a limited budget.

Hedgpeth: I think we must thank the discussant of the discussion for reminding us again that things are not simple and that they are completely interrelated. Which reminds me of the rather appalling bit of information that came out this morning: that, by 1980, one-fifth of all our runoff water will be cycled through power plants. The effect that's going to have on everything for hundreds of miles away and maybe some distance out in the ocean is rather difficult to assess at this moment, and yet we are now in 1968; that's twelve years away.

Also, these remarks about the twenty-first century: it's not so far away, either. True, I don't think we have enough time really to find out everything we're going to need. It reminds me of the rather philosophical thought, having sat through a commencement and watched another batch of Ph.D.s turned loose on the body politic. Pardon me for bringing up the Society for Prevention of Progress again, but how much of this is really worthwhile? We are going pellmell ahead on the assumption that all of these things are for the public good, and on the same assumption, that the population is going to keep on increasing to absorb them all. It has been suggested that perhaps that is not going to happen, perhaps by people who are more hopeful than circumstances really warrant.

Nevertheless, the question is, is all this worthwhile? Also I was reminded to think on this tack by this morning's San Francisco paper, which points out that perhaps the younger generations, down to the teeny-boppers (this may be one of the things that's wrong with them) are wondering if what we're doing is really for their good, when little kiddies of five and ten years old in Copenhagen get out and stop a bulldozer to protect their playground.

And seriously, there is this thought: Are all these things really the best thing we can do with our environment? Are there other alternatives? Of course, we must also admit that we have exterminated most of the whales so that we can't go back to whale-oil lamps, either.

So we are in a dilemma. But I think now and then it does help us to sober down and think of this possibility—that perhaps, as Oliver Cromwell said, "I beseech you, in the bowels of Christ, think it possible you may be mistaken."

Carlos Fetterolf: In response to the gentleman from Florida: If industry or a utility asked the ecologist if it's O.K. to place a discharge in a given location, I'm sure that I know, and I'm sure that industry knows, what attitude the biologist-ecologist would have in most cases. Therefore, industry quietly asks its own biological consultant for his opinion, weighs the cost-benefit-risk relationship, and then announces its plans and takes its chances with the regulatory agencies and the outside ecologists.

In Michigan, Consumers Power has recently undertaken a $70,000-to-$100,000 five-year pre- and post-operational study to evaluate the ecological effects one of their nuclear plants will have. American Electric Power, which plans to build a few miles south of Consumer Power on Lake Michigan, has hired well-qualified consultants to do their ecological research work for them on both radiological and thermal effects. Consumers Power and Dow are united for a combined nuclear-power project. Dow has their own division of environmental experts to evaluate what will happen. Much of the work undertaken in these evaluation projects will duplicate the work of the others. I wonder if the utility industry has an organization similar to those that the paper industry has, like the Institute of Paper Chemistry, or the National Council for Stream Improvement, which works for the industry and helps them come up with positions on environmental problems? Would the utility industry consider such an approach?

Fred A. Limpert: The utility industry does have the Edison Electric Institute. I have a copy of their catalog of research that is now under way, and much of it is in the environmental area.

Foerster: Both Dr. Hedgpeth and Dr. North alluded to the side-effects of thermal pollution. Perhaps these thermal discharges will not, in certain instances, be great enough to stimulate death of a fish or any macro-invertebrate. But what about items like shellfish poisoning with the *Gonyaulax* and the stimulation of the increase in growth with a longer warm season and longer shellfish-poisoning changes? Also, what about viral infections to the various plants, and what about the effect of temperature increasing the incubation of bacteria and viral diseases? The temperature itself may not directly kill the organism but may result in increasing the organisms that kill the organisms we call beneficial or useful.

North: You bring up what is perhaps the stickiest part of being a biological consultant. You tend to overlook some of these unexpected occurrences. Undoubtedly, if you sat down, you could make a list as long as your arm of unexpected events that might happen.

I think the only way we can guess what some of these unexpected happenings might be is to look to the south of us, where the waters are warmer, and see what the ecology of these areas are. Water there may have temperatures approximately the same as, say, location X, where a nuclear-power plant may be installed. But this is, at best, a speculative guess, because location X already has a complement of plants and animals that is quite different from the tropical situation and can be expected to react somewhat differently.

Hedgpeth: I would say substantially the same, especially if you were to watch what went on in the southern environment. However, you can't predict what some of those things might do from the side-effect basis when introduced to the more northern area.

We have in the past inadvertently spread things around the world. One of the most notorious cases is the introduction of the slipper limpet *Crepidula fornicata* to European waters, which developed there into a major competitor for oyster food, besides sitting in great piles on the oysters, whereas, in its own environment, it doesn't behave that way, since it's apparently not that numerous. It sort of exploded, so they could not get rid of it.

Sometimes, when something is moved to a new environment, it will take over. We succeeded once in this, rather remarkably, with the striped bass, in bringing out a mere handful and dumping them into San Francisco Bay in 1872 or 1876. But we haven't been able to do it with other things. We are working very hard now to get the Atlantic lobster established in British Columbia waters, after many false starts.

But quite often, it's the undesirable animal's behavior in its own environment that is more discrete and, shall we say, obscure. In other words, you don't know about some people. You get them to your house at a party and find out they are really the wrong kinds.

Charles Woelke: I have heard a number of comments about paralytic shellfish poisoning. There seems to be the attitude that this is associated only with warm water. I would suggest that Southeastern Alaska certainly does not have a great deal of warm water; yet, shellfish toxin is present. And the same thing is true in Eastern Canada. This may tend to be a little bit of a bugaboo that's being thrown up. Maybe someone else would like to say that paralytic shellfish poisoning is strictly warm-water, but I can't buy that particular point.

Hedgpeth: It was not my intention to associate *Gonyaulax* with temperature. I merely wished to point out that, because it infested mussels in the summertime, it made them less valuable to us as a food resource. The incidence of *Gonyaulax* may be more related to solar radiation or some other factor, and I trust that that information implication will be forgotten. I didn't intend that.

Loren D. Jensen: Some facts about the Edison Electric Institute are in order. Johns Hopkins University has had some research support from this institute since 1962. Currently, we have eleven research sites at operating thermal-generating plants in various kinds of meterological areas and representative types of water bodies. Since 1962, we have concentrated on studies relative to the physics of heat dissipation and, since 1965, have been collecting hydrographic and meteorological information from these eleven sites.

The data has been compiled using computer technology and will result, at the end of this year, in a rather extensive description of the sites, the meteorological and hydrographic variables that seem to be governing the ways in which heat is lost in various kinds of water. Since the first of this year, we have been doing biological research at two of these sites.

I think that it's fair to say that our support has been adequate for our needs, and we have reason to feel that our findings and recommendations will be used and passed on by the industry.

Moreover, we too have been educated by this kind of mutually beneficial association. First of all, we realize that the scientist who does consulting with industry has always had the stigma by his fellows of being somewhat of a sell-out, at best, somewhat less of a prestigious scientist. Much of the lack of realistic attitudes toward pollution problems can be traced to attitudes such as this.

There is a real need for the national experts in biology and ecology to come to the rescue of this very important problem and let their opinions be heard in discussion groups where we do what Dr. Strickland suggested this morning: let our hair down and say just how very little we know about collecting plankton and analyzing it and counting it and reporting it, and then get about the business of standardizing approaches so that we can begin to use each other's data and put it together in some kind of a meaningful fashion.

With regard to the need for such expertise, we plan to ask biologists, who heretofore have not been involved in these types of problems, to get together several times a year and assist in the development of our research project.

This certainly isn't going to be the cure-all, but I do think it is fair to say that the industry is definitely doing something in this particular field, and I might say that it plans to do considerably more in the future.

James R. Adams: Again, to comment on some of the points brought up by Dr. Jensen, I would like to say that, right now, there are about ten studies on thermal plants in the marine environment, both pre- and post-operational studies that are supported by the power companies. Now, at least four of these studies are being made by people in either the federal government or in universities with no strings attached, in which they report to private boards or to consultants.

One of these is a study by Dr. Merriman of Yale University; another is a study by Dr. Jack Pearce on the Cape Cod Canal Plant at Boston Edison; and then there are several other studies scattered around in different parts of the country. Most of these studies are financed at any-where from $75,000 to $200,000 a year, and there is going to be quite a bit of information available.

I think that the very best thing for us to do—that is, if you are con-cerned about the effects of a thermal power plant—is to go out and look at some of them. I'd say to get your feet wet, go swimming, and look around to see what you can find, because they are all over the place. If you really want to find out what is happening around these warm-water discharges, at least a gross idea, get in the water and look at it.

BANQUET ADDRESS

CERTAINLY I am greatly honored to be here tonight. It's a doubtful kind of assignment as far as any productivity is concerned. And, of course, you're all such high-powered technicians, as Dick Poston described you; I had you down as "magnificent technicians," which I think sounds a little better in the politician's ears.

But it is, first of all, a matter of pride that you have chosen Oregon for the first section of this first National Symposium on Thermal Pollution, and I hope that the relatively unpolluted surroundings will add to the productivity of your discussion. And I hope that your deliberations will point further to assisting us, the nation, and the world, in arriving at solutions to the pressing problems in this area. I certainly envy those of you like Dr. Parker and Mr. Alabaster and Dr. Krenkel, those coming to Oregon for the first time, and on a day like this.

But no matter how vital your symposium, I have to inform you that it was not the big event of the week, not the big event in Oregon in the past seven days. And Dr. Parker is brightening, I can see, because he said, "Why don't you just tell us about the Oregon Primary?" And it certainly was a wild one. We saw a fellow who was first, last time, in the Republican Primary in 1964, running a lousy third; and we saw a fellow who ran a lousy third in 1964 in the Oregon Primary, winning it handily on the Republican side.

And then we saw the tiger of the Senate, the most redoubtable and feared man in the Senate, our Senior Senator, getting less than fifty percent of the vote in his own primary. So those are some of the things that generally happen in Oregon in a preferential primary.

But there was even, I think, a more historic occasion than that: my mother, who will be 80 in August, held the first autograph party for the first book that she has ever written, just out now, called, *Ranch Under The Rimrock*. She always wanted to do everything I did. When I was on radio or a newsman for years, she wanted to be on radio; when I was on television for ten years, she wanted to be a political commentator; and then she wanted to be governor, and would have been a redoubtable one, no question about it. Finally, at age 77, she ran around end on me: she spent seven years writing this book; and by golly, they had television interviews with her, and I'll tell you she was simply magnificent, humorous, and just did a tremendous job. And so she has now surpassed me, and I think she's extremely happy.

She is a historic producer in other ways, too, I might say, because she had five children and all of them were conceived in Oregon and all of them were born in Massachusetts, and I think that was transcontinental pioneering in reverse. So we ended up, when our money ran out, on this end of the circuit and had a wonderful small ranch about which her book was written. We started a newspaper. I always wanted to be a newspaper-man, and so we started a little weekly paper up there that we circulated throughout the family all over the United States, and it was called *The Western Old World,* and we reported, very dutifully, the fights that Mother and Dad had, verbatim. And they still let it go; they didn't believe in censorship, which was pretty brave of them.

All of the assignations that took place in the pig-pens and in the dairy barns were written about; in fact, it was the most dutifully-reported ranch you've ever seen. It was really funny and there's a whole chapter on it in the book, if you ever get it. I'm not trying to sell the book, but it is good.

My Aunt Marian, who lived in Boston, sent us a check for $50, which was right before the depression officially began, although it had been going on at our ranch for quite a while. She said, "This is to pay you for past favors and to underwrite the future of your most enjoyable publication." This was the most money we had seen for a long time, so it went to our heads and we suspended publication, on the spot. That was the last issue we ever put out.

But being there on the ranch, which is a beautiful place between two rimrocks, you could look east to the Blue Mountains and west to the Three Sisters. This did give us a great background to be conservationists, and so we have always cringed at every threat to any species, and of course we cringe now, officially and personally, when we see the threat of disaster to the spring and summer chinook runs in the Columbia River. I think those of you coming from afar might like to know that the chinook runs are really regarded by scientists and by dickeybirds alike as the key to what kind of life we're having in the state of Oregon and in the Pacific Northwest.

It's not that you have to catch them; you don't have to catch them, but you want to know that they're there, that they're coursing up the river, these great fish. And when you know they're not there, if that day ever comes, then you know that we're sinking and drowning in our own pollution, because that's how important they are, symbolically. Besides, they are terrific fun to catch, when you're not so esthetic about them as I am tonight.

So surely the challenging issue that we face today is man's insatiable ability to despoil his environment. He's not foreign to this planet, but what he too frequently does is foreign to other living things on this planet. Since man is the only creature able to shape his surroundings to his needs, and the one species that possesses an absolutely incredible capacity to reproduce in violation of all of nature's laws, man must accommodate and be compatible if there is to be a tomorrow.

We always talk about what's going to happen in 1980, 1990, 2000; and Joel Hedgpeth wondered, this afternoon, if it was all worth even trying to do, because twenty years from now will find more people living on this planet than have lived on it since its entire creation.

And then I have the story that I intended to tell recalling the pilot on the jet-liner, as told by Jim Agee, in my introduction. The pilot said to the crew, "Ladies and gentlemen, this is your captain. I have two pieces of news for you tonight, one good and one bad, and I would like to tell you the bad news first: we're someplace between Portland and Seattle, and we're lost. Now for the good news: we're ten minutes ahead of schedule."

Jim stated that "this is just about where we are in our thermal-pollution symposium now. We're a little bit lost, but at least we're a little bit ahead of schedule."

It's a funny thing, the way these jokes go around. You see sort of a run on them, and I should have been a little bit suspicious, because we ran into it first when we were traveling with Governor Love in a chartered plane down the Coast and he went up to the cockpit and made this announcement in a sort of a bass voice.

But I would just like to interpolate my comment on it, that I hope we're not lost in approaching the problems of our environment. But I do think we are a bit ahead of schedule in providing people.

The program indicates my remarks are directed toward the livability in

Oregon and its future. Simply stated, that has been the thrust of my administration: enhancing Oregon's livability. It means a balanced development; it means being aware of the needs of the expanding economy, and at the same time taking care of the people, providing the jobs, and producing the items they need. But at the same time, it means giving maximum protection to our natural resources, so it has to mean conservation in the very best sense of the word.

At a time when conflicting interests of the users of our resources are at an all-time high—sports *vs.* commercial fishing, open space *vs.* industry, looking at a tree or cutting it—even defining conservation becomes most difficult. Perhaps the best definition of conservation was given by that little girl in school who was asked what conservation meant, and she replied, "Conservation is what you eat, and where you sleep, and what you wear. And if you don't, you won't!" I think that statement is most explicit for one so young.

I can't help thinking that the greatest contribution we can make to oncoming generations will be to turn away from past habits of development, production, and utilization for a single purpose; to apply our talents and knowledge to enriching human life in an improved surrounding. Past habits —let me indict them as hard as I can—have produced these problems. So let us be as imaginative as we can and not be bound by guidelines of inertia.

We had a great fellow in NIH who was out here—speaking of magnificent technicians—John Eberhard, and he told us that man's need for open space, on the basis of psychological studies made by NIH, was as great as his need for food, shelter, or sex. And he also told about a Boston urban-renewal project where the people they rooted up in that project, two years later, were suffering from the same sadness as if they had lost a very close and a very dear member of their family.

I would say that a nation which produces atomic bombs to win wars and wage peace should be capable of fashioning the atomic furnace to destroy our pollutants. A people that will put a man on the moon should be able to harness the fires of fission and not only use the heat for electric power, but also use it to enhance an environment where heat and water could conceivably turn the desert into an oasis.

And where atomic-power generation produces heat that could be damaging to aquatic life in fresh water, it might well enrich an ocean shore and add immeasurably to marine-life production, and recreational enjoyment, a la the David Black thesis, certainly a thesis that ought to be explored by you professionals before any implementation is thought about in any way.

I think ahead when an atomic plant might be located on the Oregon coast, providing still a further tool for the sea-grant work of Oregon State University. And may we be imaginative enough to make the deserts of Central Oregon and Eastern Oregon productive with the warmed waters of electric generation. And what about just thinking of this single use of atomic generation in terms of power? Perhaps warming water for 250,000 acres can be justification, just by itself, for constructing the plant, in the first place.

Oregon is moving into the atomic age, convinced we can have the needed plants without damage to our livability. Just in the last six months, I have developed a Nuclear Development Coordinating Committee, directed by Dr. Arthur Scott, of Reed College, and Dr. Chi Wang, of Oregon State University, both nationally recognized in their fields, and the committee is now preparing the state's position in a policy statement of guidelines.

Then there is my Committee of a Livable Oregon, and if there was ever a non-house committee, that one is. I put on it every person devoted to our natural resources in the sense that those resources ought never to be developed. I'm speaking of the Izaak Walton League and the Wildlife Federation and the Garden Councils and the garden clubs, and so on. This committee is working with state and federal agencies to inventory our resources, pinpointing the areas of resource conflict, and providing unbiased leadership for sound, compatible development.

The State Committee on Oceanography, which was set up during the last session of the legislature and of which I am chairman, assisted in securing sea-grant status for Oregon State University. And that, of course, received the top recognition and the top award from the National Science Foundation for all grants that it made this early spring.

We are continuing this co-operative effort between the state, the university, and industry in developing the resources of the sea; and somewhere in this equation, atomic power has a place.

In all the state agencies, boards, commissions, and departments, we have this year started a program of planning for our development. It sounds like a very dull word, and certainly planning sounds no bugles for anyone, but we're moving ahead in the most modern possible way with a great group of brains that do constitute the best "think-tank" we have ever had in our state capital of Salem.

Past, present, and continuing programs of each are being reviewed, and goals are being established. We are simply saying that a program cannot stay simply because it's been in existence; now, it has to fight for its life. Nobody ever analyzes programs in the sense that they ought to fight for their lives. They simply say, "Let's add ten percent more for this many more personnel."

Now, we're putting in new programs and the new programs are not off in the C budget that comes last, after all the other budgets are taken care of. But the C budgets, which are the new programs, are going to fight on a parity with the A budgets, to see if we can't inject some new blood and some thoughts and some new programs into state government in Oregon.

These goals and objectives will sustain and complement state objectives so we can achieve purposeful and coordinated development. Also, we have a Legislative Interim Committee proposing a bold and far-reaching program of land-use dedication, in order that our land may have its best use on the basis of the needs of agriculture, industry, cities, recreation, and commerce, without wasteful encroachment.

There is a brief analogy about a chicken and a pig, and Jim Boydston hasn't gotten onto this one. However, you will probably hear it, now, ten times in the next two weeks. The chicken and the pig were walking along a street in a small town and saw a sign that said, "Eat bacon and eggs for health." The chicken said, "Isn't it wonderful how we can contribute to the benefit of mankind?" The pig responded, "Well, to you it's a contribution, but to me it's a total commitment."

I think Dick Poston, who has lived with these problems so long and helped manage them so well in our state, will agree with me that in insuring a more livable Oregon, we are certain that each person is ready to make his contribution, whether he be an individual, an industrial employee, a businessman, or a member of some other organization or enterprize. For our happiness will come with blending our actions to our natural surroundings, because a happy people are productive, progressive, and purposeful. That's a "pollyanna" statement if I ever heard one, but I do agree with it.

I'd like to superimpose just one thought on the chairmanship of the State Sanitary Authority which I accepted for this brief period. It might sound to you, if you weren't aware of the circumstances, as if it were some kind of show-business, grandstand play by the governor. But we had nine or ten major bills before the legislature that no legislature in the history of the state has ever been able to pass. The chairman of the Sanitary Authority suddenly died of a heart attack, and I knew the chairman I wanted to have on the Sanitary Authority to succeed him: my Budget Director, John

Mosser, a brilliant Yale law-school graduate, the brightest man, by all odds —and both sides of the aisle admitted it—who had served in the Oregon State Legislature in this century. He is an absolutely fearless man, yet scientific, at the same time in keeping our environment and fighting for it and serving all his time without any pay in this job. However, he wasn't going to be available. Consequently, I took the job, to hold the chairmanship open. At the same time, my purpose was to dramatize to the legislature my commitment to the need for having this legislation passed. It worked out admirably, because we got all the legislation we were after, and we also got John Mosser, three months later. So this is one of those things in politics where there was a one-hundred-percent take, one-hundred-percent value received for the people of our great state.

I just want to say once again that I assure you of the pleasure Oregon takes in acting as host for this significant conference. I bid you welcome, especially those who are going to be hosts for the one at Vanderbilt University, when you come to the Beaver State, and you from England, too. And I hope each return will show to you that we have kept our pledge to enhance the livability of Oregon. For we really believe that in Oregon our prime selling points will always be pure air, clean waters, and bountiful natural resources. This we say, this we live, and this we state to you.

Chapter 6   Donald I. Mount

DEVELOPING THERMAL REQUIREMENTS FOR FRESHWATER FISHES

THE need for documenting water-quality requirements for various water uses, including the production of fish for sport and commercial harvesting, has been recognized for many years. The recent emphasis on water-quality standards has made apparent the importance of establishing valid and precise water-quality requirements for aquatic life. Our present state of knowledge precludes the establishing of criteria for even the common water-quality parameters, such as temperature, oxygen, and hydrogen-ion concentration. Required levels or limitations for these are as important as those for the many chemical pollutants which exist in the nation's waters, because aquatic life, and fish in particular, must get their oxygen from the water; and because water at natural temperature can hold only a small quantity of oxygen, there can be little variation in the oxygen concentrations in water if a healthy aquatic-life population is to be maintained. Higher temperatures increase oxygen requirements and reduce the oxygen-holding capacity of water. Fishes and most other aquatic animals are cold-blooded, and temperature plays an important, highly regulatory role in their physiology. Temperature intimately controls the reproductive cycles, the rate of digestion, the rate of respiration, and the chemical activities that occur in a fish's body. For these reasons, thermal requirements are among the most difficult to define and establish.

Much of the confusion that exists today concerning thermal requirements for aquatic life results from indecision as to the goals we hope to attain. For too long we have sought some "magic number" which, if met, would protect all aquatic life. Clear-thinking scientists and engineers have known for some time that the many species of fish and other aquatic animals vary greatly in individual tolerances or sensitivities. Some of the confusion can

140

be eliminated if we consider the requirements, for example, of the coho salmon, the northern pike, white bass, carp, etc., and stop searching for a single thermal requirement for "aquatic life" in general. In the latter approach, we are seeking to draw a mean for all species, leaving some underprotected and some overprotected.

Obviously, if we are to have a standard limiting temperatures in a particular water body, there can be but one standard. In determining what that value should be, however, we are confused by a number of factors, including the differing sensitivities of species and the problems of economic and social values. The biologist can simplify his job greatly by first determining for each important species its thermal requirements. Once these are determined, we can go about the task of selecting an intermediate value as a standard. If we had data for each species, we would at least know which are being fully protected, which are being partially protected, and which species are going to be eliminated.

In approaching the problem of developing thermal requirements for freshwater fishes, a basic decision must be made as to the objective to be reached. It appears that there are two different objectives which one might work toward. The first is to protect an unaltered environment, or similarly, a diversified ecology. To attain this objective, one is not concerned with protecting only those species (and food-chain organisms) which are utilized by man for sport or commercial reasons; one's interest extends to many other types of organisms found in the aquatic habitat. The second objective one might wish to achieve—quite different from the first—is to protect only those species which are desired and utilized by man and their supporting food-chain organisms, *necessary* in order to produce the crop. If there are areas in our country where man has not altered the streams or lakes, their number is small, and I shall not spend time here discussing them. Suffice it to say, there is a good case for protecting such areas for future scientific study and as a reservoir for maintaining species variety.

At the National Water-Quality Laboratory, we hold the view that we cannot protect all living organisms that normally or naturally occur in a particular water body and are now endangered or removed by man's activities. We believe, further, that the average citizen is not interested in protecting these species. Probably more than 99 percent of the people see or utilize less than 1 percent of the organisms that occur in a stream. We do need to preserve, in certain areas, a diversified ecology for scientific experimentation, but let us not kid ourselves: the public is not interested in protecting diatoms, mayflies, or water fleas, unless they know that such species are necessary to produce desirable shell- or fin-fish. Popular concern for aquatic organisms is quite different from concern for other animal groups such as, for example, birds. A rare or uncommon bird species is seen by many; people will spend time, effort, and money to see those birds and derive satisfaction and value in so doing. Such is not the case with a rare species of algae. When one views the tremendous amount of work facing those who work in water-pollution control, one is convinced that the first order of business must be to develop the tolerance limits under which the harvested species and their necessary food organisms can thrive. Even to maintain a diversity of species, the harvested ones must be studied at some time.

This paper does not seek to promote one approach over the other, but we must have one objective clearly in mind as we proceed. Thus, the remainder of this discussion will consider the development of requirements for protecting the harvested species and the necessary food organisms. Those who favor protecting an unaltered or a diversified ecology will have to pursue a different approach because they have in mind a different objective.

The first task is to determine the thermal requirements of the species that are to be harvested and then make additional studies of the food-chain organisms as needed. There is a substantial amount of evidence that fishes frequently are more sensitive to elevated temperatures than are most of the food-chain organisms. This is not to say that some varieties of invertebrates or phytoplankton are not more sensitive than the fishes, but it does imply that, under increased temperature, sufficient food organisms will be present, though they may be of different kinds, to support the harvested crop. The statement is frequently made in the literature that productivity is greater if the ecology is diversified. This is *incorrect,* if, by production, we refer only to the harvested species; it *is correct,* if, by production, we mean the total amount of tissue elaborated. If we are protecting the environment for harvested species, then it is better and more efficient to have fewer species present, concentrating on those that man wishes to harvest, plus the necessary food organisms. The work to be done is thereby lessened, because we need to determine the requirements for far fewer organisms to protect the desired species. Simplified ecosystems tend to be less adaptive, however, and the risk of a failure is higher, should a change occur.

It is the experience of those who have worked on surveys of rivers and lakes where pollution exists that the fishes usually are the first animal group to disappear and the phytoplankton is among the last to disappear. These observations are consistent with the principles of toxicology; the greater the number of species within an organism group, the greater the range of tolerances. Therefore, one would expect to have some algae growing in places where macro-invertebrates cannot survive, and one would also expect to have macro-invertebrates living in places where fishes cannot survive.

It is important to concentrate our efforts on the desirable species that are most likely to be affected, and we must not attribute scientific desires to the public. There are many places in this country where species of fishes that are very heat-tolerant are highly valued for sport or commercial purposes. Catfish and carp are two examples. We tend to think that protecting such species is accepting inferior-quality aquatic life. Such is not the case, and one would not make that statement if he were thinking of fishermen in certain areas of the nation. Many fishermen would much rather catch bullheads or channel catfish than lake trout. We must be very careful that we do not impose on the public at large the desires of a few scientists or sportsmen. Again, let me emphasize that it is a rare individual among the non-scientific people who sees more than a small fraction of the aquatic life species that occur in lake or stream. We must develop requirements that will protect the species the public wishes to harvest, so we should *begin* working with those species.

Turning now to the requirements for freshwater fishes, the information needs concerning thermal requirements are general but basic. Two functions which cannot be altered if we are to have a satisfactory crop of fish are reproduction and growth. If these are satisfactory, one would be hard pressed to justify any further restrictions on the addition of heat to a body of water, since the "crop" is acceptable. In developing thermal requirements to protect growth and reproduction of fishes, it would seem that we need to have information in three principal areas. The first is to determine maximum temperatures at which normal growth and reproduction can occur without impairment. The second area is to determine how much or to what extent temperatures may fluctuate without interfering with growth and reproduction. While most biologists are convinced that temperature fluctuations are very damaging to fishes, there is very little hard evidence to support this view. This statement is not intended to imply that fluctuating temperatures have no adverse effects, but only to point out that we do not have sufficient evidence to support such a statement fully. The third area of informational need is to determine whether cooler temperatures in the winter are necessary in order for reproduction to occur. Again, most biologists seem to feel that cooler winter temperatures are requisite, but there is little evidence to support this. Several of the fish species maintained in hatcheries at the same temperature the year round are able to spawn and reproduce normally. There is an urgent need to know which species require seasonal fluctuation for successful reproduction.

Factors other than growth and reproduction also must be considered in developing thermal requirements. Avoidance reactions, especially for migratory species, can be very important. Cases have been documented in which certain fish species attracted to hot-water areas were not able to survive for the period of time they stayed, or died when they left the area of heated water. In a similar way, fishes migrating upstream must be able to detect and avoid adverse mixing zones; if the zone of unfavorable temperature extends entirely across the stream, the migration can be completely blocked. A spawning migration blocked by a mixing zone will eliminate the population.

Temperatures different from those normally existing in a body of water can result in a change in the species composition of the fish fauna. This effect can be either adverse or beneficial, depending on the species which are to be protected and harvested. In many warm-climate areas, trout fishermen find a bonanza where cool water is discharged from the bottom of reservoirs. In other areas, trout fishermen are unhappy because the cold-water species they wish to catch have been displaced by warmer-water species, such as walleye or largemouth bass. Sometimes fine warm-water fisheries are spoiled by cool-water releases from reservoirs. Striped bass are not inferior to trout except in the minds of certain people.

Evidence suggests increased danger of disease as water temperatures are increased. The disease factor in fishes living in elevated temperatures is a significant area that has not received sufficient attention. Epidemics can easily reduce the harvest of fishes, and further information on the role of disease must be obtained in our coming research efforts.

Even though fishes are a sensitive group of aquatic organisms, it would be risky to ignore the food-chain organisms. One faces bewildering decisions in choosing the food organisms with which to work. Nearly all biologists agree that most fish species will alter their food habits considerably, depending upon the type of food organisms available. One can show, for example, that rainbow trout eat fish, frogs, insects, crayfish, leeches, snails, and many other sorts of animals, depending on the origin of the data cited.

Fishery biologists are now generally in agreement that the year-class strength of many species of fish depends on survival of the fry during their first three to four weeks of life. In many instances, the right type of food must be available at the precise time when the fry begin to feed. It appears that first priority should be given to studies of the food organisms on which newly-hatched fry feed. It is quite clear that food for the fry must be present or the crop will fail, and it is equally obvious that the kinds and types of food that newly-hatched fry can eat are much more restricted than are those eaten by fingerling or adult fishes. Therefore, of the food-chain organisms, the fry-food organism should receive our first attention.

Lakes, and reservoirs that resemble lakes, are becoming important and support substantial fisheries, both sport and commercial. In such habitats, certain of the crustacean groups play an important role in the fish-food chain. The cladocerans, for example, should receive more research effort than they have in the past, even at the expense of studies of the stonefly, mayfly, and caddis fly. Studies by the Fish and Wildlife Service, not yet published, have shown a few genera to be the principal food of the fry and fingerlings of many of the important species of fishes. These same genera are also important as food for adult fishes.

Certain of the food organisms essential to producing a fish crop must live a part of their life cycle as terrestrial forms. We need studies to determine whether these forms can thrive when pre-emergence occurs as a result of abnormally warm water in the winter. The really important consideration, and the one most difficult to evaluate, is whether these species that might be destroyed by cold air temperature actually are necesary to support a harvestable crop of fishes. It could well be that, even though the mayflies or the stoneflies are destroyed because they emerge in winter when the air is cold and they cannot mate, other organisms will suffice as the missing link in the food chain.

Most would agree that, as a water temperature is raised, the total tissue elaborated will also increase. This does not mean, however, that production of the desired species will increase. Studies are urgently needed to determine whether the organisms which thrive at the higher temperatures are suitable as food. The species composition of phytoplankton is drastically changed when the temperature is altered, and a careful evaluation is needed to determine whether this change is detrimental in terms of producing harvestable species of fishes. Primary production data are difficult to translate into fish-production values. Actually, only a very small portion of the biomass is needed to produce an adequate crop of fishes, at least in the more fertile waters.

Present research on thermal requirements at the National Water-Quality Laboratory is designed with the intent to develop information in the areas of the greatest need. We believe that one of the first and greatest needs is to determine the maximum thermal limits in which selected species of important fishes can thrive. We are presently establishing a field-study site near a power plant on the St. Croix River in Minnesota. Here we will maintain such important species as northern pike, smallmouth bass, walleye, and white bass at differing constant temperatures to determine: 1) at what constant maximum temperature the species can spawn and grow normally, and 2) if cooler temperatures are needed during a part of the year to initiate the reproductive cycle. These studies should produce useful data about individual species but will not measure the effect on the food-chain organisms.

The food-chain effects will be investigated at another location near a nuclear power plant on the Mississippi River at Monticello, Minnesota. At this site, we plan to build large outdoor ponds, small lakes, and channels adjacent to the power plant, maintaining these at elevated temperatures. The fishes will be required to live on the food organisms produced in the environment. In this way, we can determine whether the elimination of some foods (for example, the mayfly, because of winter emergence) has any significant effect on fish-crop production. I emphasize that, in these experiments, we will not feed the fishes; they must live on the organisms that grow in the ponds or channels.

Construction of these structures is planned for the spring of 1969 and actual experimentation, we hope, will begin with the summer of 1970. The Northern States Power Company has been very cooperative at both sites. They have offered the use of a building on the St. Croix River and extensive acreage at Monticello to develop these research programs.

We also expect to experiment with the best method of introducing heat. Seemingly, a catfish population would suffer little harm from warm water introduced at the surface. On the other hand, we can see where surface introduction of warm water might seriously impair the production of white bass. We hope to determine the best place and method of introducing heat into an environment of various important species.

It is the intent at the National Water-Quality Laboratory also to consider and investigate the effects of thermal discharges on uses of water other than as aquatic life habitat. One consideration suggested by those concerned is the problem of increased algal growth resulting from the warmer waters. This factor can be partially studied at the field location on the

Mississippi River. In certain lakes, such as Lake Michigan, the heat added to the water will warm the involved areas sufficiently to make them much more suitable for swimming. On the other hand, should this heat result in growths of filamentous algae or other types of aquatic plants, the advantage of warm swimming waters may be offset by the increased growth of noxious vegetation. Attention needs to be given to the effect on the bacterial quality of the water. The approaches are not very clear at this time and our planning and preparation for these studies are just beginning.

We envision that the quality of water used for industrial cooling might also be adversely affected by a rise in water temperature. It is easy to postulate that growth of slimes and other nuisance organisms in condensers would increase greatly as a result of higher temperature. We expect to evaluate this problem and develop a report outlining the results when warmer-than-normal water is used for cooling purposes. Hot water additions to sources of public water supply may also increase the cost of water treatment in municipal plants and, through the stimulation of algal growth, produce taste and odor problems.

In conclusion, there is an apparent urgent need for workers in the field to cooperate closely and pursue common objectives. The problems are with us now, and answers are needed as soon as possible, if adequate and fair water-quality standards are to be established.

DISCUSSION/ Peter Doudoroff

NOT having seen the full text of Dr. Mount's paper before preparing my formal discussion, I have decided to confine my prepared remarks to a single pertinent subject that I do not expect he will discuss fully, namely, temperature optima for the growth of fish.

Except for temperatures compatible with successful reproduction and continued survival of young and adults, fishes have no clearly definable thermal requirements. Successful populations of a fish species can persist under widely varying temperature conditions, but growth and production rates of the fish can vary greatly with these conditions, just as they can vary greatly with the abundance of food in the environment. It is with the *production* of valuable fish species that we are concerned primarily and not merely their continued existence. Meaningful temperature criteria that are to be relied upon in protecting fisheries, except where temperature increases are of brief duration only, must derive chiefly from research into the influence of temperature on fish production, rather than from information on limits of tolerance or on the distribution of fish species in nature. For this reason, major emphasis has been placed in recent years on fish bioenergetics, feeding, and growth by the Pacific Cooperative Water Pollution Laboratories of Oregon State University. In a recent paper, Warren and Davis (1967) have reviewed the state of our knowledge of this general subject, presenting some of the basic concepts and helpful background information.

If the productive capacity of fish populations is to be fully protected in regulating water quality, only temperatures that are not less favorable for growth of the fish than natural temperatures should be deemed permissible. But if the maintenance of water temperatures as nearly as possible optimal for production of fish is indeed to be our objective in controlling thermal pollution, we must ask questions concerning the constancy of the temperature optimum for growth of a given fish species. What factors or variables can influence the optimum and to what extent? Here, I want to emphasize the influence of one factor whose presumably great importance seems to have been too often entirely overlooked, namely, the abundance or availability of food in the natural environment. As I have pointed out already, the growth rates of fish in nature can vary greatly with the available food supply. The reason for this influence of food availability on growth is obvious. But how can the availability of food influence the temperature optimum for growth?

The higher the temperature of the water is, the higher is the metabolic rate of the fish, and therefore, the greater is the amount of food required for maintenance of its body weight, with no growth. The efficiency of utilization of food for growth, or its conversion to body tissue, may be reduced at high temperatures for this reason and others. Given a certain restricted daily ration, a fish living at a low temperature may be able not only to satisfy the relatively small maintenance-food requirement, but also to grow at a moderate rate. The gross food-conversion efficiency—that is, the ratio of the weight-gain or its caloric equivalent to the weight or caloric equivalent of food consumed—may be quite high. But a higher temperature that is optimal for food intake (or appetite) and for growth when rations are unrestricted, the same restricted daily ration may be barely sufficient or insufficient for maintenance of body weight. The gross food-conversion efficiency thus may fall to zero at this "optimal" temperature.

Several examples of such shifting of the thermal optimum for growth can be drawn from results of recent laboratory studies. In experiments performed at the Biological Station of the Fisheries Research Board of Canada, Nanaimo, British Columbia, and in work recently reported by Brett *et al.* (1967), young sockeye salmon kept in aquaria and given all the food that they would eat grew best at temperatures near 15° C. Their growth rates declined sharply at higher and lower temperatures. Yet, when fed a small, restricted daily ration, these fish grew best at temperatures near or below 5° C., and grew not at all at 15° C., the optimal temperature for the growth of abundantly fed fish. Intermediate temperature optima for growth were associated with intermediate rations. A similar downward shift of the temperature optimum for growth with great reduction of rations was observed in three of four experiments with juvenile coho salmon performed in our laboratory by R. C. Averett (unpublished data). When rations were unrestricted, growth was best at temperatures of 17° C. or 20° C., and was much reduced at temperatures of 8° C. or less. When rations were sufficiently reduced and uniform, the best growth occurred at the lowest temperature tested in each of the three experiments, that is, at 5° C. or 8° C. Other data (unpublished) obtained in our laboratory by R. A. Lee show that the growth rates of largemouth bass kept in laboratory aquaria on unrestricted rations of mosquito fish continued to increase with rise of temperature up to 31° C., the highest test temperature. However, their maintenance ration increased also from about 0.5 percent of their body weight per day (at 10° C.) to about 2 percent of their body weight per day. When kept at 15° C. and fed a ration (about 2 percent of body weight per day) that permitted virtually no growth or even resulted in weight losses at temperatures of 25° C. to 31° C., the bass grew at a moderate rate indicative of a very high gross efficiency of food conversion (conversion ratio near 0.40).

It is interesting to note, in this connection, that the maintenance-food requirements of fish in nature may be considerably greater than those of fish held in laboratory aquaria. Results of some preliminary experiments performed in our laboratory by R. A. Lee (unpublished data) show that the metabolic rates of largemouth bass preying on mosquito fish in small, artificial ponds were, on the average, higher, when food was scarce, than that of bass held in aquaria at the same temperature and ingesting no less food. The elevated metabolic rates of the bass must be ascribed to the activity involved in seeking and pursuing the scarce prey. Bass in aquaria had no reason to be more active when fed little, and so they conserved their energy. It can be seen that, when the level of food intake is uniform and low, temperatures compatible with growth in nature may be even lower than those compatible with growth in the laboratory.

We can now see that a successful population of fish inhabiting a cold, oligotrophic water that is very deficient in fish foods may become unable to maintain itself when the temperature is raised to a level that is optimal for growth of the species when food is unlimited or very abundant. The fish may do as well or even better at the higher temperature, of course, if the rise of temperature results in either a marked increase in abundance of food organisms or an improvement of the success of the fish in exploiting their food resources. However, materially greater availability of food at the higher temperature may not be assumed.

Naturally cold waters are inhabited by organisms that are well adapted for life at the low temperatures, and these waters can be quite productive of fish foods when there is no severely limiting deficiency of plant nutrients, light, or suitable substrate. Temperature increases accelerate life processes and thus may result in increased rates of production of fish foods; but, as we have already seen, they do not necessarily accelerate growth and production rates of aquatic organisms. When the food supply of fish is mostly of terrestrial origin, as it can be during summer months in small trout streams, it obviously cannot be much influenced by the temperature of the water.

Little is known about the influence of temperature on the feeding efficiency of fishes, and this matter certainly needs investigation. Foraging success sometimes may improve as the temperature rises, because of an increase in the activity of the fish and improvement of their swimming performance. However, any increase of the swimming speeds and agility of predatory fish with rise of temperatures may be matched or surpassed by corresponding improvement of the swimming performance or agility of their prey. In streams well populated with fish that feed predominantly on drifting organisms, most of these available food organisms may be captured at any temperature over a wide range.

We can conclude that the evaluation of temperature optima for growth and production of fishes is not easy. The abundance of fish foods or the natural growth rates of the fish in problem waters evidently must be considered in attempting to predict the effects of temperature increases on fisheries, and future research in the temperature relations of fishes must be planned accordingly. Although the growth of valuable fishes may often be improved by raising the temperature of relatively cold waters, higher temperatures that proved optimal for growth under some laboratory or field conditions should not be accepted as satisfactory until careful attention has been given to the pertinent bioenergetic considerations.

REFERENCES

Brett, J. R., J. E. Shelbourn, and C. T. Shoop. 1967. "The Relation of Temperature and Food Ration to the Growth Rate of Young Sockeye Salmon." Abstracts of Paper at the American Fisheries Society 97th Annual Meeting, September 13–15, 1967, Toronto, Ontario, Canada. pp. 9–10. (Also data presented at the above meeting and personally communicated to the author by J. R. Brett.)

Warren, C. E., and G. E. Davis. 1967. "Laboratory Studies on the Feeding, Bioenergetics, and Growth of Fish." In *The Biological Basis of Freshwater Fish Production,* edited by S. D. Gerking, pp. 175–214. Oxford: Blackwell.

DISCUSSION FROM THE FLOOR

Peter Doudoroff: I wanted to take just a minute, in addition to giving my own paper, to make one or two comments on Dr. Mount's paper, which I have now read and heard.

I think that this matter of species diversity is one about which many ecologists are quite touchy. It has been often argued that a wide diversity of species insures stability of a system. If conditions change and become unfavorable for a small number of species present in an aquatic environment, there may be almost nothing left. All the desirable fish species may be wiped out, or their food supply may be seriously impaired. On the other hand, if you have a large variety of species of fish and fish-food organisms, conditions adverse for some may not have such serious effects, because other, more resistant species will be present to take their place in the environment or in a food chain of the eliminated sensitive species.

However, I generally tend to agree with Dr. Mount in believing that we should to some extent consider our waters as crop-land and recognize that, as we seek to improve our management practices—and this is what our research should be directed towards—we should be able to learn what conditions must be maintained, and also how we can maintain these conditions. Certainly, as Dr. Mount pointed out, most of our agricultural practices are based on this idea. We increase production of desirable species through practices which reduce species diversity drastically. I think we should consider, to some extent, arguments on both sides and not go too far in one direction or the other. Furthermore, we should recognize that, as we improve our management practices and our degree of control over our environment, we may be able to sacrifice, to some extent, species diversity, which otherwise perhaps would need to be preserved.

In general, I agree with most of the things that Dr. Mount has said. Even though we have more money for research and more people working in our area than before, we should not scatter our research efforts too broadly. Without very clear practical objectives, I think that we will not make very rapid progress, even with the increased amount of effort and money that can be devoted to the work, and we haven't much time to waste. The problems facing us now will not be solved by people doing only basic research in areas in which they happen to be personally interested. Some of this is necessary, but I think, especially in our government laboratories, investigators should focus attention rather clearly on some major practical problems that can be solved in a reasonable amount of time.

Clare Idyll: Dr. Mount warned us that he was trying to be the devil's advocate, and perhaps he did not believe all he said. Rather than attack Dr. Mount, I would like to attack what I think is a concept that he put forward as a spur to discussion.

This is the concept that we should study organisms that we can see or that the public can see, to the detriment of some other organisms that perhaps are never seen by the public, but which may be of equal importance.

If we put our priority on organisms that the public sees and appreciates, we may miss the mark entirely, and I hate to think that we were setting our priority in this fashion. Instead of allowing the public to lead us to the kind of organisms that we should put our priority on, surely, we, as biologists, ought to make the decision ourselves. I doubt that Dr. Mount really meant this, but if there was any implication of it, I protest. Also, I wonder if it is valid to make a distinction between studying the whole environment or studying certain parts of it. The more I study natural systems, the more I am convinced that their parts are so closely interrelated that it is very difficult to separate segments and to study only the ones "which are of commercial or sport value." I think all of us have recognized that if you knock out any part of the food chain, you are likely to knock out the whole business, and it's going to come tumbling down on you like a house of cards.

Finally, I would question Dr. Mount's statement that the public is unaware and unappreciative of certain of the organisms that they don't see, that diatoms mean nothing to them or that clostridium means nothing to them. I think we underestimate the capacity of the public to appreciate these things, if they are properly led. I like to think that in Day County, in south Florida, we have managed to convince the public that there is, in fact, a value to organisms that they cannot see. We have tried very hard to get this idea across to them with respect to Biscayne Bay. We are trying very hard to preserve Biscayne Bay, not merely because of the organisms that can be seen, such as birds and some of the larger fishes, but because of the diatoms and some of the other things that cannot be seen. We have led them to some understanding, at least, even though it is not as much as we would like or as much as we will lead them to in the future.

We have fought hard against the idea that we are trying to preserve Biscayne Bay and Everglades National Park "for the sake of the alligators," as many people have claimed. But we make the point that the fact that the alligators are there has a value to people, and that, more than that, even organisms that they cannot see are of value.

So let us not underestimate the capacity of the public to appreciate these things, and let us not be led by them to study and protect only the organisms that they can see and can use.

John Cairns, Jr.: I would like to say a few words about Dr. Mount's comment on the simplified ecosystems, which I think will be essential in some cases. I agree with him that we obtain an increased harvestable crop from some of these, and also that they are quite unstable.

There are some hidden costs, which he didn't mention, although I am sure he is aware of them. They should be emphasized! It may be necessary to manage these simplified systems to a much greater extent than the self-regulating, complex systems. This management might include such things as adding nutrients to the system on occasion, providing hatcheries for the valuable species that might not be able to spawn under the new conditions, and introducing new species that are appropriate to this simplified system, so that the aquatic community doesn't disappear, and so that the harvestable crop is realized. This means a capital investment! The second point I want to make is that as one establishes a simplified system and stabilizes it by controlling the environment, this may preclude the introduction of new industries, since the new system has been designed for a specific set of conditions. In other words, one has, by simplifying the ecosystem and managing it to produce a specific end product, increased the rigidity of the system, and this may not always be desirable.

Donald I. Mount: I would like to comment briefly on Dr. Idyll's comment. The latter points that were made, I can agree with. In regard to identifying which organisms we are trying to protect, I certainly cannot argue that one can find justification for defending the preservation of any given species, or leading the public to appreciate the value of any given species. You must also realize that the source of the money is important. All of the money comes from the public, and in the case of agencies which support most of the research on thermal pollution, it is coming from the pocket of the public. It is not our position, in the federal agencies that are applied or service-type organizations, to tell the public what they ought to appreciate. This may be the responsibility of the university, but I clearly disagree with you that it is the responsibility of such federal agencies. If you in the universities care to shift the thinking of the public in such a way that they appreciate diatoms rather than insects, then we will adhere to the public's wishes.

At this point, I am no longer being a devil's advocate, but conveying my own philosophy, and I would be glad to have you show me where this is wrong. The public already wants more organisms that we can protect. I am sure you can keep pace with encouraging protection of other organisms faster than we can figure out how to protect them.

Ruth Patrick: I agree with Dr. Cairns's statement that there are certain hidden costs in a simplified system. On land and in water, so far as we know, but certainly, on land, simplified systems are relatively unstable systems. One only has to look at the costs of insecticides and fungicides to see what happens when one simplifies a system. Also, as Dr. Idyll pointed out, we often do not understand, certainly at this stage of knowledge, the interrelationships of species to each other, and what may happen when we eliminate predators, such as those on a disease-causing organism. Indeed, if we so simplify these systems, we may subject the species we wish to preserve to severe diseases and plagues.

All of us know that man must use his surface waters, and man must learn to use and not abuse these waters. I would like to suggest that, rather than allowing simplified systems to develop as a result of pollution, we should seek to learn how to manage our rivers so that conditions will be present that will support diversified aquatic life. This probably can be done by planning the use of the watershed; by structurally altering currents or substrates; and perhaps by adding minute amounts of some nutrients. The diversity of a pristine stream is not necessarily needed; rather, a high-enough diversity should be present to provide the necessary predator-prey relationships to maintain a stable community.

Secondly, I would like to comment on the fact that in many of the natural eastern streams that I have studied very thoroughly, very few Cladocerans have been found. Furthermore, the caddis fly larvae and mayfly nymphs are in the streams all year long. One of my colleagues, Mr. J. Richardson, who is getting his doctor's degree from the University of Utah, has recently completed very extensive studies that show that these organisms are present in the streams all year long, and that different species emerge at different times of the year. So let us not belittle the food value of these organisms.

Mount: I appreciate these comments very much, because I think this is exactly what we need to discuss. I can't disagree with you, particularly, but continuing to play the devil's advocate role, I must mention our observations in a lake near our laboratory. There are more man-hours of fishing pleasure derived from that one lake than from all of the trout streams in the state. The point I am making is that reservoir and lake fishing is becoming greatly expanded, and the amount of pleasure derived from it is fantastic in parts of the country, and yet more effort (as near as I can tell, from looking at what is going on) is being spent on stream organisms which are, from the standpoint of public use, much less important than some of the ones that are the basis of the food chain for important fisheries.

Also, another point made by Dr. Patrick, and again a broad one, is that we are taking chances if we go to a more simplified ecology. I have two comments. You imply that if we protect a simplified ecology we will not be protecting many other species. I think that the species that man desires to harvest cover a broad range of tolerances and that many other species will also be protected. But even if that were not true, I think that one of the greatest faults that we as biologists have had (and perhaps here the engineers have been ahead of us in the game, although it hurts me to admit it) is that we have not been willing to take risks and base our efforts and work on probabilities. Surely, every time engineers build a bridge they know that there is a chance that it will fail. I think the same is true in biology. We must recognize that we are taking chances with the ecology and with the "crop land" if we predict effect.

No matter what we do, we're taking chances, and I am pleading that somehow we must try to get together and use the information we have. I think if NASA had not decided they were going to go to the moon, but rather that they would go to any old planet that they could reach, they wouldn't be nearly as far along as they are now. If we should protect other than a harvestable crop, I'll go along with it, because I am more interested in getting cooperation to complete a job than riding off in twenty different directions at once, which is what we are doing right now.

Patrick: I simply say that I do feel that we must do research to understand the hazards of the simplified system. In other words, how far can we push diversity by decreasing it before we get into trouble? I certainly don't think we ought to ride off on white horses now, simplifying every environment.

Mount: The coho experiment in Lake Michigan ought to be an excellent opportunity to do the very sort of thing that you are suggesting. Those of you who were at the Eleventh Great Lakes Research Conference heard some discussion about the hazards involved in trying to manage a coho fishery in Lake Michigan where only a few species are involved, especially the lake trout, and the coho. I agree with you wholeheartedly, and this is one area of research which I very clearly left out in my paper. I appreciate your comment.

Donald R. Johnson: I can't help comparing the biologists's challenge with that of the engineer. When the engineer designs his bridge, he is dealing with steel that he pretends is about one-fourth or one-fifth as strong as it actually is, for his design purposes. Therefore, he has a safety factor of about four or five in his design. The biologists, on the other hand, don't always include this in their particular challenge.

Max Katz: I am sitting here and I am watching the biologists blow their opportunity again; in fact, we are blowing it completely. The trouble with most of us biologists is that we are in a dream-world. There is one thing that we are forgetting. We have created the world population explosion. The population of the world is doubling every thirty years. The wastes we are producing are doubling. Our demands for goods are doubling and tripling. Let us not forget that these goods are going to be produced. These thermal power plants are going to go in. The engineers are going to build them, and the engineers are going to come to the biologists for answers.

We are going to have to supply these answers and we have to get out of the ivory tower, forget about impressing ourselves with our academic purity, get our hands dirty, and supply some useful answers.

Yesterday, we sat through a session that was completely negative. The speakers, all of whom have a lot of good information, tried to compete with each other to show how pure they were and presented

dozens of reasons why they could not come out with any firm information. It was very amusing. It was very well done. Nothing was presented
and the engineers in the audience, the guys who will build the thermal
power plants, will say, "Well, the hell with biologists. We're going to
do what we want."

Don Mount and Peter Doudoroff are presenting a positive approach.
Now, I say that we biologists had better get on the cotton-picking wagon,
or we'll lose out even more than we have.

Marc J. Imlay: This idea that in agriculture a simplified system works best
has a lot of sting; but I happened to attend a lecture on agriculture
around the world, which described attempts to import the American
system, and it turns out that, in warm-weather areas, they have gone back
to diversity. They have gone back to planting one potato plant, one
squash plant, and one pea plant next to each other, because they found
that the disease problem is just too great with the American system im-
ported from the colder areas.

Dr. Mount, I thought that it might be possible in the artificial ponds
to tackle this question of competitive species, thereby narrowing the
actual range of temperature that an animal can tolerate. Perhaps one
could have a non-harvested type of fish along with a harvested type of
fish. It might turn out that the non-harvested fish, under the warmer
conditions, could better utilize all the types of food organisms that occur
under warm conditions when the stoneflies and mayflies are gone. This
would demonstrate that, although the valuable fish might be able to exist
in isolation on the warm-water food organisms, the non-harvestable fish
might do even better in competition, and eventually become the only fish
present.

Idyll: I was amazed to find myself defending an academic and basic research position. I spend my life as a fishery biologist on the other side
of the fence.

I think Dr. Mount missed my point. I think that we have the responsibility to tell the public what they like, and what they want. I think we
have this responsibility in the most urgent sense, but not because we want
them to look at diatoms and appreciate them. But if diatoms happen to
be part of the ecosystem which is essential to the well-being of the
harvestable animals that I am interested in as a fishery biologist, then
we had better lead them to the appreciation of diatoms. Actually, I am
after the same objectives as Dr. Mount, Dr. Doudoroff, and the other
people here; namely, to save the environment for the sake of the useful

organisms, useful whether they are providing recreation and money or simply for aesthetic values. But I am on the side of Dr. Patrick. I think I see a great danger in attempting to study a "simplified" community as opposed to the whole community, because there is the possibility that if we don't understand the whole of it, we won't save any of it.

I was struck by the discussion in which it was said, many times, that there is a great danger in studying organisms in the laboratory, because when you are outside the laboratory, the answers that you obtain don't apply. I would like to put in a plea for more observation of the natural community. This is the approach that we're taking at the University of Miami in our ecological work. We don't decry the necessity or the usefulness of laboratory experiments on thermal tolerance of organisms. Yet, organisms react in the field in nature in a different way, and we are spending a great deal of time looking at the community as it exists in nature. We are also studying the kinds and the quantities of animals as they exist in relatively unchanged environments compared to what happens when the environment does change. I believe that this has to be done, eventually, because when you are through studying the physiology of the organisms in the laboratory, you still must look outside. Therefore, we had better get outside now, as well as making these studies in the laboratory.

Mount: I agree completely with you on that point. However, I think that it is worth pointing out that if one of our objectives is to understand fully what is going on in the out-of-doors and in natural habitats, field investigations often fail to answer *how.* I think what you really find is *what has happened,* but there is little likelihood of understanding *why* it happened.

Given enough time, you could devise experiments to gain an understanding, and I presume this is what you are proposing. But the reason I make this point is, simply, that much of what I said this morning was not just in regard to thermal work, but of our bigger responsibility of water quality in regard to all types of pollution. It is quite obvious to those who are familiar with waste-disposal problems in this country that there isn't now and will not be enough time or enough money to do such studies for all the important kinds of pollution. We will be forced to use quick laboratory studies to give an estimate to a plant either in existence or going into operation. We are taking great risks, but I think the risks are even greater if we refuse to estimate. We biologists have been entirely too reluctant to use our best judgment in trying to predict what will happen and accept the damage if it occurs.

I don't think there is any other way to keep up with the problems if we don't do this. There has been a terrific amount of criticism of the standards that the states have developed and that have been, or have not been, approved by the federal government. But at least we're establishing *some* kinds of levels of quality, which is better, even though they may not be adequate, than not having any at all. We, as scientists and biologists, are not applying what we know, and we are not ready to take the risk of being wrong. We must realize that we are not just working on thermal pollution, and we are going to fail if we don't spread our effort correctly over the whole problem. Therefore, eliminating every non-essential project is mandatory to effective progress.

Chapter 7   Ruth Patrick

SOME EFFECTS OF TEMPERATURE ON FRESHWATER ALGAE

THE algae of freshwater lakes, ponds, and streams are found growing on or near the shores and bed of the body of water, attached to various kinds of debris, to the substrate of the bed, or to aquatic plants. Round (1965) has classified these algae as being epilithic communities which are encrusting or basally attached to rocks; epiphytic communities which are species that are attached to larger algae, bryophytes, and angiosperms; and epipelic communities which may be motile species living on sediments ranging in size from coarse sand to fine silt, and rich organic deposits.

In lakes and some ponds, reservoirs, and estuaries, one often finds algae which spend their vegetative life afloat and are known as plankton, although they differ from the true plankton of the sea, which spend their vegetative and reproductive stages afloat.

The kinds of algae found in fresh water commonly belong to the Cyanophyta (blue-green algae), Chlorophyta (green algae), Chrysophyta (yellow-green algae), and Bacillariophyta (diatoms). A few species belonging to the Rhodophyta (red algae) are also found.

These various major groups of algae represent many species. Each species has its range of temperature tolerance and its range of optimum growth, photosynthesis, and reproduction. In general, the diatoms are represented by the largest number of species with relatively low temperature tolerances —that is, below 30° C.—although some of them have ranges that are higher. The species of the green algae form a group whose tolerances cover a wider temperature span. The blue-green algae have more species that are tolerant of very high temperatures, although some of them are characteristic of cool to cold water.

Hustedt (1956) classifies diatoms according to the following temperature ranges:

Stenotherms

    Cold-water stenotherms   15° C.

    Temperate stenotherms   15°–25° C.

    Warm-water stenotherms   above 25° C.

Meso-stenotherms: those forms that can withstand 10° C. variation in temperature

    Tropical cold-water forms   10°–20° C.

    Temperate forms   15°–25° C and 20°–30° C.

    Warm-water forms   25°–35° C. and 30°–40° C.

Meso-eurytherms: those forms that can withstand 15° C. variation in temperature

    Cold-water to temperate forms   10°–25° C.

    Temperate forms   15°–30° C.

    Temperate to warm-water forms   30°–45° C.

Eu-eurytherms: those forms that can live in 20° C. or more variation in temperature.

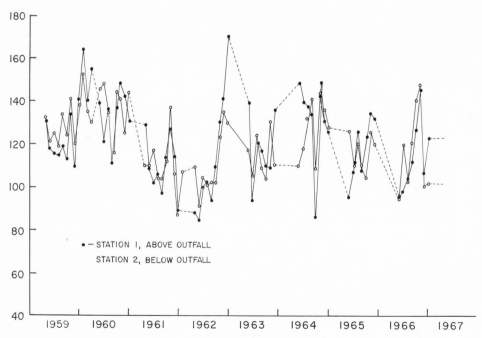

Fig. 1—.Diatometer studies—Potomac River. Number of species of Diatoms at Station 1 and 2, 1959–1966.

Under natural conditions, one of the main factors in seasonal succession of species is the different temperature optimums of species; and, as a result, as the temperature increases or decreases, replacement takes place. Since, in nature, the population size of a species depends upon the ability of a species to compete successfully against other species and to survive predator pressure, we often find the range in tolerance and of optimum development to be different from that in the laboratory.

Under natural conditions, in contrast to laboratory conditions, increases in temperature may have various indirect effects on algae. For example, increase in temperature may bring about an increase in bacteria which mineralize organic material that is a source of nutrients and thus increase the nutritive value of the water. Increased temperature and/or bacterial activity may also reduce the presence of oxygen which, if lacking near the bed of the river or lake, may bring about the release of trace nutrient elements or toxic substances previously held by the oxidized microzone.

Changes in temperature from warm to cold and vice versa are also important in causing the overturn of lakes and the circulation of nutrients. The melting of ice and snow in the watershed during the spring often causes nutrient enrichment in lakes and streams and thus stimulates algal growth.

Under natural conditions, algae may be classed as those with optimum ranges for growth above 30° C.; intermediate species, which prefer temperatures between 10° C. and 30° C.; and cold-water forms. Thermaphilic algae are commonly found in thermal springs and streams resulting from geysers.

Although there have been some excessively high temperatures reported for algae, Castenholz (1967) reports that recently there has been no incontestable evidence of chlorophyll-containing organisms at constant temperatures above 74°–75° C. Schwabe (1936), in his work on island thermal algae, records the ranges of temperatures in which he found various algae (Table 1). These ranges are for blue-green algae and diatoms. As one might expect, the temperature ranges for diatoms were generally lower than the blue-greens, except for *Achnanthes marginulata,* which had a range of 26°–41.5° C., and *Diploneis interrupta,* which was found at 48.2° C. Petersen (1946) lists 39 species of thermal algae. Sprenger (1930) found an apparently unialgal culture of *Pinnularia appendiculata* var. *nave-ana* at about 60° C. Brock and Brock (1966) report that the approximate upper limit for growth of *Mastigocladius laminosus* (a blue-green) in Iceland was 60° C. The maximum temperature for this same blue-green alga in New Zealand was 61° C. (Nash, 1938). Various species

TABLE 1. Algae Occurrence

Achnanthes marginulata	26°–41.5° C.	Phormidium valderianum	48° C.
Cocconeis schlettum	36°–34° C.	P. corium	48° C.
Diploneis interrupta	48° C.	P. angustissima	47° C.
D. oculata	20°–30° C.	Lyngbya sp.	48.2° C.
Eunotia tenella	25° C.	Oscillatoria chalybaea	47° C.
Mastogloia smithii	20°–25° C.	O. proboscidea var. westii	41.5° C.
Navicula variostriata	11°–14° C.	O. tenuis	44° C.
Nitzschia filiformis	31°–35° C.	O. sancta	41.5° C.
Nitzschia tryblionella	10°–25° C.	Scytonema varium	47.7° C.

Source: Data from Schwabe, 1936.

of the blue-green genus *Synechococcus* also grow well at high temperatures and have been found in North America, Japan, and Greece. The genus *Phormidium* also is found at these high temperatures. In general, algae do not grow as well, nor are there as many species, at these extreme upper ranges as there are at slightly lower temperatures.

Mann and Schlichting (1966) studied two groups of springs in Yellowstone Park which they designated as the White Creek Group and the Zomar Bench Group. In all the springs, the benthos was composed of *Phormidium bijahensis,* which was first or second in frequency between 60°–80° C., and five species and varieties of *Synechococcus. S. elongatus* var. *amphigranulatus* was the most variable in response to temperature change. The growth maximum for the species occurred between 66.5° and 68° C. in the Zomar Bench Group, and between 74.5° and 74.8° C. in the White Creek Group. The other species occurred at somewhat lower temperatures and were more similar in their maximum-growth temperatures in the two groups of springs. All were characteristic of water with temperatures greater than 50° C.

Stockner (1967) studied two thermal streams; one, Stream A, originated from two boiling springs at the base of Castle Geyser and the other, Stream B, was runoff water from a number of hot springs behind Old Faithful. At 56° C., *Schizothrix calcicola* was the dominant form and replaced the filamentous bacteria which were present at higher temperatures. At 50° C., it reached its greatest density and it disappeared when the water cooled below 42° C. *Mougeotia* sp., a green alga, became the conspicuous alga when the water temperature fell below 35° C.

In Stream B, samples were taken at intervals along a temperature gradient from 45° to 33° C. during January. The flora consisted of filamen-

TABLE 2. Diatom Species Distribution Relative to Temperature in Stream B.

Temperature	Species
45° C.	Pinnularia microstauron
	Gomphonema parvulum
	Achnanthes grimmei
39° C.	Pinnularia microstauron
	Gomphonema parvulum
	Achnanthes grimmei
	Navicula cincta
36° C.	Pinnularia microstauron
	Gomphonema parvulum
	Achnanthes grimmei
	Navicula cincta
	Rhopalodia gibberula
	Amphora coffeaeformis
33° C.	Pinnularia microstauron
	Gomphonema parvulum
	Achnanthes grimmei
	Navicula cincta
	Rhopalodia gibberula
	Amphora coffeaeformis
	Denticula elegans

Source: Data from Stockner, 1967.

tous greens, blue-greens, and diatoms. The diatoms found at the various temperatures are listed in Table 2. As the temperature decreased, there was a marked increase in species which were represented by smaller populations. Stockner also lists the species of diatoms commonly occurring above 35° C. in Yellowstone Park and at Mt. Rainier (Table 3).

Just as there are thermophylic species, there are also kryophylic species—that is, species which grow best at low temperatures and typically have low temperature ranges. These species are found at high altitudes, in very cold water at various altitudes, and in the Arctic and Antarctic regions. Many such species which have only been found in cold water are known among the diatoms (Cleve & Grunow, 1880; Petersen, 1946; Patrick and Freese, 1961) and they are also recorded for other groups of algae (Hutchinson, 1967). Rodhe (1948) found that *Melosira islandica* var. *helvitica* could be reared at 5° C., 10° C., and 20° C. Although the best growth was at first produced by the culture at 20° C., it soon died; and, over time, the culture grown at 5° C. was the best. It is interesting to note that, although the blue-green algae are represented as a group by more

TABLE 3. Diatom Species Commonly Occurring at Temperatures Above 35° C.

Yellowstone	Mount Rainier
	Biraphidineae
Rhopalodia gibberula	Rhopalodia gibberula
Denticula elegans	Denticula elegans
Mastogloia elliptica	Mastogloia elliptica
Amphora coffeaeformis	Amphora coffeaeformis
Navicula cincta	Navicula cincta
Nitzschia parvula	Nitzschia frustulum
Nitzschia ignorata	Nitzschia obtusa
Pinnularia microstauron	Nitzschia thermalis
Gomphonema parvulum	Caloneis bacillum
	Diploneis interrupta
	Navicula cuspidata var. ambigua
	Monoraphidineae
Achnanthes grimmei	Achnanthes grimmei
Achnanthes lanceolata	Achnanthes lanceolata var. dubia
Achnanthes gibberula	Achnanthes exigua
	Achnanthes pinnata
	Achnanthes minutissima

Source: Data from Stockner, 1967.

thermophylic species, there are species in this group, such as *Oscillatoria rubiscens,* which would be classed as a kryophil (optimum growth is 4°–10° C.). Ruttner (1953), Hutchinson (1967), Sorokin (1967), and others have noted that a single species may have strains with very different temperature optimums.

By far the largest number of species of algae are those that live at intermediate temperatures. They are the species found in most temperate and tropical lakes, ponds, and streams. They may also be found in the Arctic and Antarctic regions in warm water, particularly when the weather is warm. These freshwater algae are often able to endure temperature ranges adverse for growth by ceasing to divide or by forming resting stages of various kinds. An example of this ability was seen in our laboratory in experiments carried out under my direction and published by Cairns (1956). In these experiments, we seeded slides in the Sabine River at temperatures about 18°–20° C. These slides, at the time the experiment started, appeared to be composed mostly of diatoms, although a very few greens and blue-greens were present. The temperature of the culture media was gradually raised to 40° C. Between the temperatures of 20°–30° C., diatoms were the dominant species; between the temperatures of 30°–35° C., green algae became dominant; and above 35° C., blue-green algae were dominant. The temperature was then gradually lowered and, at the respective temperature ranges, green algae and diatoms reappeared. This indicated that all of the organisms of a given group were not killed when its optimum range was exceeded, but the species were not able to compete successfully with those species better suited to a given temperature range.

The group of species which live in intermediate temperature ranges are by far the most important to the ecosystem of the fresh waters, because they are so widespread. To date, little is known of the relative food value of various species of algae, but many field observations have recorded very different predator pressures on certain species, such as species of blue-green algae and species of diatoms. In these natural ecosystems, we typically find many species of algae. As yet, we do not know the importance of maintaining diversity of species for food sources at various stages in the food web.

In the intermediate-temperature range, as in the other temperature ranges, there are species which are stenothermic as well as eurythermic. It is the species that have distinct temperature preferences living with those that have broader temperature tolerances that allow seasonal succession to occur but assure a maintenance of species as primary producers in the ecosystem. Many floristic papers have been written which discuss the change of algal species at various seasons of the year and their apparent temperature preferences. Diatoms are usually a dominant group in the fall, spring, and winter. However, under some natural conditions in early spring, certain green algae, such as *Ulothrix zonata,* may become very common and also some blue-greens may occur. In late summer, one may find, in natural conditions in the temperate zones, blue-green algae becoming more common, although often this is not the case, and greens and diatoms remain as the more common summer forms. Although temperature is an important factor in seasonal succession, many other ecological factors, such as light and nutrients, are very important in determining what species are present.

Hopkins (1963) has found that the movement of diatoms through the top several millimeters of mud in the Ouse estuary in Sussex is related to temperature and also to light. He found that the smaller species of diatoms, such as *Stauroneis salina, Nitzschia closterium,* and *Navicula* sp., had a greater increase in velocity when the temperature was raised from 5°– 15° C. than do larger diatoms (*Trophidoneis vitrae* and *Pleurosigma* sp.).

This horizontal velocity did not increase above 17.5° C. and decreased at 20° C. Only two diatoms, *Nitzschia closterium* and *Stauroneis salina,* were moving at 0° C. On bright days, the rate of vertical movement of the small diatoms was 1 mm. in 6 minutes at 10° C. and 1 mm. in 2 minutes at 15° C. After being exposed to light, it took varying lengths of time before the diatoms stopped moving in the dark. He concluded that these diatoms could reach the mud surface when the temperature was between 7.5°– 20° C., although the greatest movement is between 10°–17.5° C.

This paper is cited because it probably indicates what happens to the diatoms in freshwater wetlands. The photosynthesis and resultant oxygen production in wetlands is largely dependent on diatoms. It is very important that the optimum temperature regimes be preserved in such areas if oxygen production is to be maintained.

Many laboratory experiments have been carried out to determine the effect of increased temperature on cell growth, photosynthesis, and various metabolic functions.

Dutrochet (1937) reports the maximum temperature for *Nitella flexilis* as 45° C.; Sachs (1864) gives 45°–60° C. for *Cladophora;* and deVries (1870) reports 44° C. as death point for *Spirogyra* and *Oedogonium,* and 46° C. for *Hydrodictyon.* Ayres (cited by Oltmans, 1923) found that *Ceratium tenuissimum,* when exposed to 28° C. for 30 minutes and then returned to normal lower temperatures, died after 320 minutes.

Miquel (1892) conducted a series of experiments on mixed algae cultures of diatoms and green algae. All algae were in good condition at 30°–31° C. At 37°–38° C., all large species of diatoms were killed, but some specimens of smaller species persisted. At 42° C., the green algae lived; but at 44.5° C., they were dead, with the possible exception of *Protococcus.* Klebs (1896), in his studies on the conditions for reproduction of algae, found that *Vaucheria repens* was unable to live at about 33° C.; *Oedogonium diplandrum,* above 39°–40° C.; and that the maximum temperature for *Stigoclonium tenuis* was 27° C.

The studies with *Chlorella* strains have shown that, at lower temperatures, optimum growth requires lower light intensity. The greater productivity of *Chlorella* at higher temperatures is due to the fact that these high-temperature strains can utilize more light (Sorokin, 1959, 1967). Furthermore, there is evidence that light saturation for growth is slightly higher than that for photosynthesis at low temperatures and the reverse is true at high temperatures. High-temperature algae are higher-efficiency algae, as measured by the absorption of incident light and the conversion of the absorbed energy into a product.

Sorokin (1960) has shown that the accumulation of cell material and the amount of cell division of *Chlorella* influenced the values of activation energy for growth.

Daletzkaya and Chulanovskaya (1964), working with medium-temperature and high-temperature strains of *Chlorella,* found that, with increased rate of growth at high temperatures, the dry weight and size of the cells decreased. Furthermore, as found by other workers, the optimum temperature for growth and photosynthesis were not the same.

Margalef (1954) found that cultures of *Scenedesmus obliquus* grown at 23° C. had smaller cells than those grown at 13.5° C. He also found evidence that the dry-weight content is not as different as the size of the cell would suggest. The larger cells contained more water. The smaller cells were shorter and broader and had higher rates of oxygen consumption.

Munda (1960) found that, with increase of temperature from 5°–15° C., the desmid, *Closterium leibleinii,* became more resistant to hypertonic solutions such as higher chloride concentrations.

The growth pattern of *Chara zeylandica* is affected by increase in temperature (Anderson and Lommasson, 1958). The most abundant growth of the main shoots and the rhizoidal system occurred at 24° C. At higher temperatures, the main shoots ceased to grow. At 32° C., the lateral branches were limited to small outgrowths.

Barker (1935) found the maximum rate for photosynthesis for *Nitzschia palea* to be 33° C., and at 40° C., the rate was irreversibly lowered.

The diatoms *Nitzschia filiformis, N. lineraris,* and *Gomphonema parvulum* seem to have their optimum growth between 22°–26° C. (Wallace, 1955). Growth is greatly reduced or inhibited for *N. lineraris* at 30° C. and for *N. filiformis* at 34° C. In contrast, *G. parvulum* continued to grow fairly well at 34° C., although the rate was not as great as at 20° C.

The optimum temperature for growth and photosynthesis for *Aphanocapsa thermalis* was found to be 40° C. Phycocyanin does not seem to be as adversely affected by high temperatures as chlorophyll A (Moyse and Guyon, 1963). Feldfoldy (1961) worked with various unialgal strains of some green algae and found the following optimum temperatures for photosynthesis: *Coelastrum microsporum,* 30° C.; *Scenedesmus obtusiuscula,* 35° C.; *Ankistrodesmus falcatus,* 25° C. Löwenstein (1903) found that when *Mastigocladus laminosus* was taken from thermal waters it could live indefinitely at 49° C. and up to three days at 52° C.; but if it was grown at room temperature for any length of time, it lost its ability to live at higher temperatures.

Aruga (1965) studied in the laboratory the optimum temperature for photosynthesis of algae in Shinjiike Pond, which is on the campus of the University of Tokyo. He found that the optimum temperature for photosynthesis for the various algae increased as the season progressed from cold to warm weather. The highest optimum temperature was about 30° C., in August. From December to March, the optimum temperature for photosynthesis remained about 20° C., except for January, when it was 15° C. This change in optimum temperature for growth did not seem to be correlated with changes in major types of algae—i.e., changes from diatoms to green algae—but rather, in change in species or physiological changes.

Studies on the effect of increased light on pigments have been made with the blue-green alga, *Oscillatoria brevis*. The formation of chlorophyll A is increased between 20°–35° C. at low light intensities. At 45° C., the amount of chlorophyll A is practically the same as at 35° C., but the total and specific activity of chlorophyll A is less at 45° C. The quantity of phycocyanin does not vary much between 35°–45° C. but is greater at 30° C. (Garnier, 1962). Under high light intensity, the concentration of these pigments is greater at 40° C. than at 20° C. (Garnier, 1958). Carotene B does not seem to be greatly affected in formation or activity by these changes in temperature. In *Euglena*, Brown (1960) has found that heat affects the various chlorophylls in varying degrees. When he heated the cells for five minutes at 45° C., little change in the absorption spectrum occurred. When he heated the cells for fifteen minutes, the amount of chlorophyll absorbing at 695 mμ was greatly reduced and the chlorophyll absorbing at 673 mμ increased.

The effect of temperature on reproduction by zoospores has been studied for several green algae. The lower temperature limits for production of zoospores for *Vaucheria repens* and *V. clavata* was 3° C.; for *Oedogonium diplandrum*, 0°–0.5° C.; and *Tetraspora gelatinosa*, 3°–11° C. The upper temperature limits for zoospore production of *V. repens, Bumilleria,* and *Ulothrix zonata* was 25°–26° C., in *Oedogonium diplandrum*, 35° C.; and for *Tetraspora gelatinosa*, it was not determined (Rhodes and Herndon, 1967). According to Klebs (1928), zoospore production in *Bumilleria* was completed in less time (12–24 hours) at 24° C. than at 13°–17° C. (20–24 hours). League and Greulach (1955) found the reverse to be true for *Vaucheria* and this was also found to be the case in *Tetraspora gelatinosa*.

The ability of algae to tolerate abrupt changes in temperature has been investigated by Peary and Castenholz (1964) for the blue-green alga, *Synechoccus lividus*. They found that this alga could withstand, without apparent damage, sudden changes in temperature from 70° C. to room temperature. On the other hand, Marre and Servettaz (1958) found that *Aphanocapsa thermalis* was very sensitive to changes in temperature from that to which it had been acclimated. Such changes affected both growth and respiration. Löwenstein (1903) and Allen (1959) found that growing thermophilic algae at lower temperatures for long periods of time caused them to lower their temperature for optimum growth.

The effect of increasing temperature on the sorption of radionuclides by *Phormidium boryanum* was studied by Harvey (in press): [57]CO had the highest sorption at 25° C. and decreased with temperatures up to 40° C.; [54]Mn had increased sorption with an increase in temperatures up to 35° C.; [65] Zn, during the first 14 days, had more sorption by *P. boryanum* at 40° C. than at lower temperatures; but at 21 days, the amount of sorption was similar for all temperatures. Although the sorption of ^{59}Fe, [85]Sr, and ^{137}Cs was slightly higher at 40° C. than at lower temperatures, there was no positive relationship between temperature and sorption levels.

Several efforts have been made to examine the effect of increased temperature and/or light on the algae community by the use of various kinds of artificial streams. Kevern and Ball (1965) studied two streams—one maintained at 20° C. and the other at 25.6° C. They found that there was an increase in gross productivity with increase in temperature, but due to increased respiration there was no significant net increase in productivity. They state, as Phinney and McIntire (1965) found, that had the temperature differences been greater, the differences between the two streams would have been more marked.

Phinny and McIntire found that increasing the temperature from 11.9° to 20° C. produced an increase in oxygen evolution from 337 to 447 mg O_2 per m^2/hr. when the intensity of illumination was approximately 22,000 lux. However, the increase in temperature did not affect oxygen evolution if the light was below 11,100 lux.

Studies on the effects of various environmental changes, including temperature, on diatom communities in the Philadelphia area have been made by me in a series of experiments extending over the last few years with the assistance of a grant from the FWPCA. The diatom communities are developed on glass slides mounted in plastic boxes through which natural stream-water flows. The experiments are set up in a greenhouse so that as near natural light conditions as possible prevail. The results of these experiments show that cell division and, hence, population sizes, of most diatoms are severely limited by temperatures less than 15° C. Increasing temperatures over the naturally occurring temperature increased the biomass until temperatures of 29°–30° C. were reached. At 30° C. and 33.8° C., there was a decided decrease in biomass. The greatest diversity as measured by numbers of species and similar sizes of populations occurred at about 24° C. At 18° C. and at 30° C., in the October-November experiments, the communities were dominated by one or two extremely common species. At 33.8° C., in the September-October studies, this tended to be true, but the degree of dominance was not as great. More species were very common. The dominant species in these experiments at 30° C. and 33.8° C. was *Nitzschia palea,* which is not necessarily a thermophil but is a broadly tolerant species of many ecologically variable factors. There was no marked reduction in numbers of species. These conclusions support laboratory findings that high temperatures near the limits of tolerance repress cell division.

Comparatively few studies have been made on the effect of thermal discharge on the algae in rivers, lakes, or ponds. In general, the effects of thermal discharges are to alter the current pattern in the receiving body of water and to cause a change in the natural existing temperature regimes. Usually, a gradient of decreasing temperature from the outfall occurs which varies in pattern mainly because of variation in morphology and current patterns existing in the receiving body of water and the effect of wind. Of course, other ecological variables may affect the pattern of the temperature gradient. The method of discharge also affects this pattern. As shown from the above studies, increases in temperature within the range tolerated of the existing species, if light and nutrients are sufficient, may increase productivity. However, if the natural ranges are exceeded, species composition will change; and if the change is great enough, shifts in the flora from diatoms to a mainly green algal or blue-green algal flora may occur.

Dryer and Benson (1957) studied the effect of temperature increases due to discharges from the New Johnsonville steam plant on Kentucky Lake. They found no significant increase in plankton could be associated with the steam-plant discharges. There was a 10° F. rise in the steam-plant discharge harbor between October and March, and a 5°–6° F. rise in April and May. The temperature in the lake proper was only 1°–2° F. above that of the lake above the plant.

The effects of warm water from a power station in Poland (Stangenberg and Pawlaczyk, 1961) decreased the number of species when the temperature was above 30° C. High temperatures favored the growth of the diatoms *Navicula cuspidata, N. ambigua,* and *Tabellaria* sp., but blue-green algae were not abundant.

Trembley (1960, 1965) studied the effects of thermal discharges from the Martin's Creek plant of the Pennsylvania Power and Light Company on the algae on slides placed in the Delaware River. He studied areas above the discharge (B), at the discharge (A), 650 feet below the discharge (C), and 4,500 feet downstream (E). The temperature at the discharge during the period of study ranged between 36°–42° C. from June to September; and at the point 650 feet downstream, the range was 34.5°–40° C. for the same period. The point 4,500 feet downstream had nearly normal river temperatures.

The results of these studies show that there were fewer species and they were represented by more individuals in the heated areas. During the periods when the temperature exceeded 34.5° C., there was an increase in blue-green algal species in the areas at the discharge and 650 feet downstream. This was probably due to the fact that the temperature in these areas was high for a long period of time. In the natural river (Station B), the temperature only exceeded 27° C. on one day, June 6, when it reached 27.2° C.

The effect of chlorination of the heated water at the point of discharge (A) seemed to reduce the sizes of populations—that is, the total number of individuals, but not the number of species. This was suggested by the data from Station B, but Trembley states, due to the small amount of difference in number of individuals and the amount of data, no definite conclusions can be drawn.

At about the same time that these studies were being made on the Delaware by Lehigh University, studies were also being made in the vicinity of the Dickerson Plant of the Potomac Electric Power Company on the Potomac River. This plant has an 8° C. rise of temperature of water going through the condensers. The water, as it leaves the discharge channel, conforms to the Maryland state law which, at that time, was that the water could not exceed 38.7° C. beyond a distance of 50 feet from the mouth of the effluent channel. Studies were made in 1956–57, before plant operations started. Periphyton studies have been made continuously since before operations started. These have been made by the use of the diatometers described by Patrick, *et al.* (1954). Each year, the conditions of the aquatic life at certain selected stations have been studied. These studies have been made during late August or early September, when the flow is

TABLE 4. Common Algae in Potomac River

	Before plant operation 1957			1963			1965		
	Sta. 1 above plant	Sta. 2 below plant	Sta. 3 below plant	Sta. 1 above plant	Sta. 2 below plant	Sta. 3 below plant	Sta. 1 above plant	Sta. 2 below plant	Sta. 3 below plant
Blue-green Algae									
Anabaena inaequalis	—	×	—	—	—	—	—	—	—
Anacystis montana	—	×	—	—	—	—	—	—	—
Calothrix juliana	—	—	—	×	—	—	×	×	—
Entophysalis lemaniae	×	×	—	—	—	—	—	—	×
E. rivularis	×	—	—	×	—	—	×	×	—
Lyngbya aestuarii	—	—	—	—	—	—	×	×	—
Microcoleus vaginatus	×	×	—	—	—	—	—	×	—
Nostoc verrucosum	×	—	—	—	—	—	—	—	—
Oscillatoria amphibia	×	—	—	—	—	—	—	—	—
O. brevis	—	—	—	—	×	—	—	×	—
O. princeps	—	—	—	×	—	—	—	×	—
Schizothrix calcicola	—	—	—	—	—	×	×	×	—
Spirulina subsalsa	—	—	—	—	—	—	—	—	—
Symploca muscorum	—	—	—	—	—	—	—	—	×

Green Algae	1	2	3	4	5	6	7	8	9
Chlorotylium cataractarum						×			
Cladophora glomerata	×	×	×		×				
Congrosira debaryana			×						
Hydrodictyon reticulatum		×	×	×				×	×
Oedogonium sp. 1		×	×	×			×	×	×
O. sp. 2							×		
O. sp. 3							×		
O. grande									×
Pediastrum tetras							×		
P. tetras var. tetraodon							×		
Pithophora sp.		×							
Rhizoclonium hierglyphicum		×							
Scenedesmus bijuga							×		

(Continued on p. 176)

Table 4, continued

	1957 Before plant operation			1963			1965		
	Sta. 1 above plant	Sta. 2 below plant	Sta. 3 below plant	Sta. 1 above plant	Sta. 2 below plant	Sta. 3 below plant	Sta. 1 above plant	Sta. 2 below plant	Sta. 3 below plant
Green Algae (continued)									
S. dimorphus	—	—	×	—	—	—	—	—	—
S. opoliensis	—	—	×	—	—	—	—	—	—
S. quadricanda	—	—	×	—	—	—	—	—	—
Spirogyra sp.	×	×	—	—	—	×	×	—	—
Tetraspora gelatinosa	—	—	×	—	—	—	—	—	—
T. lubrica	×	—	×	—	—	—	—	—	—
T. sp.	—	×	—	—	—	—	—	—	—
Diatoms									
Achnanthes lanceolata	—	—	—	—	—	—	—	×	×
A. sp.	—	×	—	—	—	—	—	—	—
Amphora veneta	×	—	—	—	—	—	—	—	—

Species	103	100	102	87	110	81	78	102	73
Cocconeis placentula var. euglypta	X		X						
Cyclotella atomus				X					
C. meneghiniana	X								
Cymbella tumida	X								
Fragilaria capucina var. mesolepta	X								
Gomphonema parvulum	X							X	
G. sp.	X								
Navicula cryptocephala var. intermedia	X								
N. lanceolata	X								
N. minima		X	X					X	
N. radiosa var. tenella			X						
N. symetrica			X						
N. tripunctata var. schizonemoides							X		
N. sp.	X	X	X						
Nitzschia amphibia	X								
N. intermedia		X							
N. palea	X	X	X	X	X		X	X	
Stephanodiscus astraea var. minutula						X			
Total number of species	103	100	102	87	110	81	78	102	73

lowest and the water reaches its highest temperature. During most of the years, these studies have also been made in late spring and early summer. These studies are made by a team of scientists who are specialists in each of the various major groups of aquatic organisms.

The plant is located a short distance downstream from the Monocacy River, which receives the treated sewage from the town of Frederick. Because the Potomac River is so shallow, most of the flow of the Monocacy stays close to the left bank. The intake water of the plant is composed of Monocacy water plus Potomac River water.

Diatometers were positioned in an effort to determine the conditions of the river before the water passes through the power plant and they are also placed downstream from the power plant, approximately 1200 feet below the mouth of the discharge channel. Slides are examined every two weeks except for the months of January through April, when, due to irregular flow of the river, it is impossible to maintain the diatometers.

The results of these studies indicate that there has not been any significant difference in the number of species of diatoms nor in the numbers of individuals between the upstream and downstream areas. The temperature at the downstream station rarely reaches 34.5° C., according to temperature studies, whereas, in the area studied by Trembley, the temperature was between 34.5° and 40° C., from June to September, whenever samples were taken.

Studies were also made by a phycologist examining the area between the plant outfall and an area approximately 0.8 miles downstream. Most of the studies were made at the area 0.8 miles downstream, as this was the region most ecologically similar to the upstream station. A station farther downstream at Whites Ferry was also studied.

The effects of thermal discharges were hard to separate from the effects of the organic enrichment due to sewage effluent, as no typical thermal species developed. The common algae in August were blue-greens, greens, and diatoms at all stations, before and after plant operations started (Table 4). There has been an increase in the organic load in the river, over time. This is particularly true downstream from the Monocacy River.

The diversity of species has increased a bit at Station 2 since plant operations started. At Station 1, there has been a slight, but not significant, decrease in species. Although a diversified algal flora was present, there has been a tendency for blue-green algae to increase a little on the left bank below the outfall. The increase was more noticeable in the first 100 feet below the outfall. Blue-green algae will increase due to increased organic

TABLE 5. Number of Species

	Stations			
	1	2	3	4
1961	89	51	74	83
1965	65	52	56	64

load and/or to rise in temperature. The temperature in the area 0.8 to 1 mile below the outfall rarely reached 34.5° C. and was usually well below the temperature that increases thermophilic blue-green algae. Most blue-green algae develop well in high-nutrient warm water and probably both contributed to the small increase noted in 1965. The chemical analyses of the water in 1956 as compared with 1965 shows an average P as PO_4 increase from 0.007 to 0.60 mg/1, which is about an eighty-five-fold increase.

A few studies were made to determine the effect on phytoplankton of passing water through the condenser of the Dickerson plant. These studies were made by Dr. Griffith of Hood College and the Limnology Department of the Academy of Natural Sciences. The Academy's studies were made in August. The results of these studies indicated that although a few algal cells showed some apparent change in observable conditions before and after passing through the condensers, the difference was not significant. These studies were made under highpowered water-immersion lenses. No attempt was made to culture these cells by the Academy. Similar results were observed by Dr. Griffith.

Recently, we have been studying the depth at which an intake should be placed to avoid as much as possible the plankton organisms. The results of these studies indicate that a few feet below the lower limit of the photosynthetic zone supports the least plankton.

Another series of studies was carried out by the Limnology Department of the Academy of Natural Sciences of Philadelphia in the Green River in 1961 and were repeated in 1965, after the addition of the new unit to the Paradise Plant. The river in the area we studied receives some runoff from strip mines in the area, as well as heated water from the power plant. The algae at Station 2 during the 1961 survey seemed to be most adversely affected (Table 5).

There was a diversified algal flora of blue-greens, diatoms, and green algae in August and September, consisting of 47 species at this station, which is a considerable reduction from the 83 species found at Station 1,

TABLE 6. Green River

Common Algae and Species Numbers

	1961 Stations				1965 Stations			
	1	2	3	4	1	2	3	4
Lyngbia bicolor	—	—	—	—	X	—	—	—
L. putealis	X	X	X	—	—	X	X	—
Entophysalis lemaniae	X	X	X	X	—	—	—	—
Plectonema wollei	X	—	—	—	—	—	—	—
Schizothrix calcicola	—	—	—	—	—	—	—	X
Microcoleus vaginatus	X	—	X	X	—	X	—	X
Amphithrix janthina	—	X	—	—	—	—	—	—
Oscillatoria splendida	—	X	X	—	—	—	—	—
Schizothrix friesii	—	—	X	—	—	—	—	—
S. fasciculata	—	—	—	X	—	—	—	—
Symploca muscorum	—	—	—	—	X	—	—	—
Spirogyra sp.	—	—	—	—	—	—	—	—
Oedogonium sp.	—	—	X	—	—	—	X	X
Cladophora glomerata	—	X	—	—	—	—	—	—
Rhizoclonium fontanum	X	X	X	—	—	—	—	—
R. hieroglyphicum	—	—	—	—	—	—	X	—
Schizomeris leibleinii	—	—	—	—	—	—	X	X
Vaucheria sp.	X	—	—	X	—	—	—	—
Amphora submontana	—	X	—	—	—	—	—	—
Achnanthes minutissima	X	—	X	X	—	X	—	X
A. nollii	X	X	X	X	—	—	—	—
Cocconeis sp.	X	—	—	—	—	—	—	—
Cymbella turgida	X	—	—	—	—	—	—	—
Navicula symmetrica	X	—	—	—	X	—	—	—
Nitzschia gracilis	X	—	—	—	—	—	—	—
N. amphibia	—	—	X	—	—	—	—	—
Gyrosigma sp.	X	—	—	—	—	—	—	—
Navicula tripunctata var. schizonemoides	—	—	—	—	X	—	—	—
Gomphonema sp.	X	—	—	—	—	—	—	—
Nitzschia filiformis	—	X	—	X	—	X	—	—
N. frustulum	—	—	—	—	—	—	X	X
Synedra fasciculata	—	X	X	X	—	—	—	—
Navicula germainii	—	—	—	—	X	—	—	—
Cymbella cistula	—	—	—	—	X	—	—	—
Cymbella tumida	—	—	—	—	—	—	—	X
Batrachospermum sp.	X	—	—	—	—	—	—	—
Compsopogon sp.	X	—	—	—	—	X	X	X

which was not affected by increases in temperature. The common algae are listed in Table 6. In shallow water and in the splash zone, the very common blue-green algae were *Lyngbya putealis, Symploca muscorum, Am-*

phithrix janthina, and *Oscillatoria splendida,* which were associated with the diatoms *Nitzschia filiformis* and *Amphora submontana;* and the green algae *Rhizoclonium fontanum* and *Cladophora glomerata.* In deeper water, the red alga, *Batrachospermum* sp., and the blue-green, *Entophysalis lemaniae,* were very common. During this period of study (August and September 1961), the temperature range between 7:00 A.M. to 10:00 A.M. was 26° to 27° C., and reached 33° to 35° C. or higher in the middle of the day.

In 1965, the river survey was repeated in September and October. As seen in Table 5, there is a great reduction in species at both Stations 2 and 3. At Station 2, the temperature for the surface water at the time of the study ranged between 26° and 28° C. The only very common algae were the blue-green algae, *Lyngbya putealis* and *Microcoleus vaginatus;* the red alga, *Compsopogon coeruleus;* and the diatom, *Nitzschia filiformis.* During the summer months, the temperature at this station ranged between 24° and 36° C.

At Station 3, at the time of the study, the temperature was 23° to 25° C. in the surface waters. The most common algae were the green algae *Rhizoclonium hieroglyphicum, Schizomeris leibleinii,* and *Oedogonium* sp.; and the most common diatom was *Nitzschia frustulum. Lyngbya putealis* was fairly common, and *Compsopogon coeruleus* was common. Other fairly common species are given in Table 6. During the summer, the temperature in this area ranged between 23° and 32° C. and was usually above 28° C. This is an example of how dominant algae shift from blue-greens to greens as the temperature is reduced.

The results of studies on the Susquehanna River showed that heated water reduced the kinds of species of algae present (Slack and Clarke, 1965).

CONCLUSIONS

From these studies, it is evident that the common freshwater algae belong to the Chlorophyta (green algae), Bacillariophyta (diatoms), Rhodophyta (red algae), and Cyanophyta (blue-green algae). In each of these major groups, there are many species. These species cover a wide range of temperatures, from those that prefer cool-water conditions to those that prefer very warm water conditions. Each species has a range in temperature that it can tolerate and a range in temperature in which its growth is optimum.

In general, the blue-green algae have more species that prefer temperatures from 35° C. upward, whereas the green algae have a relatively large number of species that grow best in temperatures ranging up to 35° C., although some can grow at higher temperatures. Most of the diatom species prefer lower temperatures—that is, temperatures below 30° C. The natural seasonal succession of species which we find is largely due to the fact that species can out-compete each other under varying temperature conditions. Of course, other ecological conditions also control the kinds of species which we find present at various seasons of the year. These conditions are light, nutrients, and so forth.

The effect of artificially increasing the temperature regime of a species tends to increase growth and photosynthesis so long as light is sufficient for these functions and the limits of temperature tolerance are not reached. As one approaches the limits of temperature tolerance for a species, cell division is repressed, as is photosynthesis, and the formation of reproductive cells may be repressed. The cell size is often reduced and the oxygen required for respiration is increased. The pattern of growth may be greatly altered.

The temperatures at which species may successfully grow in the laboratory may be higher than those which occur in nature because of lack of competition with other species under natural conditions and the effect of predator pressure. Relatively few studies have attempted to understand the effect of sudden changes on algal growth. Those that have been done with the blue-green algae indicate that they may or may not be resistant to sudden changes in temperature.

Studies made concerning the effect on algae of passing them through a condenser indicate that if the temperature does not exceed 34°–34.5° C., little, if any, harm is done.

Today, we know very little about the effect of thermal discharges upon algal communities. Much more study needs to be done to understand these effects. Will increasing the winter temperatures—that is, those below the optimum range of growth—be beneficial? Are other ecological conditions, such as nutrients and light during the winter months, suitable for increasing optimal growth with temperature rises? What are the effects of high temperature on algal communities under different environmental conditions? To date, the evidence we have indicates that when the temperatures rise above 35° C., blue-green algae often become dominant, if this high temperature is maintained for fairly long periods of time. Since these algae are a poor food source, the ecosystem may be severely damaged. Likewise, we often find an increase in green algae if the temperature is between 32.5°

and 35.0° C. for several weeks, whereas, below these temperatures, diatoms usually seem to be dominant in streams that have not been adversely affected by other types of pollution.

We know that natural systems are characterized by a high diversity of species and that raising the temperature beyond the optimum for these species reduces diversity. The question is, how much and in what way can this diversity be altered and not reduce the energy flow and productivity of the system, and more important, the stability of the system through time?

REFERENCES

Allen, M. B. 1959. "Studies with *Cyanidium caldarium*, an Anomalously Pigmented Chlorophyte." *Arch. Mikrobiol.* 32:270–277.

Anderson, R., and R. Lommasson. 1958. "Some Effects of Temperature on the Growth of *Chara zeylanica* Willd." *Butler University Botanical Studies* 13(2):113–120.

Aruga, Yusho. 1965. "Ecological Studies of Photosynthesis and Matter Production of Phytoplankton. I. Seasonal Changes in Photosynthesis of Natural Phytoplankton." *Bot. Mag. Tokyo* 78(926–927):280–288.

Barker, H. A. 1935. "Photosynthesis in Diatoms." *Arch. Mikrobiol.* 6:141–156.

Brock, T. D., and M. L. Brock. 1966. "Temperature Optima for Algal Development in Yellowstone and Iceland Hot Springs." *Nature* 209(5024):733–734.

Brown, Jeanette S. 1960. "Factors Influencing the Proportion of the Forms of Chlorophyll in Algae." *Carnegie Inst. Ann. Rep. Dir. Dept. Plant Biol.* 1959–1960:330–333.

Cairns, J., Jr. 1956. "Effects of Increased Temperatures on Aquatic Organisms." *Indus. Wastes* 1(4):150.

Castenholz, Richard W. 1967. "Environmental Requirements of Thermophilic Blue-Green Algae." In *Environmental Requirements of Blue-Green Algae,* pp. 55–79. Richland, Wash.:Pacific Northwest Water Laboratory.

Cleve, P. T., and A. Grunow. 1880. "Kenntniss der Arctischen Diatomeen." *K. Svenska Vet.-Akad. Handl.* 17(2):1–121.

Daletzkaya, I. A., and V. Chulanovskaya. 1964. "Vlinyanie temperatury na rost i fotosintez Khlorelly." *Bot. Zhurn.* 49(8):1147–1159.

Dryer, W., and N. G. Benson. 1957. "Observations on the Influence of the New Johnsonville Steam Plant on Fish and Plankton Populations." *Proc. Ann. Conf. S.E. Assoc. Game and Fish Comm.* 10:85–91.

Dutrochet. 1837. "Observation sur le *Chara flexilis*. Modifications dans la Circulation de cette Plante sous l'Influence d'un Changement de Temperature, d'une Irritation Mecanique, de l'Action de Sels, etc." *C. R. Acad. Sci.* 5:775–784.

Felfoldy, L. J. M. 1961. "Effect of Temperature on Photosynthesis in three Unicellular Green Algal Strains." *Acta. Biol. Acad. Sci. Hung.* 12(2):153–159.

Garnier, J. 1958. "Influence de la Temperature et de 1'Eclairment sur la Teneur en Pigments d'*Oscillatoria subbrevis* Schmidle (Cyanophycees)." *C. R. Acad. Sci., Paris* 246(4):630–631.

————. 1962. "Action de la Temperature sur le Renouvellement des Differents Pigments d'*Oscillatoria subbrevis* Schmidle (Cyanophycees), a la Lumiere." *C. R. Acad. Sci., Paris* 254(12):2218–2220.

Hopkins, J. T. 1963. "A Study of the Diatoms of the Ouse Estuary, Sussex. I. The

Movement of the Mud-Flat Diatoms in Response to Some Chemical and Physical Changes." *J. Mar. Biol. Ass. U. K.* 43(3):653–663.

Hustedt, F. 1965. *Kieselalgen (Diatomeen).* Stuttgart: n.p.

Hutchinson, G. E. 1967. *A Treatise on Limnology, vol. 2.* New York: John Wiley and Sons, Inc.

Kevern, N. R., and R. C. Ball. 1965. "Primary Productivity and Energy Relationships in Artificial Streams." *Limnol. Oceanogr.* 10:74–87.

Klebs, G. 1928. "Die Bedingungen der Fortpflanzung bei eninigen Algen und Pilzen." *Sweite, unveranderte Auflage.* Jena: Gustav-Fischer.

League, E. A., and V. Greulach. 1955. "Effects of Day Length and Temperature on the Reproduction of *Vaucheria sessilis.*" *Bot. Gas.* 117:45–51.

Löwenstein, A. 1903. "Uber die Temperaturgrenzen del Lebens bei der Thermalalge *Mastigocladus laminosus* Cohn." *Ber Deutch. Bot. Gesell.* 21:317–323.

Mann, J. E., and H. E. Schlichting, Jr. 1966. "Benthic Algae of Selected Thermal Springs in Yellowstone National Park." *Trans. Amer. Micr. Soc.* 86(1):2–9.

Margalef, Ramon. 1954. "Modifications Induced by Different Temperatures on the Cells of *Scenedesmus obliquus* (Chlorophyceae)." *Hydrobiologia* 6(1–2):83–94.

Marre, E., and O. Servettaz. 1956. "Ricerche sull'adattamento proteico in organismi termoresistenti. I. Sul limite di resistenza all'inattivazione di alghe di acque termali." *Atti Accad. Naz. Lincei, Rend. Cl. Sci., Fiz., Mat. e Nat.,* ser. 8 20(1): 72–77.

Miquel, P. 1892. "Recherches Experimentales sur la Physiologie, la Morphologie et la Pathologie des Diatomees." *Ann. Microgr.* 4:273–287.

Moyse, A., and D. Guyon. 1963. "Effet de la Temperature sur l'Efficatie de la phycocyanine et de la Chlorophylle chez *Aphanocapsa* (Cyanophycee)." In *Studies on Microalgae and Photosynthetic Bacteria* (a special issue of *Plant and Cell Physiology*), pp. 253–270. Tokyo:Univ. of Tokyo Press.

Munda, Ivka. 1960. "Vpliv temperature na rezistenco vrste *Closterium leibleinii* Kutzing pri razlicnih koncertracijah kloridov." *Bioloski Vestnik Ljubljana* 7:3–9.

Nash, A. 1938. "The Cyanophyceae of the Thermal Regions of Yellowstone National Park, U.S.A., and of Rotorua Whakarewarewa, New Zealand; with some Ecological Data." Thesis, Univ. Minnesota, n.d.

Oltmanns, F. 1923. *Morphologie und Biologie der Algen.* 2d Aufl., vol. 3 Jena:G. Fischer.

Ostrup, E. 1918. "The Freshwater Diatoms of Iceland." *Bot. Iceland* 2:88.

Patrick, R., and L. R. Freese. 1961. "Diatoms (Bacillariophyceae) from Northern Alaska." *Proc. Acad. Nat. Sci. Philadelphia* 112(6):129–293.

Patrick, R., M. H. Hohn, and J. H. Wallace. 1954. "A New Method for Determining the Pattern of the Diatom Flora." *Not. Nat. Acad. Nat. Sci. Phila.* 259:12.

Peary, J. A., and R. W. Castenholz. 1964. "The Effect of Temperature on the Morphology of a Thermophilic Alga." *Amer. Jour. Bot.* 51(6):680.

Peterson, J. B. 1946. "Algae Collected by Eric Hulton on the Swedish Kamchatka Expedition, 1920–22, Especially from Hot Springs." Kgl. Danske Vid. Selsk., *Biol. Medd.* 20:3–122.

Phinney, H. K., and C. D. McIntire. 1965. "Effect of Temperature on Metabolism of Periphyton Communities Developed in Laboratory Streams." *Limnol. Oceanogr.* 10(3):341–344.

Rhodes, R. G., and W. Herndon. 1967. "Relationship of Temperature to Zoospore Production in *Tetraspora gelatinosa.*" *J. Phycol.* 3(1):1–3.

Rohde, Wilhelm. 1948. "Environmental Requirements of Freshwater Plankton Algae.

Experimental Studies in the Ecology of Phytoplankton." *Symbolae Botanicae Upsaliensis* 10(1):149.

Round, F. E. 1965. *The Biology of Algae*. New York:St. Martins Press.

Ruttner, F. 1953. *Fundamentals of Limnology*. Toronto:Toronto Univ. Press.

Sachs, 1864. "Uber de obere Temperaturgrenze der Vegetation." *Flora* 47:4–12.

Schwabe, G. H. 1936. "Beitrage zur Kenntinis islandischer Thermal biotope. (Contribution to the Understanding of the Iceland Thermal Biotope)." *Arch. Hydrobiol., Suppl. Bd.* 6(2):161–352.

Slack, K. V., and F. E. Clarke. 1965. "Patterns of Dissolved Oxygen in a Thermally Loaded Reach of the Susquehanna River, Pennsylvania." Geol. Surv., Prof. Pap. 525–C:C193–195.

Sorokin, Constantine. 1959. "Tabular Comparative Data for the Low and High Temperature Strains of *Chlorella*." *Nature* 184:613–614.

———. 1960. "Kinetic Studies of Temperature Effects on the Cellular Level." *Biochem. Biophys.* 38:197–204.

———. 1967. "A New High-Temperature *Chlorella*." *Science* 158(3805):1204–1205.

Sprenger, E. 1930. "Bacillariales aus den Thermen und der Ungebung von Karlsbad." *Arch. f. Protistenkunde* 71:502–542.

Strangenberg, M., and M. Z. Pawlaczyk. 1961. Nauk. Pol. Wr. Wroclaw No. 40, *Inzyn. Sanit. Water Poll. Abst.* 1:67–106.

Stockner, J. G. 1967. "Observations of Thermophilic Algal Communities in Mount Rainier and Yellowstone National Parks." *Limnol. Oceanogr.* 12(1):13–17.

Trembley, F. J. 1960. "Research Project on Effects of Condenser Discharge Water on Aquatic Life." *Progress Report, 1959–1960*. Bethlehem, Pa.: Lehigh University Institute of Research.

———. 1965. "Effects of Cooling Water from Steam-Electric Power Plants on Stream Biota." In *Biological Problems in Water Pollution,* pp. 334–335. U.S. Dept. Health, Education, and Welfare, 999WP–25. Washington, D.C.:U.S. Gov't. Printing Office.

Vires, H. de. 1870. "Materiaux Pour la Connaissance de l'Influence de la Temperature sur les Plantes." *Arch. Neerl. Sci. Exact. Nat.* 5:385–401.

Wallace, Natalie. 1955. "The Effect of Temperature on the Growth of some Freshwater Diatoms." *Not. Nat. Acad. Nat Sci. Philadelphia* 280:11.

DISCUSSION/ L. A. Whitford

THE excellent presentation by Dr. Patrick of some effects of temperature on algae makes a formal discussion a little difficult. There are few things which she leaves, either by omission or incomplete coverage, for adverse criticism; and I do not want merely to repeat to you her major points.

There is one point I should like to make, however, which may not be exactly pertinent to this discussion but which is of importance in this symposium. I'm sure it has occurred to all the biologists and to many non-biologists. This is the fact that there is little direct or actual pollution of water by either a rise or a fall in temperature. (This was written, of course, before Under Secretary Black mentioned it in his address.) I believe we should regard a significant *lowering* of temperature as much a "thermal pollution" as a *rise* in temperature. Some day this also may be of practical importance. Thermal pollution affects water chiefly through its effect on the biota and not directly. We must, therefore, study thermal pollution by studying its effect on the biota.

If water contains organic compounds, a rise in temperature could, as pointed out by Dr. Patrick, bring about a more rapid degradation by bacteria and fungi of these compounds. If volatile or labile compounds are present, a rise in temperature might increase or decrease the possibility of tastes and odors in potential drinking water. A rise in temperature would normally bring about a more rapid loss of dissolved gases, volatile solutes, and such materials as chlorine, which might have been added to it.

Dr. Patrick briefly touched on the semi-annual turn-over of water in deep lakes. It is possible that where thermal-water-holding reservoirs are used, warm water, if properly introduced deep in the reservoir, might bring about an artificial turn-over and, therefore, circulation of nutrients. This might be desirable, if the reservoir were used for fish production.

In her listing of the kinds of algae found in fresh water, Dr. Patrick

186

omitted the Chrysophyta, or golden algae. This group is certainly of considerable importance in the plankton of soft-water areas. In the Southeast, species of *Mallomonas, Synura,* and *Dinobryon* are more abundant than diatoms for some weeks in early spring. *Synura* is a notorious odor-producing organism.

I most certainly agree with Dr. Patrick that field studies or observations should always supplement laboratory work. Predator pressure is only *one* of the factors which careful field work could bring out.

Her classification of organisms according to temperature tolerances is a good one. I should like to point out, however, that some of our work (Whitford & Schumacher, in press) and that of others indicate that 15° C. is a very critical temperature. Many microthermal or cold-water organisms rapidly disappear at this temperature and mesothermal or mid-range organisms begin to increase at this temperature. We would prefer to place the lower limit of effective growth of mesothermal algae at 15° C., rather than 10° C. (When I mention "our work," I refer to recent work by Dr. George Schumacher of New York State University, Binghamton, and me, and our students.)

Dr. Patrick's discussion of the thermal or high-temperature algae is excellent, but I should like to note that some investigators report that most, if not all, thermal species grow as well at room temperature as at elevated temperature, while others report the reverse. This point should be further investigated.

The cryophillic, or low-temperature species, which occur not only in the Arctic and Antarctic but also in alpine regions, are a very interesting group. Some have as narrow a range of temperature for growth and reproduction as any organisms. Chodat (1918), in his interesting old book, *Les Plants Aquatiques,* states that they grow just at the temperature of melting ice and, in alpine regions, are frozen solid each night. Some, he says, fail to grow at a temperature still too low for barley to germinate. Our limnologist, Dr. John Hobbie, recently investigated species from Lapland. All his plants, which include several species of *Chlamydomonas,* soon die out at room temperature. These species should be further investigated. There have been only a few recent studies, including some by Dr. Janet Stein of the University of British Columbia. In North Carolina, we have a number of winter species which should probably be classified as cryophills. They belong to the Chrysophyta mentioned earlier, and all grow and reproduce at a temperature below 10° C. One of them, *Tetrasporopsis reticulata,* was first described from Lake Baikal in Siberia, but it also occurs in winter in both

America and Europe. Another, apparently an undescribed species, is found in water only a few degrees above freezing.

Investigations have already been made of the possibility of using heat derived from water as a source of power. Such investigations might eventually lead to the production of large quantities of low-temperature water in which cryophills would be important.

The literature abounds, as pointed out by Dr. Patrick, in references to the effect of temperature on algae. It should be noted, however, that it is possible that the effects of light and temperature can be confused. Adequate records of neither the intensity nor the duration of light are easily or frequently made. There is some evidence that light intensity, quality, and duration may be more important than present data indicate. Many morphogenic responses in plants are triggered by phytochrome, a pigment-absorbing light in the red (660 mμ) and far-red (730 mμ). Variation in light quality and intensity during seasonal changes may be equally as important as temperature. Day-length and total incident light are now known to be important to many species, particularly marine algae. Holmgren recently reported (unpublished data) that *Cyclotella comta,* a small planktonic diatom, grows only in very diffuse light, and other studies indicate that many photosynthetic organisms reach their greatest abundance at some depth, almost certainly a response to low light. Some of our own studies (Whitford, Schumacher, Dillard 1965, 1966) indicate that light, rather than higher summer temperatures, inhibits a number of species. We first conceived this idea from field studies in spring and autumn in North Carolina, where spring and autumn are fairly long and extended observations are therefore possible. In springtime, certain green algae may become more abundant as temperature is rising, but the same species may increase in autumn as temperature is decreasing. This seems to be a response to favorable light conditions, rather than to temperature. Laboratory studies confirm these observations. More careful and detailed studies of light should be made, at least under field conditions, along with studies of temperature.

One other observation should be made on the relation between temperature and another factor. This is the effect of a current of water on attached or benthic algae. Long ago Ruttner (1926) noted that running water is physiologically richer than still water. A current will negate, at least in part, the effect of an otherwise harmful rise in temperature. A current of water produces a steeper diffusion gradient around an attached organism and can therefore partly overcome the effect of a lower gas content or even mineral content in running water. Ambul (1959), in studies of animals, indicates that this is true, as do some of our studies on algae. The construction of rapids where warm water is discharged into a stream might serve two useful purposes where growth of algae is desired: the rapids would more rapidly cool the water and, at the same time, promote growth of algae due to the direct effect of a current.

Dr. Patrick mentions that recent studies of warm-water discharges indicate that elevated temperature may increase the growth of algae during cooler seasons. With proper temperature control, this might prove to be a beneficial effect of thermal pollution; but more careful and detailed studies are needed, both to prove this and to devise methods for taking advantage of it, if it be true. The same studies would lead, of course, to methods of repressing the growth of algae, in case pure water is wanted for drinking or industrial uses.

Where growth of algae is desired to increase fish production, studies are badly needed on the importance of various species. This has been adequately stressed by Dr. Patrick and I also should like to point out the great need for such investigation. Here is where a team of investigators including a phycologist, an invertebrate zoologist, and an ichthyologist should work together. Prowse (1960) in Malaysia recently found that an important freshwater fish there (*Tilapia*) can digest and use blue-green algae much better than it can *Euglena*. We know little about the food chain, or web, leading to any important freshwater fish.

As I pointed out in the beginning of my discussion, thermal pollution has its effect through the biota; nevertheless, the services of physical scientists and engineers are needed in the study of thermal pollution. For instance, a study of the diffusion gradient in a current was made by chemical engineers at my own university some years ago (Ferrell, Beatty & Richardson, 1955). We were able to use their data in some of our work. Physical scientists and engineers could not only give valuable assistance in devising and constructing instruments for the measurement of light, temperature, *et cetera,* but they could render valuable assistance in setting up and conducting actual investigations, also.

In summary, I can largely only echo what Dr. Patrick has indicated and adequately discussed; the need for many carefully planned and executed studies of thermal pollution. Only such studies can indicate to operators and engineers the correct design and operation of facilities to reduce the effects, or possibly, in some cases, to increase the effects, of changes in water temperature produced by man.

REFERENCES

Ambuhl, H. 1959. "Die Bedeutung der Ströming als Ökologischer Faktor." *Schweiz, Zeitschr. Für Hydrobiol.* 21:133–264.

Chodat, R. 1918. *La Biologie des Plantes. I. Les Plantes Aquatiques.* Geneva: n.p.

Dillard, G. E. 1966. "The Seasonal Perodicity of *Batrachospermum macrosporum* Mont. and Audouinella violacea (Kütz.) Hamel in Turkey Creek, North Carolina." *Jour. Elisha Mitch. Sci. Soc.* 82:204–207.

Ferrell, J. K., K. O. Beatty, and F. M. Richardson. 1955. "Dye Displacement Technique for Velocity Distribution Measurements." *Industr. Eng. Chem.* 47:29–33.

Prowse, G. A. 1960. "Digestion of Fresh-Water Algae by *Tilapia mossambica.*" Report, 1960–61: Tropical Fish Culture Inst. Malacca.

Ruttner, F. 1926. "Bemerkungen über den Sauerstoffgehalt der Gewässer und dessen Respiratorischen Wert." *Naturwiss.* 14.

Whitford, L. A., and G. J. Schumacher. 1964. "Effect of a Current on Respiration and Mineral Uptake in Spirogyra and Oedogonium." *Ecology* 45(1):168–170.

Whitford, L. A., and G. J. Schumacher. "Notes on the Ecology of Some Species of Fresh-Water Algae." *Hydrobiologia* (in press).

DISCUSSION FROM THE FLOOR

John W. Foerster: We are conducting an FWPCA-sponsored before-and-after study on the Connecticut River to determine the effects of thermal waters discharged by the Yankee Atomic Power Project. Therefore, I would like to offer some data reinforcing Dr. Patrick's information. Samples recently analyzed from the upstream station for plankton showed that 79% to 82% of the population were diatom flora. These samples were taken in March and April, when water conditions upstream from the intake were 50° F. Of course, there is leeway between samples.

At the effluent point, where the thermal discharge occurs into the small receiving stream which empties into the Connecticut River, the water temperature was approximately 73° F. With this difference of approximately 23° F., the diatom flora dropped to about 69% of the population. When we looked at these percentages in counts, one did not see the exact impact; however, when we looked at the species distribution, we found that there was considerable change in the over-all composition, particularly in the different species of Melosira. There was a change from the typical type of Melosira that we found in the spring, to the type of Melosira that one would find later on in the summer months. About a mile downstream from the plant, where the temperature dropped back to about 51°–52° F., there appeared to be a shift back to the diatom flora, comprising about 77 percent of the total population of the plankton.

There is one interesting point that I want to bring out, about grazing and fish and this relationship. In some earlier studies that I have done on Periphyton, I noted that *Keratella cochlearis* ingested diatoms, and this organism, a rotifer, was then ingested by blunt-nosed minnows. *Keratella cochlearis* was present in water above the thermal discharge into the stream; but in the thermal discharge, this organism was absent,

191

and other rotifers, which didn't seem to be present above the thermal discharge, then became prevalent.

Also, could Dr. Patrick elaborate on the Green River project? I noticed that, between 1956 and 1965, there appeared to be a drought.

Ruth Patrick: That was the Potomac River; the Green River studies were in 1961 and 1965.

Foerster: In the slide, you demonstrated 89 species in 1956 and in 1951; and then, in 1965, there were only 64 or 65 species. Do you feel that this was due to an encroachment upstream by the thermal discharge, or was this just naturally-occurring phenomena?

Patrick: No, this was not due to the thermal discharge, and as I said, this area is in an area of strip mining, and there was much more strip mining going on in 1965 because it was more profitable. Also, there was more shipping as a result of the strip mining of the coal, and this combination of factors caused considerable addition of silt and solids. I believe these factors were partially, if not largely, responsible for the decrease in the species upstream.

In regard to your *Keratella* statement, we have been looking at what happens to *Keratella* as it goes through the condensers. In this plant on the Patuxent River, there is only a 12° F. rise, and we found that at this time of year (in the spring), when there is no temperature problem, there is no deleterious effect. There are just as many live animals coming out as going in.

John Strickland: My paper yesterday was supposed to be solely concerned with the marine plankton. I could have given an erudite 45 minutes on what we know about the effect of temperature on the phytoplankton and the zooplankton, which isn't much, but we do know something about it. My negative attitude, and my plea for a lot more fundamental work, and the lag that is inevitable until we obtain competent people, was obviously associated with the difficulty of prediction with an interrelated ecosystem. This has been stressed by others.

I have been a government servant for more years than I have been a university professor, so I do realize that people must make decisions. All I can say at the moment is that I am profoundly thankful that I am not being faced with some of the decisions that must be made. But I don't think we do any justice to the problem by kidding ourselves by stating that we know more than we do.

I would like to conclude with a rather sad thought. Half the profession is spending its time running around trying to save the starving masses by fertilizing water and increasing productivity. When this happens, as it does when we discharge sewage into our rivers, lakes, and estuaries, we produce a new condition, known as "eutrophication."

The other half of the profession spends its time trying to stop this condition. The reason is, of course, that the things that get produced in fertilized waters are not necessarily the things we want. We simply still do not understand properly the processes occurring in a complex aquatic ecosystem.

Marc Imlay: I know there are many individual cases that are well documented of the lethal affects from blue-green algae. What is the general situation when a lake, or even a river lake, becomes thermally polluted and there is mainly blue-green algae? Are the blue-green algae toxic to the fish in the lake, and if so, why isn't blue-green toxicity more common in natural warm-water lakes where fish are prevalent?

Patrick: Whether they are toxic or not depends upon the species which becomes abundant, and even on strains of the species. Dr. Gorham, in Canada, has done extensive research in this field, and I believe you would find it most interesting to write to him for his reprints. It is, in general, the strain of the species that becomes abundant under these normal conditions that may or may not be toxic.

Imlay: You would feel that, in cases of thermal pollution, the types of blue-green algae that would generally appear would not be toxic?

Patrick: I would rather think this would be true, but I would hate to make any positive statement on this without knowing the conditions.

Walter Glooschenko: Most of the work that has been discussed today is rather pertinent to studies done in temperate regions. I myself am from a subtropical region down in Northern Florida. However, other people I talk to are alert more to a tropical system. Unfortunately, work has been done that tends to indicate that perhaps subtropical and tropical systems are much nearer the upper lethal thermal limit than are systems in temperate regions. So I would like to make a plea for more research to be done, both by the federal government and by other institutions, on some of the effects of thermal pollution in a warm area, not a temperate area—say, a subtropical or tropical region. Some of this work is now being done.

Unfortunately, the reference that is quoted all the time is a study by Mayer, done at the Dry Tortugas Laboratory near the Florida Keys, in 1914. Also, Dr. Otto Kinne's paper is an excellent review-paper on temperature effects on organisms. I would like to know if there has been more work done since 1914 on whether tropical organisms are at their upper limit?

If this is true, then the major problem we will have will be in power plants set up in the southeastern United States or in tropical regions. I think we may be kidding ourselves to do studies up in Minnesota, when maybe we should be doing these same studies with federal support down in Florida, Alabama, and Georgia.

Also, when we get into a warm-weather area, a subtropical or tropical region, we ought to consider primary production in a little different light. In more temperate regions, it is true that most primary production will be by benthic diatoms, blue-green algae, and phytoplankton; however, when we get into some of our subtropical and tropical estuaries, we now must deal with the marine angiosperms, grasses, such as *Thalassia*, (commonly called turtle-grass), and other marine grasses. There has been work done by the Florida Board of Conservation that shows that once you get up to about 36° or 37° F., a range of normal temperatures, these grasses are injured by heat. This is entirely natural and has nothing to do with heated effluents. Perhaps one of the things that we should worry about in some of our work is what the effect will be on marine grasses when we get into these warm areas and not just strictly the "traditional" temperate primary-producers, such as the phytoplankton.

With respect to work on Cryophilic algae, I think Oregon State can take a little bit of credit. There is a program now going on at Oregon State University under Dr. Herbert Curl and his research assistant, Jack Hardy, that I hope will get into print fairly soon. Mr. Hardy now is in Palau in the Micronesian group, working on warm-water algal forms with the Peace Corps. Therefore, some of this work will be coming out on some of the Cryophilic algae of the Pacific Northwestern peaks, from Mt. Rainier down into Oregon.

Another interesting fact is that some of the same species apparently occur in the thermal hot springs of Yellowstone, and may occur growing on the snow surface, caused by some of these thermal hot springs. Apparently, there is no difference, from observation. This is a rather interesting effect. I would like to again plead for more research and more comments on the tropical and subtropical water where most of the pollution may be seriously damaging, not just in the nice cool waters of Minnesota and Maine.

Eugene B. Welch: I am really at a loss to interpret the species-diversity concept. I think I appreciate the concept, but I am a little familiar with information from the Green River and find the changes in diversity difficult to interpret in the light of other data. We have studied periphyton growth on slides, and we find that, in the heated zone, we observe a significant increase in the rate of growth. Copepods are probably an important link in the food chain leading to fish in this area. The copepods could well be benefiting from such increased periphyton growth, even though species diversity may be less, and a large part of it may be bacteria. I think there are some areas where copepods probably feed quite intensively on bacteria and detritus. I am not sure of the direct evidence, but I have some associations in mind.

Patrick: You say that you appreciate the concept of diversity, but you want it explained in regard to the Green River results, and that you have found that copepods feed on bacteria?

Welch: I have not actually found that, except that we have observed an increase in the rate of periphyton growth in the heated zone along with a decrease in the algal component. This increase in growth is probably due largely to bacteria and other heterotrophs. If the zooplankton possibly increase due to this increased heterotrophic or bacterial growth, then, looking at the whole food chain, we actually may get an over-all increase in the fish production. How do you interpret species diversity in this regard?

Patrick: Well, you see we do not use only one group of organisms, and we look at sizes of populations as well as numbers of species. If we find that the fish are increasing, then obviously they have a food chain that does not involve some of the species of algae that have been eliminated. In no case did we find an elimination of algae. We did get a shift, particularly at Station 2, to blue-green, so we do look at the whole food chain to see what is going on.

Welch: This is, in a way, my point. I have seen fish data from this area and there is no discernible decrease in fish crop that can be directly associated with the heated water. However, an increase in the periphyton growth was observed, which could indicate an increased fish-food supply, even though you observed a shift in species. So how can one interpret such a shift in species diversity?

Patrick: The measure of diversity is to tell you whether or not the area has changed from a control area of the stream in regard to the various groups of aquatic life. As I said in my paper, we do not know at the present time how diversity is related to energy flow. It is for this reason that we look at all groups of organisms in the ecosystem. In our studies, as I recall—and you may have more information about fish than I do— we did get a decrease in fish species at Station 2.

Welch: I have looked at the 1961 and 1963 results in reports by Hunter Hancock, and there was a decrease in fish crop between these years, but there was also a decrease at controls and at downstream stations as well. However, results by the fisheries group in TVA have shown that, in the winter, the game-fish species have increased more than the rough-fish species in vicinity of the heated-water discharge. All I am saying is that, to determine how far one should go with reducing heated-water discharge, based on species diversity of food-chain organisms, to me is very difficult. Cooling towers are going in there; and, in a way, it is very difficult for me to justify the need, based on the biological data at hand.

Patrick: I agree with you. There should have been much more thorough studies made. We were not given the money to make those studies. We did only two studies. I think Milo Churchill has a group of biologists down there making studies, so that I would agree with you that just the two studies we made would not tell much about productivity. Different types of studies would have to be made. What they do show is that there was a change, and the change was more severe after the introduction of the second unit.

Donald R. Johnson: I suggest that, for the group at large, the conceptual scrutiny in terms of the diversity point of view is the most important thing. The group at large may lose track of some of the detail here.

Patrick: Typically, a natural ecosystem is composed of many species which function as primary producers, such as herbivores, carnivores, and secondary carnivores. Some 550 of the many studies we have made, scattered through the country, indicate that when a stream is in a natural condition and stays in that condition and is not adversely effected by pollution, the number of species do not change very much. I am not saying no pollution is entering the stream, but that the stream is perfectly capable of assimilating that pollution, and by all criteria (chemical, B.O.D., etc.), the stream seems to be natural. Under these conditions, we find a high species diversity.

Now, the effect of pollution is to reduce this diversity in varying ways. By going into a stream and looking at all groups of organisms and seeing what kind of species and how many there are, and which ones have abnormally large populations, one is able to make some judgments as to whether organic population has occurred, and if the organic load was very heavy or whether toxic conditions have occurred. The degree of severity or of the degradation can be determined by this measure.

Of course, there are other people that would measure the pollution by B.O.D. measurements. Others might measure pollution by temperature, or by oxygen. The diversity concept is a measure whereby one can determine the degree of degradation of aquatic life against the characteristics of a so-called natural system.

Steven Carson: It appears to me, from a number of the comments that have been made, that the federal people and the state people are faced with the need to make decisions. They must find some answers.

Two basic issues were raised: one, the question of an interruption in the food chain; and the other, the question of possible toxic activity, or toxic by-products as a result of a modification that takes place. What level of reproducibility have you found in the studies that you have carried out where you observe each year or every other year, to follow a system that has been under modification or change due to thermal pollution, or even natural thermal modifications? Also, what level of extrapobility do you have out of this data that would allow federal people to make the kind of judgments that they obviously must come up with? We don't have five years to investigate the potential. We must take the data that is currently available, make the extrapolation, and come up with answers that are going to stand up.

Patrick: In the studies which we have made relating to thermal pollution, we continually look at the stream and study it. In other words, it isn't just now and two or three years later. Examples include our Potomac River studies and our Patuxent River studies, which I didn't mention because they are in brackish waters. We do make very frequent studies on many different parameters, such as plankton, fish, crabs, other invertebrates, and so on. What can be concluded from these studies is the trend of change, if it is occurring. We would like to be able to measure stages in the food web. One of the great problems encountered in trying to measure standing crop accurately is the fact that organisms are not randomly distributed in a stream; and, to get a statistically accurate estimate of standing crop, either by weight or numbers of organisms, is extremely difficult. Such measures are difficult in lakes, but they are much more difficult in streams. Therefore, we try to determine relative abundance—that is, whether a given species is common, whether it is frequent, or whether it is rare. Such determinations are made by scientists who are specialists in particular groups of organisms. They go into an area over and over again, and they determine relative abundance from many collections from substrates, by dredging, and by any other suitable collecting method. By shifts in numbers of species, kinds of species, and relative abundance, one can estimate the degree of degradation.

We also have developed an instrument—a diatometer—on which diatom communities develop. These communities of diatoms are very similar to those growing in the stream on the natural substrates. A number of papers of other people have verified this: that these slides develop typical communities. That is, the number of species and the relative sizes of the populations of species are similar to those of communities in the stream. Now, by watching the trend of the shift of the truncated normal curve developed from these data, one can determine the degree of change. This is quite a sensitive method for determining change.

Krystyna Mrozinska: In Poland, we pursue such investigations in Cracow. Dr. Patrick was so kind as to mention this. We carry on such investigations on five aspects: microbiology, hydrology, limnology, ichthyology, and water quality. Detailed information on these studies can be obtained by contacting Dr. Krenkel, who can direct you to the proper investigator in Poland.

Chapter **8**   Charles B. Wurtz

THE EFFECTS OF HEATED DISCHARGES ON FRESHWATER BENTHOS

THE title of this symposium refers to *thermal pollution,* but the several papers, for the most part, include in their respective titles the expression "the effects of heated discharges." This latter is more meaningful to me, since I do not know what *thermal pollution* means. This past spring, I participated in a program at Rutgers, where I was asked to speak on thermal pollution. I felt compelled to make an observation to the effect that I assumed the organizers of the meetings meant by thermal pollution the raising or lowering of water temperatures to levels that left the water unusable by other water-use interests. I make the same assumption here. If the definition of thermal pollution is highly restrictive, the national problem is aggravated; if the definition is broad enough, the problem is inconsequential. However, I have not been asked to speak on thermal pollution, but only on the effects of heated discharges. As to whether or not such effects constitute pollution in any given area, I cannot say. Pollution is a product of legislation; and in any particular body of water, its identity will be established by prevailing definitions in legislative acts.

I am particularly pleased to be asked to discuss the benthos. For many years, I have been in agreement with Ward (1919) who said:

As a means for determining the suitability of the water body for the existence of fish life and its favorableness for the multiplication of fish species, it is better to study the small organisms rather than the fish themselves. The fish are evidently less subject than are the smaller organisms to the control of the immediate environment and better able to change continuously their position, as well as to undergo for a limited period unfavorable conditions without really adapting themselves to the situation.

In my opinion, a half-century ago Ward recognized the most powerful biological tool for pollution studies.

With very few exceptions, all the concern demonstrated about temperature changes in our surface waters relates to increases in temperature. The artificial cooling of water by such things as the discharge of hypolimnetic waters has been but a minor problem, to date. Unquestionably, heated discharges are of much greater concern. Such discharges have received so much attention in the press that the public believes many of our surface waters have been irreparably destroyed. This is far from the actual situation. To the best of my knowledge, the Mahoning River passing through the steel complex in the vicinity of Youngstown, Ohio, was at one time the only stream in the nation so heated that biological activity could be considered completely destroyed. Historically, this river had temperatures as high as 140° F. The construction of dams with controlled releases of water has considerably alleviated this problem. The stream, nevertheless, is still a long way removed from complete recovery. In a study made in 1960, only eight macroinvertebrate species were present. Based on a comparison with other streams in the vicinity, the Mahoning River, at that time, was calculated to be reflecting 87 percent biological depression. No measure of other forms of aquatic life was made, but algae were present. It would be logical to assume that bacteria and other micro-organisms were present. This stream, at one time the most affected by heat of all the streams in the nation, is today biologically active. Its potential is real and its recovery can be predicted.

When the North American continent was first settled by Europeans, it was exploited for an agricultural economy. This initial society made intensive efforts to open the land to farming. In time, this led to the destruction of vast stands of vegetative canopy and exposed the land to the effects of solar radiation. An obvious corollary was the warming of our waters. In my opinion, the temperature regime of our surface waters has been more deeply affected by agricultural and forestry pursuits than it has been by industrial discharges.

These environmental alterations, with the concomitant temperature increases, must have altered the biological structure of our streams. Subtle temperature changes must have occurred, and such changes will produce biological changes. If the temperature change is permanent, the biological change will be permanent. Lake Erie is a typical example of this. In the decade 1918 to 1928, the average annual temperature of Lake Erie was 50.0° F. Between 1925 and 1930, there was an increase in this average annual temperature, and the mean annual temperature is now about two degrees warmer than previously. Before the years 1918 to 1928, the bottom fauna was dominated by mayflies; currently, it is dominated by known pollution-tolerant midge larvae and oligochaete worms. The current concern of the pollution of Lake Erie appears to be, in part, supported by what appears to have been a natural climatic change in the temperature regime of the lake. The midge larvae and oligochaete worms are better adapted to the slightly warmer temperatures than the mayflies of yesteryear. This situation, combined with the "discovery" of a thermocline in the lake, by some enlightened limnological dilettante, seems to have supported the political concept of a dead sea. In spite of the talk about Lake Erie being a dead sea, it has always been, and still is, the most productive of the Great Lakes for commercial fishing. For example, Lake Erie yielded 54 million pounds of fish to the commercial catch in 1966. This was 46.5 percent of the total for all the lakes. In 1967, the Lake Erie catch was 48 million pounds, which is about the average annual catch figure for the past 50 years. I do not deny that pollutional materials are being introduced into the lake; I do deny that irreparable damage has been done

THE THERMAL ENVIRONMENT

Rate of Temperature Change

When we present or read a figure for average annual temperature, or maximum summer temperature, or average daily temperature, or some similar derived value for surface waters, we commonly fix that figure in mind as *the* temperature of that particular body of water. It is surprisingly easy to overlook the daily fluctuations of water-quality characteristics, such as temperature.

Most temperate-zone species of freshwater benthic organisms are eurythermal. In the course of a year, they live at temperatures somewhere between 32° F. and 90° + F. This is not to say these organisms can casually plunge from one extreme to the other; but at any given time, most species can tolerate relatively wide temperature fluctuations.

Sprules (1947), for example, worked on a small stream in Algonquin Park, Ontario. He observed that, during July and August, there were some instances where diurnal temperature fluctuation was as much as 17.5° F. This would be a rate of change of 0.7° F. per hour for 24 hours, if the change was constant and uni-directional. If the fluctuation was cyclic, as would be expected, this would be 1.4° F. per hour for 24 hours.

The stream studied by Sprules was eight miles long, located at 45°33′ north latitude, and ranged in elevation from 1400 to 1282 feet. The maximum recorded water temperature during the study was 85° F. In spite of location and elevation, this sounds like a warm-water stream.

Sprules's study was devoted primarily to the mayflies, caddis flies, and stoneflies. Detailed data on the species present were secured at each of four stations along the stream. Considering the uppermost station as Station 1, and the lowermost as Station 4, there was a heat increment in the average summer water temperature at each succeeding station. Station 2 had an average summer temperature 4.4° F. higher than Station 1; Station 3 was 1.7° F. greater than Station 2; and Station 4 was 5.4° F. greater than Station 3. Total temperature difference in the average summer temperature between Stations 1 and 4 (a distance of about eight miles) was 11.5° F. Sprules collected 108 species of mayflies, caddis flies, and stoneflies at these four stations. Table 1 presents these insect data.

TABLE 1. Insect Distribution Data

Station	Mayflies	Caddis flies	Stoneflies	Total
1	10	15	12	37
2	17	19	11	47
3	22	21	11	54
4	28	30	7	65

During the past few years, I have used some very simple descriptive statistics to interpret biological conditions in an aquatic environment. I would like to present these statistics for Sprules's insect data. The statistics are standard deviation $(S = \sqrt{\Sigma(d2)/N-1})$, variance (s^2), and the coefficient of variation $(V = 100s/X)$. Table 2 presents these statistics for the data of Table 1.

TABLE 2. Descriptive Statistics

Statistic	Mayflies	Caddis flies	Stoneflies	Total
s	7.64	6.35	2.24	11.79
s^2	58.33	40.33	5.00	139.00
V	40.2%	30.2%	22.4%	23.1%

Variance is a measure of the variability among the groups and increases as the variability increases. The coefficient of variation is independent of variance and reflects variation about the mean value of the group. Low values (less than 20 to 25 percent) of V reflect a high degree of biological stability and indicate that the biological materials present are in equilibrium with their environment. From the statistics presented in Table 2, it is apparent that the stoneflies in the study area had little variability in species composition among the stations and reflected considerable biological stability. I interpret this as the insect group with the best adaptive tolerance to a broad thermal environment (of the three groups studied, of course). The caddis flies were a less stable element and the mayflies the least stable element, reflecting decreasing thermal tolerance. When all species of all groups are considered, there is a very high variance, but also a high degree of stability.

Deriving chi square by using Dayhaw's k × 1 table for the four stations and three groups of insects, no significant difference was found for the data (x^2 = 7.043; df = 6). In spite of temperature differences in this small stream, the structure of the stream insect fauna as represented by these three groups was in equilibrium with the thermal environment.

I would like to pursue the concept of a fluctuating thermal environment further. My interest in this stems from the limiting rate of temperature change included in many of the new state criteria. In general, these newly adopted regulations stipulate that temperature rate of change shall not exceed 2 or 3 degrees Fahrenheit per hour. I can find no basis for this figure and believe it is unnecessarily restrictive.

At four stations studied in Ozark headwater streams in 1965, April temperatures ranged from 50° F. to 64°F., with an average of 57.6° F. August temperatures ranged from 70° F. to 80° F., with an average of 73.5° F. At one of these stations in August, the temperature at 12:30 P.M. was 58° F. and at 3:15 P.M. it was 61° F. This was a rise of slightly more than one degree per hour, if the rise occurred at a constant rate. In May 1968, the temperature range at these stations was 57° F. to 63° F. with an average of 60° F. In one instance, the temperature rose from 60° F. to 63° F. in two hours: a rate of 1.5 degrees per hour. Such rate changes in headwater streams must be common phenomena. These headwater streams were highly productive and had a very diverse macroinvertebrate fauna. During the 1965 work, a total of 154 species was taken at these four stations.

At six stations in headwater streams in Puerto Rico, temperatures varied from 71.5° F. to 78° F. in one August, and from 68° F. to 73° F. in December. At one station in August, the stream temperature rose from 74° F. to 78° F. in one hour and twenty minutes. This is a rate of change of three degrees per hour in a region with a relatively narrow annual temperature range. In the same region in Puerto Rico, three stations were located on impounded waters. These waters ranged from 74° F. to 83° F. in August and 69° F. to 86° F. in December. Some macroinvertebrate species were common to both stream and impoundment. On any given day, such species (though probably not the same individuals) were living in a temperature spectrum of as much as 18 degrees.

Dorris and others (1962, 1963) made some studies on a backwater channel of the Mississippi River. The channel locally is called a chute and is located behind an island of the river. This channel was 11,880 feet long, 90 to 240 feet wide, and had a maximum depth of 12 feet. Dorris *et al.* (1963) said: "A wide annual range in temperature was observed. The river temperature ranged from 0° to 33.2° C. [32° to 91.8° F.], while the upper end of the chute [the channel described above] had a temperature range of 0° to 34.2° C. [32° to 93.6° F.] and the lower end of the chute had a range of 0° to 33.9° C. [32 to 93.0° F.]." It is apparent that a different temperature regime prevailed in the river when compared to this backwater channel. Further, different regimes prevailed within the channel itself. A further observation by the authors stated: "Stagnation of the water current at the lower end of the chute was sufficient to permit the development of thermal stratification in midsummer. On 2 July [1955] a temperature difference of 3.7° C. [66.6° F.] between the surface and the 4-foot depth was observed." Both fish and macroinvertebrate organisms must have been moving back and forth throughout this temperature gradient. Emerging insects would, of course, pass through the gradient, but the time of passage would probably be so short that stress conditions would not develop. However, insect larvae, such as mosquito larvae, that normally move vertically from top to bottom must have been able to withstand the changes in temperature associated with their movement through the water.

South Branch Forked River, a tributary of Barnegat Bay in New Jersey, reflects pronounced thermal stratification associated with water density. On June 15, 1965, surface temperatures of this stream ranged from 65° F. to 70° F., while bottom temperatures ranged from 70° F. to 78° F. Water depth varied from three to four feet. At one station, in four feet of water, there was an 11-degree difference between surface and bottom. The surface temperature was 65° F., the bottom was 76° F. These very strong thermal gradients were natural phenomena associated with the layering of fresh water over salt water. The bottom fauna of the stream included a mixture of both freshwater and estuarine species. For example, errant polychaete worms were present, along with midge larvae. It follows that the resident fauna must have been able to withstand pronounced and rapid temperature changes associated with the tidal oscillation of the river.

This entire area of tolerable rate of temperature change is badly in need of investigation, and this is especially true for the macroinvertebrate element in biological systems.

Maximum Tolerable Temperature

No doubt one of the areas of interest to this group is the maximum temperature that can be tolerated by aquatic life without injury. For this reason, I would like to refer to this subject briefly.

During the summer of 1967, I visited a river in Alabama, where historic records from preceding years showed surface-water temperatures as high as 93° F. with temperatures at the bottom of 90° F. Interestingly, the bottom was 45 feet below the surface. This river was not receiving heated discharges; these were normal temperatures. At the time of my visit, the river temperatures at the surface were 81.5° F. to 82.0° F. In a bayou off the river, the surface water was 84.4° F. This bayou had a very rich fauna associated with mats of water pepper (*Polygonum* sp). The fauna included mosquito larvae, water striders, giant water bugs, water scorpions, midge larvae, mayfly nymphs, dragonfly and damselfly nymphs, aquatic beetles, sponge, amphipids, a fingernail clam (*Pisidium* sp.) and snails from the genera *Helisoma* and *Lymaea*. Obviously, these waters were not biologically depressed.

Our surface waters can, of course, be subjected to biological depression by heated discharges. A 1959 study (Wurtz and Dolan, 1960) on the Schuylkill River near Philadelphia showed this. The river was also suffering from biological depression from pollutants introduced above the heated discharge. The effect of these pollutants was not measured. In any event, one-fourth mile below the heated discharge, the 1959 recorded temperatures showed 33 days with temperatures at or above 95° F. and 13 days with temperatures at or above 100° F. The maximum recorded temperature at this site was 106.5° F. Table 3 summarizes the data from four stations of this survey and presents the descriptive statistics.

It is apparent from Table 3 that the river had reduced fauna at all stations. Variability was slight and the biological community was quite stable. This stability reflects biological equilibrium between the benthic fauna and the environment. It is self-evident that the macroinvertebrate organisms include species tolerant of elevated temperatures.

TABLE 3. Schuylkill River Macroinvertebrate Data

2.5 miles above discharge	32 species
1.0 miles above discharge	24 species
0.25 miles below discharge	20 species
3.0 miles below discharge	29 species

$\overline{x}$ = 28 species
s = 5.68
s^2 = 32.33
V = 20.3%

Biological Responses and Temperature

Biological rhythms have evolved as adaptive responses to environmental influences. The simplest categorization of such rhythms is the dichotomy that recognizes genetically impressed endogenous rhythms on the one hand and environmentally controlled exogenous rhythms on the other hand. Many rhythmic phenomena are known where the adaptive association with some environmental influence cannot be identified. Conversely, many are known where the adaptive association is clear-cut. Of all the environmental influences that exert controls over biological rhythms, photoperiod appears to be the most important. At the same time, however, temperature is also a very important influence on rhythmic behavior in and by organisms.

Sprules (1947) observed: "The maxima of temperature and insect emergence did not coincide in any twenty-four hour period." He found, in general, that maximum temperatures occurred at 11:00 P.M. Insect emergence began to build up between 8:00 P.M. and 9:00 P.M., corresponding with a decrease in light intensity.

Although temperature exerts a powerful control on metabolic rates, at least some organisms have an inherent capacity to physiologically circumvent the influence of temperature. Some midge larvae, for example, may complete all the metabolism associated with pupation and be physiologically ready for emergence during mid-afternoon peak-temperature hours. Nevertheless, these organisms do not emerge until nightfall. The influence of photoperiod in such cases is more compelling than the influence of temperature.

Frenling (1960) studied the emergence of the mayfly *Hexagenia bilineata* along the Mississippi River. He said:

Particular attention was given to *H. bilineata* emergences at Keokuk, Iowa, because field operations were centered there. Daily air temperatures were made available by the U.S. Army Corps of Engineers, and daily water temperatures and pool levels were provided by the Union Electric Power Company. No correlation was found between air temperature and emergence in the Keokuk area. . . . Similarly, no correlation was evident between emergence and water temperature, water level, or phase of the moon.

It is very easy to say that, as temperatures increase, metabolic rates increase. The beautiful simplicity of this statement is most appealing; but, unhappily, *it ain't necessarily so*. For example, Rao and Bullock (1954) observed: "It . . . was already well documented years ago that at least some species, when cold-acclimated or -adapted, show higher metabolic rates (or other rate functions) at any given temperature, within limits, than when warm-acclimated or -adapted." In the same article, these authors also said: "Temperature response is a complex function and in many respects varies among animals . . . so that we are recognizing common trends rather than rules."

There is, in Georgia, a large lake at the head of which a heated discharge enters. A well-defined thermal gradient is developed here. This is a life-size experiment already established and simply waiting for someone to record the available data. In July 1967, I had an opportunity to pay a brief visit to this locale. At the time, discharge temperatures were 95° F. to 102° F.; surface water below the discharge was 97° F.; and a short distance farther along the shoreline, the surface water was 91° F. to 92° F. In another arm of the lake that was not influenced by the discharge, surface-water temperatures were 85° F. to 86° F. In the warmest water (97° F.), the macroinvertebrate fauna included flatworms, bryozoans, snails and their eggs, mosquito pupae, and midge larvae. A single mayfly nymph was also observed. These species must represent forms with a broad thermal tolerance.

Walshe (1948) studied heat resistance in a series of midge larvae and clearly demonstrated a temperature gradient along which the species were arrayed. The gradient was established on the basis of a thermal index, which was that temperature where 50 percent of the experimental animals died within 22 hours. Survival temperatures for all specimens would, of course, be lower than the 22-hour thermal index. Table 4 presents the gradient developed by Walshe.

The distribution of mayflies, caddis flies, and stoneflies as found by Sprules in his Algonquin Park study clearly shows such a gradient existing in nature. With each temperature increment at the successive stations, there was a loss of some species from the fauna with a gain of others. This is clearly evident from the fact that, of the 108 species in these three

TABLE 4. Walshe Midge-Larvae Data

22-hour Thermal	Species
84.2°F	Tanytarsus brunnipes
86.0°F	Prodiamesa olivacea
86.9°F	Anatopynia nebulosa
94.1°F	Chironomus riparius
95.0°F	Chironomus albimanus
95.9°F	Chironomus longistylus
101.8°F	Anatopynia varia

groups collected by Sprules, only six species were common to both Stations 1 and 4.

Biological responses to water temperatures must be as complex as responses to dissolved-oxygen concentrations. Fox *et al.* (1937) recognized three patterns in rates of oxygen consumption among mayflies. These included: 1) oxygen-consumption rates that are independent of oxygen tension over a wide range of concentrations, but with abrupt thresholds above and below this range; 2) oxygen-consumption constant over an intermediate range of oxygen concentrations and with the rate of use increasing or decreasing gradually above and below this range; and 3) a rate of oxygen consumption that is dependent on oxygen content. I believe the same three patterns can be recognized among the macroinvertebrate organisms in relation to temperature. The intimate relationship between temperature, oxygen consumption, and metabolism is well known to biologists. The most direct measure of metabolic rates is through the measurement of oxygen consumed. This is the purpose of such instruments as the Warburg respirometer.

Within a heated-discharge area, I would expect to find species of macroinvertebrates that were characterized by a pattern where metabolic rates are independent of temperature over a broad range, but where abrupt threshold responses occur. In effect, such species would have a very narrow stress zone beyond which they would be killed. Those 20 species found in the Schuylkill River one-fourth mile below a heated discharge probably represent such species. The organisms found included three flatworm species, one bryozoan, three oligochaete species, one leech, two sphaeriid clams, four pulmonate snails, one damselfly nymph, one beetle, and four midge larvae.

Nearly 40 years ago, Běléhradek (1930) said: "None of the temperature formulae proposed up to the present in biology can be said to hold good in every case, nor is there a rational temperature law in biology." Běléh-radek discussed the Q_{10} rule and concluded that: ". . . a serious analysis by means of the Q_{10} is not possible, and the practical use of this constant for comparative study will not therefore be very significant." Still, today, many biologists lean on this concept of biological responses to temperature as though it had originally been included in the text of the Sermon on the Mount. Rao and Bullock, cited above, observed: ". . . temperature coefficients commonly increase with the adaptation of the protoplasm to higher temperatures. . . . Q_{10} varies commonly, not invariably, with this factor."

Běléhradek explains the persistence of the Q_{10} rule thus: "It may be explained only by the well-known mathematical timidity of biologists, that the Q_{10} still figures with some importance in many biological publications." Běléhradek believed that thermal summation was nearer to reality than is the Q_{10} rule. Thermal summation states that the time necessary to reach a particular stage of development, multiplied by the temperature, gives a constant. The temperature is counted from "biological zero," which is the temperature where cold stops development. This rule, according to Běléh-radek, has been shown to hold good for many plants, as well as in the development of insects. Thermal summation is of particular interest in agriculture, where it is expressed as length of growing season.

Brock (1967) recently published on organisms living in hot springs. One of his conclusions was to the effect that we do not yet know the upper temperature for life. He does state that the upper temperature for multicellular animals is less than 122° F.; however, various protistans live at temperatures up to 167° F. Brock meant living, not simply surviving. Survival would occur at even higher temperatures, as in the case of the African midge larva that, when dessicated, can withstand 215° F. for a minute or two.

The carbon compounds that are characteristic of biological materials can function over a range of temperature from a little below freezing to about 170° F. Macroinvertebrates, as well as all other living organisms, must either live in a thermal environment where their chemical systems retain stability or maintain their own internal thermal environment. Poikilothermic animals, which have a relatively variable body temperature, choose the environment. The macroinvertebrates are, of course, among these forms. However, many macroinvertebrates are heterothermic. Such organisms are facultative endotherms in which body heat is derived from the individual's oxidative activity. Among the insects, for example, many of the Lepidoptera are heterothermic. I am not aware that heterothermy has been investigated among aquatic macroinvertebrates. Heterothermy does give an organism some independence from environmental temperatures.

A recent book (Gordon, 1968) contains some comments on animal function of particular interest to the ecologist concerned with the effect of heated discharges. I quote:

> It is important to recognize that, aside from certain marine situations, for example, the abyssal depths below the photic zone, the physical environment is thermally complex. Many animals behaviorly exploit this thermal complexity and often demonstrate impressive thermoregulatory capacities that may go unnoted in strictly physiological treatments of the topic of the control of body temperature.
>
> The physiology of animals living under natural conditions frequently violates the conventions that have gradually developed from laboratory investigations.

Gordon, who wrote the book I have quoted from, is of the opinion that, aside from extremely cold or extremely hot situations, a hypothetical ectothermal poikilotherm has a body temperature that passively follows and is indistinguishable from that of the environment. No energy, therefore, would be expended on thermoregulation. He also feels that heat exchange in aquatic animals is largely by means of conduction, in which he is probably correct.

Shallow waters, like terrestrial habitats, represent physically complex systems, and this results in a thermally complex environment. In view of this, it is not possible to describe the thermal environment in terms of ambient temperature alone. The energy flux in streams can be represented by some standard formula such as that presented by Gordon, where:

$$S \pm R - LE \pm G \pm A + M = T$$

In this formula, as calories/cm²/minute, S = solar radiation, R = thermal radiation from and to objects in the habitat, L = latent heat of evaporation, E = rate of evaporation, G = conduction to or from the channel substrate, A = convectional exchange with the atmosphere, M = metabolic heat, and T = the sum of gains and losses of energy and this is related to body temperature. In a stream, animals can behaviorly control the magnitude of some of these components. Many insect nymphs are cryptic and avoid S; conversely, many snails, such as some of the Pleuroceridae, actively seek exposure to solar radiation. In a stream L, E, and A must represent but indirect factors as regards individual organisms, but they do influence the total habitat. It would appear that G is directly involved with organisms at the individual level, and of course, M would be also.

The variability among the components of the formula as they relate to the members of any phylogenetic group reflects amazing complexity. I have hypothesized a situation among the midge larvae of three life forms commonly found in the family Tendipedidae. This table is purely a figment of my imagination. I do not know whether or not it approaches reality at all. However, for what it's worth as a thought, Table 5 presents an instantaneous energy picture at 3:00 P.M. and 3:00 A.M. on a sunny day and clear night in mid-summer in, let us say, Mud Creek, U.S.A. Energy gains are indicated by plus, losses by minus, and where I did not care to guess, I inserted a query. Only S, R, G, and M of the standard formula are included.

TABLE 5. Hypothetical Energy Exchange in a Thermal Environment for Midge Larvae

Life Form	3 PM	3 AM
	S R G M	S R G M
Burrowers	− + − +	− + + ?
Case builders	− + − +	− − + ?
Clamberers	+ + − +	− − + ?

Each of the 24 cells of Table 5 represents a continuous variable during any 24-hour period. There is no evidence to indicate whether any rate of change is constant or variable. In the Ozark streams mentioned earlier, 154 species of macroinvertebrate species were taken; of these, 21 were midge larvae. The complexities of energy exchange at the species level are obviously enormous. In addition, other factors such as water chemistry influence metabolic rates and, therefore, temperature. Physical phenomena of the environment also affect metabolism. Fox and Simmonds (1933) found the oxygen consumption of the aquatic Sowbug *Asellus aquaticus* to be 1.5 times greater in swift water than in still water. Similar results have been obtained for mayfly nymphs where differences as great as 3 or 4 to 1 have been found for specimens from swift water, as compared to those from still water. Whitney (1939) found that mayfly nymphs from slow waters had greater heat tolerance than those from fast water. Walshe (1948) found that, among ten midge species, those from still waters were much more resistant to anaerobic conditions than were those from streams, and that the stream forms had a lower thermal resistance. Habitat does influence the thermal regime of macroinvertebrate species.

If the hypothetical situation presented in Table 5 could be developed for all the species of a community it could form a base for predicting the biological effect of a heated discharge. Of course, all the physical phenomena of heat exchange would have to be incorporated into such a study. But, eventually, studies of this type will have to be done. How else can we learn how to control our environment? When I was invited to present this paper I was also asked to suggest areas where research was needed. This is certainly one of the critical areas.

CONCLUSION

Throughout this presentation I have deliberately sought to use field observations. This was done because, in my opinion, we do not yet know the various patterns of thermal adaptations or responses that exist among benthic organisms. Much good experimental work has been done, but much of this cannot be incorporated into the body of utilitarian biological knowledge until basic response patterns are developed.

We do know that today no surface water of the nation is so heated that a "biological desert" exists. At the same time, there are heated discharges that, at least seasonally, drive fish, and probably other organisms, from certain areas. Here I again agree with a remark made by Ward (1919). In speaking of the region of Big Bay, several miles above Glens Falls on the Hudson River, Ward referred to the probability of environmental alteration in spawning grounds for fish because of impending construction of a dam. He said: "This loss may be inseparably connected with the adequate utilization of the stream for industrial purposes, and if so, must be faced as one of the features in natural conditions that must be sacrificed in order to allow the profitable utilization of natural resources by the human race."

I, for one, believe in stream classification. Most recently I discussed this at the 17th Southern Water Resources and Pollution Control Conference (Wurtz, 1968). At that conference I said:

The management of our waters must include recognition of stream classes: in effect, zoning in the same sense we now have municipal land zoning. Some waters must be recognized as angling waters, others as municipal water supply sources, others as primarily for navigation, etc. Within this classification some waters should be recognized as industrial. . . . Water classification is already a fact. To ignore it simply aggravates our difficulties in developing regulatory controls.

REFERENCES

Běléhradek, Jr. 1930. "Temperature Coefficients in Biology." *Biol. Reviews* V(1): 30–58.

Brock, T. D. 1967. "Life at High Temperatures." *Science* 158:1012–1019.

Dorris, T. C., and B. J. Copeland. 1962. "Limnology of the Middle Mississippi River. III. Mayfly Populations in Relation to Navigation Water-Level Control." *Limnology and Oceanography* 7(2):240–247.

Dorris, T. C., B. J. Copeland, and G. J. Lauer. 1963. "Limnology of the Middle Mississippi River. IV. Physical and Chemical Limnology of River and Chute." *Limnology and Oceanography* 8(1):79–88.

Fox, M., and I. G. Simmonds. 1933. *Journal. Exp. Biol.* 10:67.

Fox, M., C. A. Wingfield, and I. G. Simmonds. 1937. "The Oxygen Consumption of Ephemerid Nymphs from Flowing and from Still Waters in Relation to the Concentrations of Oxygen in the Water." *Journal Exp. Biol.* 14:210.

Fremling, C. R. 1960. "Biology of a Large Mayfly, *Hexagenia bilineata* (Say), of the Upper Mississippi River." *Iowa State Univ., Research Bull.* 482.

Gordon, M. S. 1968. *Animal Function: Principles and Adaptations.* New York: MacMillan.

Rao, K. P., and T. H. Bullock, 1954. "Q_{10} as a Function of Size and Habitat Temperature in Poikilotherms." *Amer. Nat.* 88:33–44.

Sprules, W. M. 1947. "An Ecological Investigation of Stream Insects in Algonquin Park, Ontario." *Publ. Ont. Fish. Res. Lab. No. 56.*

Walshe, B. M. 1948. "The Oxygen Requirements and Thermal Resistance of Chironomid Larvae from Flowing and from Still Waters." *Journal Exp. Biol.* 25:35.

Ward, H. B. 1919. *Stream Pollution in New York State.* Albany, New York: New York Conservation Commission.

Whitney, R. F. 1939. "The Thermal Resistance of Mayfly Nymphs from Ponds and Streams." *Journal Exp. Biol.* 16:374.

Wurtz, C. B. 1968. "The Realities of Thermal Pollution-Environmental Limitations and Ecological Adaptations." Paper read at the 17th Southern Water Resources and Pollution Control Conference.

Wurtz, C. B., and T. Dolan. 1960. "A Biological Method Used in the Evaluations of Effects of Thermal Discharge in the Schuylkill River." In *Proceedings, 15th Industrial Waste Conf.*, pp. 461–472. Purdue.

DISCUSSION/ John Cairns, Jr.

DR. WURTZ'S main point seems to be that each species is programmed to exist in a highly variable environment, and that no two species are alike. When there is a fairly good match between the capabilities of an aggregation of species and environmental parameters, a state of dynamic stability exists. Previous speakers voiced concern about our ignorance about most of the species on this planet, which hampers our ability to protect them from the impact of technological advances (i.e., power generation, insecticides, etc.). This is quite true, but it does not leave us in a hopeless position until more is known. We do know enough about the responses of aggregations of species to stress, or pollution, for management purposes. Although each species is the end product of a lengthy and complex evolutionary process and should be protected and studied, we don't have either the time or personnel to do this now. As a result, we must rely on group responses. Fortunately, it is possible to determine when the environment has become distinctly unfavorable to an aggregation of species, even though the requirements of the species comprising the aggregation are not known. Since ecosystems, including resident species, consist of interdependent and interacting parts, it is desirable to place primary emphasis upon the functioning of the entire system and secondary emphasis upon the functioning of individual species. This is simple and more economical than protecting species individually, and, for the time being, the only option really available, when one considers the vast array of species to be protected. However, certain species of special importance to society should receive particular attention in both research and management. This is a dangerous game which our rapid technological advances have forced us to play. It is virtually certain that there are many species vitally important to human survival whose roles are not now recognized. One of the major purposes of ecological systems analysis is to protect these species. Thus, ecologists have a dual role with respect to heated discharges and other wastes: (1) protection of entire ecological systems in viable form, and (2) protection and nurture of species particularly important to man. I suspect Dr. Wurtz and I agree on these goals, although our means of achieving them might differ. However, a clear statement of our broad general goals seems essential to a harmonious relationship between a technological society and the environment upon which its survival depends.

Dr. Wurtz correctly points out that organisms can exist in Alabama and other areas at impressively high temperatures. He also gives several examples of rather striking temperature changes through both time and space which are tolerated by aquatic organisms. I have no quarrel with these statements, although they may be a bit misleading to the uninitiated. For the average human, a 14° F. increase in temperature in August is generally easier to take in Oregon than in Kansas. In common with other organisms, we respond more favorably to temperature increases when we are not already close to our maximum tolerance. Unfortunately, from a regulatory standpoint, ranges of temperature tolerance vary enormously from species to species and are, in addition, modified by other environmental qualities.

I hasten to add that I fully recognize the need for industrial use of water, provided the environment is not abused. I have worked for a number of years in two situations where restrained use of river water for coolant purposes has not degraded the biological communities in the receiving streams. In both cases, there was a pre-operation ecological survey, followed by continual ecological monitoring of river conditions at several points below the outfall and at an upstream control station. Before and after each increase in thermal loading, the amount of data gathering was increased, so that ecological changes could be detected relatively soon. I support industrial use of our natural resources when it is accompanied by and *influenced* by ecological monitoring programs of this type.

I disagree with Dr. Wurtz in several important respects. First, thermal pollution can be defined in a variety of ways. None are perfect, but many are usable! My own operational definition of pollution is: any environmental change which alters the species diversity more than 20 percent from the empirically-determined level for that particular locale (Cairns, 1967). I would be willing to negotiate on the percentage, but not on the principle. Ecological stress, typically, results in simplification of biological communities, and this can be assessed. It is true that this may be followed by colonization of new species more fitted to the new environmental conditions and former levels of complexity may be reached, but this biological adaptability is not a sufficient justification for the disruption of ecosystems, any more than the ability of societies to survive revolution is a justification for that form of stress. Moreover, establishment of new species takes time; Dr. North indicated a 5-to-8-year lag period, assuming his efforts are successful, to re-establish a new community of heat-tolerant organisms following a heat kill of resident species. Undoubtedly this restoration process will begin with a simplified system and evolve into one of considerable biological complexity. However, if heated discharge of water were to cease at a critical period, it is quite likely that this new "warm-adapted" community would suffer. The effects might well be as severe as was the initial discharge of warm water. The responsibility of an industry to maintain heated discharges of water, once a biological community has become adapted to and perhaps dependent upon these new conditions, has not been clearly stated. I refer, of course, only to those cases where heated discharges resulted in displacement of existing species by those adapted to warmer waters and not native to the area. Presumably, these can only exist and function as long as the power plant continues operation. Cessation of operation for even a brief period in midwinter would probably kill most of the species. Situations of this sort are inevitable, and the consequences should be considered before an emergency occurs. This is particularly true of the freshwater benthic organisms, which are often less mobile than fish and slower to reinvade than diatoms or protozoans. In nature, complex biological systems are usually considerably more stable than simple systems. If we wish to stress the environment to serve our needs, simplification means increased management and control. There is a price to pay for both waste treatment and the disruption of ecosystems.

My second major point of disagreement with Dr. Wurtz stems from his agreement with Ward (p. 199). When a person or an organization purchases a piece of property, this involves more than land alone; one is becoming part of an ecosystem—generally because certain environmental qualities (i.e., recreation, potable water, process water, etc.) are considered desirable. To justify sacrifice of an ecosystem, or some of its qualities, on the grounds that this is necessary for "the profitable utilization of natural resources by the human race" suggests that man exists independently, rather than as part of the environment. We want the advantages of industrialization, as well as a vigorous, diverse environment. I also believe in stream classification, but as a series of goals, rather than as a means of legalizing mismanagement. However, I suppose even a classification which merely acknowledges mismanagement has its uses. An industrial-use zone would, then, be that area of a stream in which incremental environmental disasters had made the stream unfit for other uses. A recreational or multiple-use zone would be one in which these events had not yet occurred. But the word *management* appears in Dr. Wurtz's statement and this implies goals, as well as recognition of present conditions. If this is stream classification, then I agree.

As Dr. North mentioned earlier, species may be present but not functioning well. Dr. Wurtz's data and much of the other data available, including my own, emphasizes presence or absence of species and the structure of aggregations of species, but has little information on relationships and performance. If we are to have meaningful stream classification and management, some rudimentary knowledge of species and community function is essential.

One interesting possibility in species relationships is that areas with thermal discharges which attract fish may have markedly different species structure and size distribution of the invertebrate populations than adjacent areas, even if the heat does not directly affect the invertebrates. This, of course, is due to increased predator pressure on a single area. Brooks and Dodson (1965) have shown the effects of alewives on the invertebrate population of a lake, compared to a comparable, adjacent lake lacking alewives. Although the number of species remained about the same, the kinds of species and size distribution were markedly altered. We should not hasten to label all changes *pollution* without an attempt to understand the process involved!

The effects of thermal loading are so dependent upon the types and relationships of the organisms inhabiting the receiving stream or lake, as well as the chemical and physical environmental characteristics, that prediction of the consequences of a particular heated-water discharge can only be made in a most general way, from evidence gathered in other areas. We are thus faced with the very real danger of overshooting the thermal-loading capacity of an area and thereby damaging it for considerable periods of time or of restricting the power industry from making full, reasonable use of receiving capacity. It therefore seems essential that each area scheduled to receive heated discharges of water should have its own extensive chemical, physical, and biological monitoring system and base-line survey, before operations begin. In order to be meaningful, these studies will be quite expensive in terms of the total operating cost of the plant, and in terms of the alternative ways of disposing of the waste heat, such costs are ludicrously small. Since many power-generating plants and others with thermal-loading problems now have such surveys and have had them for many years and are, apparently, still paying dividends to stockholders and operating quite well, it would seem that these costs are not the disastrous burden that plants not now sponsoring such studies would have us believe. It is quite evident that a private industry cannot expect state or federal agencies to carry out its data-collecting, particularly when such evidence may be necessary for court hearings and other due processes of law. Such data-gathering on receiving-stream conditions should be considered a regular part of operating costs.

One of the major problems regarding the effects of heated discharges on freshwater benthic organisms, as well as on other aquatic organisms, such as fish, is the comparatively enormous time lag between the collection of data, the detection of any changes that may have occurred, and the determination of the significance of these changes. This point has already been mentioned by Dr. Hedgepeth. In short, even when considerable data-gathering on biological, chemical, and physical characteristics is in progress below and above a thermal-loading zone, the time required for feedback of biological information is relatively enormous, compared to that now possible for chemical and physical characteristics. Really precise information on community structure of aquatic organisms based on the determination of species by specialists in each of the major groups may require months from the time of collection to the final report and will usually require several days per station for collecting alone. Even relatively rapid biological monitoring systems based on the number of species and the number of individuals per species usually require several days to a week from the time the collection occurs. Since, in many areas of the United States and other parts of the world, sudden increases in temperature are likely to occur, one would hope that, eventually, a biological data-gathering system could be developed, with feedback time comparable to that now existing for chemical and physical data. This would enable power industries to make use of a receiving stream to its fullest capacity and enable them to detect dangerous conditions, not weeks or months after they have existed, but, rather, hours, or perhaps ultimately even minutes after they have begun to exist. The utilization of a receiving stream by an electric power-generating plant should involve the obligation to prove that at all times its use is not endangering other beneficial uses. It would be hoped that each plant would be designed and operated to achieve this goal. Since power-shunting from one area to another is now a fairly well-developed practice, it should be possible to refrain from use of a receiving stream when normal environmental conditions approach dangerous levels for the aquatic organisms inhabiting it. However, present methods of prediction and monitoring usually do not provide information on the biological consequences of environmental changes until long after they have occurred. *It would seem, therefore, that one of our urgent needs is the development of relatively rapid biological assessment techniques.*

An equally important factor is that, once we do establish the effects of heated discharges on freshwater benthic organisms, we are not able to relate these in any meaningful way to the economic value of the loss or impairment of other beneficial uses. Only when we develop resource economics to a far more sophisticated stage than now exists will we be able to assess the economic impact of environmental degradation. At present, we can crudely document the economic results and consequences of severe degradation, but it is quite difficult to relate moderate changes in the structure and composition of the aquatic community to particular changes in the economic value of a drainage basin. Nor is it possible to predict or even crudely estimate the rate at which elimination of pollution and restoration of a river will restore maximum achievable economic values. None of these things will be easily calculated, but the attempt must be made if we are to progress from our rather precarious environmental relationships to ones that are more harmonious.

The coefficient of variation appears to be a simple and useful means of assessing the distribution of benthic organisms. The groups (such as the Table 2 stoneflies) with little variability in species composition among the stations may indeed reflect considerable biological stability and the best adaptive tolerance (of the three groups studied) to a broad thermal environment. It might also reflect other characteristics, such as the degree to which each group indulges in downstream migration as described by Waters (1961). It might also reflect the amount of speciation which has occurred within the group and the degree to which the environment has been partitioned among these species.

Dr. Wurtz states that "newly-adopted regulations stipulate that temperature rate of change shall not exceed two or three degrees Fahrenheit per hour." He further states that he can find no basis for this figure and believes that it is unnecessarily restrictive. Perhaps it is; but that is not the point! The main issue is: who has the burden of proof! Before a new drug is placed on the market, the manufacturer must demonstrate that it will not be unduly harmful. The burden of proof falls on the producer. I feel this should also be true for heated discharge water and other potential pollutants. However, merely to state that no harm will result, as has often been done, is not enough. There should be conclusive evidence, based on local conditions, including the particular species and communities involved. A literature search to show what happened in comparable situations is useful, but it is not enough. Each body of water has its own unique characteristics, and these will influence the effects of a heated discharge, so that generalizations are not sufficient.

Dr. Wurtz mentions the destruction of our forests and consequent warming of surface waters. It is quite likely that the effects he suggests occurred. However, it is the sum total of environmental stresses to which organisms respond, and it is this that should determine our course of action, rather than past events.

Other speakers have mentioned effects upon organisms swept up from the benthos into a condenser cooling system. Although both damage and passage without harm have been mentioned, no one has mentioned the possibility of benefits from this exposure to increased temperatures. There is a possibility that organisms exposed to lethal temperatures may have increased nutritive value to other aquatic organisms in downstream areas. One such case, in which *Tetrahymena pyriformis* could not survive on live *Sarcina lutea* cells alone but could, on killed *S. lutea* cells, has been reported by Johnson (1936). Even if this were generally true, the ecological consequences are not clear. Certainly, more information is needed about the consequences of passing aquatic organisms through a condenser system, both to the organisms themselves and to the ecosystem, including the biota, to which they are returned.

REFERENCES

Brooks, J. L., and S. I. Dodson. 1965. "Predation, body size, and composition of Plankton." *Science* 150(3692):28–35.

Cairns, J., Jr. 1967. "The Use of Quality-Control Techniques in the Management of Aquatic Ecosystems." *Water Resources Bull.* 3(4):47–53.

Johnson, D. F. 1936. "Growth of *Glaucoma ficaria* Kahl in Cultures with Single Species of Other Micro-organisms." *Arch. f. Protistenk.,* 86:359–378.

Waters, T. F. 1961. "Standing Crop and Drift of Stream-Bottom Organisms." *Ecology,* 42:532–537.

DISCUSSION FROM THE FLOOR

J. A. Roy Hamilton: Dr. Wurtz would like to say something in rebuttal.

Charles B. Wurtz: I don't think there's anybody here that knows me that thinks that I would hesitate to approach the microphone after that. The thing that is a little bit embarrassing is that Dr. Cairns and I really don't disagree too much. I don't think he's thinking clearly, naturally, but basically we really don't disagree.

If we go back into the history of biological studies—and I mean go far enough back to that day in 1948, when John Cairns and Charles Wurtz joined Ruth Patrick, back in Philadelphia, bought their first pair of waders, and went to work together. Back in those days, we had the idea that somehow, someway, sometime, we would be able to express biological information that would be meaningful in the sense that Max Katz wants it, and all of his associated engineers that pay his bill. What they want is numbers. Are we going to build this bridge, and is it going to collapse? They don't want someone saying, "Well, the probabilities are ninety-nine times out of a hundred it will hold up till you drive that truck over; then it will fall in the creek." Now, what we wanted was to be able to stand up and have figures and the fortitude to say, "This is the way it's going to be." Every one of us, even though we have since divided and moved along in our paths in other areas, is still working for the same thing.

Another item that I would like to bring up, before we go into hysteria, is Dr. Cairns's reference to this concept of stream classification. I am not advocating that we give a license to anybody to ruin any portion of water. I am saying that every portion of water has a use, and that you do have to think about costs. How much is a particular piece of water worth to you? I don't say give dischargers a license and let them pour whatever they want into it. You can't do this; but I think we must be realistic. It's a very simple, pragmatic world, and a real one; let's move

221

it to suit us, because we do own it. Perhaps, a hundred years from now, the ants are going to take it over; however, I'm not worried about a hundred years from now: how long can you live? There was a time when I though I might live forever, but I've given up this idea. If I can sweat through another forty or fifty years, I'll be older than Peter Doudoroff.

We must be realistic, because we cannot save every piece of water. Let's use the water. The Nile River valley has been a human artifact for over six thousand years. Should we let this revert to the sands? We are not thinking clearly if this is our answer, and this is going to happen here.

Carlos Fetterolf: I'm not going to comment directly to either Dr. Cairns or Dr. Wurtz, but to the rest of the group that's here, along sort of the same line. This is the "Heated-Discharge Symposium," where the biologist is king. In attendance are three interest groups. There are those with an academic interest, those with an economic interest, and those with enforcement interest. I fall into the latter group, as a state water-pollution biologist. Later this summer, there is going to be a "Future" meeting, on the engineering and economic aspects of heated discharges. At this second symposium, the engineer will be king. Part of the formal program will involve the discussion of water-quality standards for various water uses, including aquatic life. I am disappointed that there is no session here on this same subject. By this omission, we, in effect, are abandoning a part of our responsibility. I keep waiting to hear discussion on numerical temperature limitations in the interstate water-quality standards that the state and federal agencies are attempting to enforce, or are about to enforce.

So far, we've talked all round the subject as if it didn't exist, but it does exist, and if you don't believe it, just ask the power people that are here. Perhaps this afternoon or tomorrow, or sometime before this symposium is over, we could squeeze in a comment or two by a number of Clarence Tarzwell's aquatic-life subcommittee, which was advisory to the FWPCA in their task of recommending temperature limitations for trout streams, warm-water streams, et cetera. Their deliberations must have been monumental. They have my admiration, as well as my sympathy. But we could gain from a discussion of all, or only one, of their recommendations: for example, "3° F. increase allowable in the epilimnion of lakes." In Michigan, we must base our enforcement on injury to a use or threat of injury. We cannot deprive a riparian of his right to use the water without demonstrating that such use will be or may become injurious to another use.

A second point of vital interest to all of us is how the states will establish mixing zones for heated discharges in their receiving waters. Definition of mixing zones appears to me to be of much greater significance to the environment than the temperature limitations established. Would a representative of the FWPCA enforcement group present care to comment on how these mixing zones will be delineated?

Ruth Patrick: In response to Dr. Fetterolf's request about why didn't the aquatic-life subcommittee have some specific recommendations, I think that it is very difficult, so far as algae are concerned, to make one general rule for the whole country. But I would like to say this: I believe that many bodies of water in the temperate parts of the United States will have blue-green algae blooms if the temperature of the water is maintained above 95° F., day in and day out, for relatively long periods of time. These blooms will, nine chances out of ten, reduce the energy flow in the ecosystem. Is that concrete enough?

Fetterolf: My point is that Tarzwell's committee was able to be concrete, and I question on what basis they felt they were able to be concrete on recommending a three-degree temperature rise for the epilimnion of lakes. Is there someone on that committee here that would care to comment on the 3° F.?

Hamilton: If there is anybody here, apparently he is not prepared to comment.

George E. Burdick: I believe there is an error, because I do not believe in the committee report there is a two-degree temperature rise at the present time. I will say, however, that there is a very rational basis for a two-degree temperature rise, when you are approaching the upper thermal limit or the lower-temperature limit for a species. It is well indicated, in experimental laboratory work, that with acclimatization at the rate of 1 degree an hour, a much higher temperature can be reached before mortality occurs, than if acclimatization is at 2 degrees an hour. I would also have to disagree with the statement of Dr. Wurtz that no one here is concerned with a falling temperature. Frankly, there is a great deal more danger to the fisheries resources of the country in dropping down from a relatively high temperature as a result of a thermal discharge, by reason of closing down the plant, to a 33½- or 33-degree temperature. The acclimatization to an increasing temperature is much more rapid than it is to a decreasing temperature. If, in the wintertime, at low stream temperatures, the plants are suddenly shut down and the fish have to rather rapidly adjust themselves to a temperature of around 33°, you are going to have fish killed.

Wurtz: Dr. Burdick did bring up a point that I glossed over, and I don't deny this. I did not say that a falling temperature was not of importance. Instead, I said that it was of minor importance to the extent that you read about these things in the public press. All the public are concerned about increasing temperatures. I am well aware of the fact that falling temperatures can indeed be critical. If you will recall, I did give testimony on some of these hypolymnetic waters in the upper Delaware system. I am aware of this, but I think the critical problem here relates to increased temperatures, rather than lowered temperatures, and I think Carlos Fetterolf's questions relate to this area, rather than a lowering of temperatures. But you are quite right, of course. Indeed, lowering temperatures can be critical, and you are again right that organisms will adapt to a rising temperature much quicker than they will to a lowering temperature.

James R. Adams: Dr. Burdick stated that a declining temperature can cause a fish-kill when a power station goes off the line. Have you had examples of this any place?

Burdick: No, I haven't known of any. I was looking at the literature. This concerns temperature in a laboratory.

Adams: I am talking about at a power station.

Burdick: No, I haven't had any examples at a power station. But I am anticipating that there will be. Work done by Brett in regard to thermal tolerance definitely shows that, with acclimatization which will be well within the range where the fish will be adjusted to temperatures below some power stations, a temperature drop within the period of time you would expect, if the plant had to shut down completely, will produce a kill at the temperature level of the natural stream.

There is one additional comment: there will be a sharp delay in the fish trying to hold in the warmer water. Not in all cases will that be sufficient protection.

Adams: I direct my question to Dr. Alabaster, because in Great Britain there are far more peaking operations at thermal power plants than we have in the United States.

John Alabaster: Our peak demand comes in the wintertime, and stations using estuarine or seawater carry a fairly constant load, while most of the peak demand within a 24-hour period is met by stations using fresh water for cooling purposes. We don't have any instances, that I know of, of mortality of fish in the wintertime, when power stations come off load.

I once observed a mortality of fish which had been exposed, first, to 20° C. in experimental conditions in the spring, and were then transferred to a temperature of about 10° C. At first, the fish seemed to be all right, but they died after about two weeks. If such a drop of temperature were to occur in a river in Britain, its harmful effect might be offset by the power station coming back on load again, as it normally would, probably within 24 hours.

George Eicher: Dr. Cairns mentioned or alluded to a fact that has been somewhat well established in biological circles, that of the fish being almost irresistibly attracted to these discharges from thermal plants. Do you have any idea why this occurs? I have heard about it and seen it in actual cases for a long time, but no one, as yet, that I have heard, has given any explanation. Do you have one?

John Cairns, Jr.: No. It occurred to me in a discussion last night that this attraction of fish to warm water might have influences on the invertebrate population other than the direct effects of heat, and one of the things that suggests this is the paper by Brooks and Dodson, which appeared in *Science* in October 1965.[1] This was a study of two lakes. One had an alewife (fish that feed on zooplankton) population, and the other did not. Brooks and Dodson studied *Bosmina, Daphnia,* and the other planktonic invertebrates in these lakes. The presence or absence of the alewives affected the structure and size distribution of the invertebrate community. It occurred to me that this might easily occur in an area receiving heated waste water, and one might think that the invertebrate populations were changing as a direct result of the heated discharge, whereas, in fact, it would be changed predator pressure due to the attraction or repulsion of the fish to this area. Such things should be studied in more detail, so that they might be better understood. But I'm sorry, I have no actual information on it.

Leon Verhoeven: I want to make certain that George Eicher's comment is not misunderstood; because certainly, salmon, in my opinion, are not actually attracted to warm water. In fact, we know that in their migrations they avoid warm water, find places of cool water, and lie there until the water temperature in the area through which they wish to pass drops before they proceed.

1. J. L. Brooks and S. I. Dodson. 1965. "Predation, Body Size, and Composition of Plankton." *Science,* 150:28–35.

Eicher: I didn't have salmon in mind. To my recollection, I don't recall salmon having been attracted to discharges.

Peter Doudoroff: I want to ask John Cairns a question concerning his definition of the word *pollution.* Suppose that you have two cases under comparison, one in which you have a small discharge of untreated sewage contributed by a typhoid-carrier, but causing no detectable change in number of species present in the receiving stream; and another case where you have an industrial discharge of organic matter that does result in a 20- or 25-percent decrease in the number of species present, but the receiving stream is still quite suitable for swimming or other recreational use, no health hazard being involved. The fish in the latter stream happen to be growing faster than they did before, and fishermen are happy. Which, would you say, is the more polluted stream?

Cairns: I would select the one in which the population profile had changed to the greatest extent.

I have been working recently with cluster analysis, using dendrograms, much in the same way that the computer taxonomists use them. One can compare species aggregations and community profiles easily, using this technique. I would say that, when the profile changes substantially, due to introduced heat or toxicants, this represents biological evidence of pollution. I recognize that there are many other ways to describe pollution, but I think this is one valid way to do it.

If you are willing to put forth some other definition, then I am willing to consider that, as well. Protecting one quality does not obviate the protection of others.

Doudoroff: Of course, the point that I was trying to make is that I am inclined, myself, to define pollution in terms of interference with use. As you probably know, I have questioned the validity of definitions which are so strictly biological that they do not agree with the meaning of the word for most of the people concerned with pollution. I think there is too much danger that, when you say that a water is polluted, in your special sense of the word, the public will misunderstand. They will think that you have evidence of an impairment for use, because, to most people, pollution of water does mean defilement resulting in some interference of their use of the water, rather than merely a biological change. The layman may well interpret your statements as meaning that you have found a biological measure or index of what he thinks pollution of water is, which is its impairment for use. I think that this is the danger of using the word *pollution* in such a very special biological sense.

Cairns: I don't think we disagree on that. On the other hand, the biological community represents an information system about the environment; and when there is a substantial reduction in species, this means that the old information is not adequate, and the aquatic community is probably not operating as well as it should. One may then have restricted or lowered the options for future uses. Since we don't know what these will be, why not maintain the water resource in the best possible condition? I think your point is a good one. However, it is dangerous to go on wiping out species that have no immediate benefit, because there may be some future use for these species. There has to be a balance between your point of view and mine! I think we have to take both points of view into consideration in making decisions.

John Spindler: With regard to Carlos Fetterolf's question concerning the two-degree temperature rise, Montana has adopted this for what we call D–1 fishing waters. We have three classes of fishing waters in Montana, D–1, D–2, and D–3. D–1 is a reproducing trout fishery; D–2 is a trout fishery which has only marginal propagation; and D–3 is what we call son-salmonid fisheries.

The two-degree temperature change that we adopted in our standards was deliberate for one reason, and one reason only: it is an attempt to restrict water use in D–1 waters. Quite obviously, I am an advocate of Dr. Wurtz's previous proposal, i.e., zoning. I feel that our headwater streams in Montana have no place in the industrial complex. I am talking about heavy industry, and in this I must include heated effluents. Therefore, there is a question in my mind that we should admit that the establishment of the D–2 or the D–1 streams, with their two-degree temperature-rise allowance, is a deliberate attempt to keep industry out of those areas. The D–2 stream is a little bit more lenient, and the D–3 stream is more lenient yet.

For three years, I was a part of a subcommittee here in the Pacific Northwest to develop what we thought might become model water-quality criteria for at least the Pacific Northwestern states. If I learned nothing else in the process of coming up with numbers and figures for water-quality criteria that later became part of standards, it was that we shouldn't have them. I think that each case should be treated on its own merits on the basis of water use; and this is, of course, where we are parting company with the federal people and Secretary Udall and his anti-degradation clause. I am sure that history will prove that we must apply our findings, biological, physical, or what have you, on the basis of each particular case, and that these over-all guide lines that we are developing now for criteria are worthless. I'm not talking about standards which include compliance, and so forth, I am talking about criteria.

Chapter 9   Donald P. de Sylva

THEORETICAL CONSIDERATIONS OF THE EFFECTS OF HEATED EFFLUENTS ON MARINE FISHES

HEATED effluents from several types of industrial discharges may produce thermal pollution. The effects of above-normal temperatures are difficult to evaluate because they are long-term, sublethal, and interrelated with other complex environmental factors. This discussion revolves about the theoretical effect of high temperatures on marine fishes at all stages of their life histories, including reproduction, development and growth, food and feeding, physiology, behavior, and ecology. Emphasis is placed upon the effects that effluent high temperatures, both alone and in combination with other substances, may have upon the physico-chemical environment. The need is stressed for careful surveys and a reasonably detailed knowledge of the environment *prior* to the selection of sites where thermal pollution may occur.

Effects of artificially heated water on marine fishes *in situ* have been little studied, yet the literature contains much information on the direct and indirect effects of a temperature increase upon fishes and upon the varied aspects of their physiology and metabolism, and on the ecology of their environment. Because temperature must be viewed "as a subtle, all-pervading, complex environmental factor" (Brett, 1960), this paper is intended to elicit the subtleties of thermal effects upon aquatic life, specifically marine fishes, and to summarize these effects by discussing the extremely complex and interrelated events resulting from even a slight temperature increase. Yet it will be seen that there are no simple rules for assessing these effects upon the fish and its environment. Using information based essentially upon laboratory experiments, it is hoped we may

Contribution No. 1027 from the Institute of Marine Sciences, University of Miami.

hypothesize on the effects of artificially increased temperatures upon marine fishes under natural conditions.

Although marine fishes are more exacting in their environmental requirements than brackish-water or freshwater forms (Kinne, 1963; Hedgpeth, 1957a; Naylor, 1965), because marine fishes do not differ greatly from freshwater fishes in their basic life histories, I thus have used those examples of the effects of temperature increases on freshwater fishes which seem especially pertinent to the over-all problem.

Marine environments are extremely complex, much more so than freshwater habitats. We still do not understand many biological and physical processes in the sea, and we still know relatively little of marine ecology and the dynamics of physical and chemical processes which affect marine organisms, especially in the coastal environment. In many instances, we do not even know the identity of the organisms with which we are working, or of their population dynamics, genetics, or physiology. We are, in most cases, quite ignorant of the sea and its inhabitants.

The Problem

Thus, it should not be surprising to find a dearth of literature dealing with the problem that confronts us: the effects of artificially heated effluents upon marine fishes. The literature discloses numerous historical experiments in which organisms are subjected in the laboratory to increased temperatures, but relatively few researchers have examined the effects of large quantities of artificially heated water *in situ*. This is because the problem of thermal pollution—be it thermal addition, thermal enrichment, thermal loading, or whatever euphemism the question assumes—has become a potential problem only recently. Power needs increase geometrically with an expanding population. While only 62 million kilowatts were required for power production in 1950, the demand will be for 400 million kilowatts by 1975 (Laberge, 1959). The amount of fresh water required for cooling condensers is estimated to increase from the 1955 usage of 60 billion gallons a day to 200 billion gallons—one-sixth of the fresh water in the United States—by 1988 (Picton, 1960). Future estimates of electrical requirements range from 30 times our present usage to possibly 256 times by the year 2010 (Hoak, 1961, 1963; Cadwallader, 1964; Trembley, 1965; Mihursky and Kennedy, 1967). One of the chief users of water is the steam-electric station (S.E.S.), which returns cooled water to the environment at from 6° to 9° C. above ambient water temperatures (Jones, 1964; Alabaster, 1964; Alabaster and Downing, 1966). In some cases, receiving water may be heated to 60° C., as occurred in Ohio (Cairns, 1956a). It is anticipated that, in the thermally less efficient nuclear-power plants, which also require up to 40 percent more water than conventional plants, the discharge temperatures will be in excess of 11° C. above ambient, with an expected discharge by a single plant of fresh or salt water of up to 1,250,000 gallons per minute (Cadwallader, 1964; Naylor, 1965; Talbot, 1966; Davidson and Bradshaw, 1967). Dr. Peter Krenkel (personal communication) states that, in the United States, as of December 1968, there were 13 nuclear reactors in operation, and 86 more being built, ordered, or planned. Thus, we are faced with a potentially huge volume of warmed water being added to the freshwater, estuarine, and marine environments.

Where power stations were once located in places remote from civilization or in already polluted areas (Klein, 1957; Hynes, 1963) which for decades were too septic to permit an evaluation of thermal loading *per se,* increased usage will place new power plants in the vicinity of suburban, often unpolluted, regions. The inevitable conflict which will increase in seriousness between industrial needs and recreational desires will require intelligent, imaginative planning in the selection of power sites, their proper operation, and a mutual, rational understanding between industrial and recreational interests (Hoak, 1961; Douglas, 1968; Stroud and Douglas, 1968). This problem will require the planned deliberation of environmental bioengineering.

Survey and analysis of the problems of increased temperature on aquatic life has been greatly facilitated by the bibliographies on temperature effects compiled by Raney and Menzel (1967) and Kennedy and Mihursky (1967). The scope, background, and future of thermal pollution have been discussed in papers by Cairns (1956a, 1956b), Klein (1957), Cottam and Tarzwell (1960), Hoak (1961, 1963), Simpson and Garlow (1960), Wurtz (1961), Arnold (1962), Markovski (1962), Hynes (1963), Alabaster (1963, 1964), Jones (1964), Naylor (1965), Cronin (1967), Olson and Burgess (1967), Douglas (1968), *Sewage Industrial Wastes* (1956), *Wastes Engineering* (1961), and the U.S. Senate Committee on Public Works (1968). Power-generating stations are the prime potential sources of thermal pollution in the United States (Trembley, 1965; Mihursky and Kennedy, 1967). Gameson *et al.* (1957) list, in addition to power stations, as sources of heated water: (1) industrial effluents from gas works, paper mills, and sugar refineries; (2) sewage effluents, primarily industrial discharges to the sewers; (3) natural freshwater discharges from tributaries;

and (4) biochemical activity, i.e., exothermic processes from the oxidation of organic matter. To this, Jones (1964:155) adds as sources the cooling of gas at gas works, manufacture of sulphuric acid, sugar beet refining, preparation of dried milk, pickling of steel, and washing of wool.

In spite of the source of the increased temperatures, it is conceded that temperature is one of the single most important factors controlling the distribution and survival of aquatic organisms (Bĕléhradek, 1931; Hela and Laevastu, 1960, 1962). It will be shown subsequently that temperature increases affect fishes directly, and indirectly and sublethally, as well, by changing physiological and behavioral processes, or by changing some aspect of the environment on which fishes depend (Jones, 1964; Kinne, 1963; Naylor, 1965).

The Direct Effect of Temperature on Marine Fishes

The general effects of heated effluents upon fishes have been discussed by Cairns (1956b), Brett (1956, 1960), Doudoroff (1957), Tarzwell and Gaufin (1958), Hoak (1961), Trembley (1961), Kinne (1963), Alabaster (1963, 1964), Alabaster and Downing (1966), Jones (1964), Naylor (1965), Mihursky and Kennedy (1967), and de Sylva (1968).

Numerous experiments have been designed to determine the upper temperature limit that a fish will tolerate before it dies. The very fact that a laboratory test animal can survive being collected and the subsequent trip from the ocean to the laboratory indicates in itself that the test animal indeed must be a hardy species to begin with, and one must question the value of generalizing on thermal or any other environmental requirements in nature for such hardy fishes. Researchers seem to have a propensity for testing tolerance of killifish (*Fundulus*), carp (*Cyprinus carpio*), goldfish (*Carassius auratus*), and fathead minnows (*Pimephales promelas*), all of which appear to be virtually immortal, then drawing profound conclusions on the ability of laboratory animals to survive, and finally projecting their results as thermal or physiological requirements for all fishes. The need for long-term experiments was stressed by Kinne (1964b) when he stated that "numerous experiments seem to have been conducted on sick or dying specimens."

Much work must be done at *all levels* of the life cycle (Brett, 1960) on the less hardy fishes which are typical of the marine environment and whose temperature tolerance is limited (Cairns, 1956b; Hedgpeth, 1957b; Kinne, 1963; Naylor, 1965). For example, Waede (1954) showed that, of two closely related flounders (*Pleuronectes*), the brackish-water species could tolerate higher temperatures for longer periods, even at high salinities, than the marine species. Herrings (Clupeidae), silversides (Atherinidae), cods (Gadidae), sea basses (Serranidae), flounders (Pleuronectidae), sea trouts (Sciaenidae), and other common fishes of the coastal waters, even though difficult to maintain in captivity, should be studied because they are the important elements of the biomass, either as primary or secondary predators.

Heat death is seldom observed in nature, yet many organisms are killed at temperatures only slightly above those at which they normally live (Gunter, 1957; Kinne, 1963; Naylor, 1965). The cause of heat death in fishes has been extensively discussed by Brett (1956) and Fry (1957). It has been attributed by Fisher (1958) to synapse failure occurring in the pacemaker, myoneural junctions, and during smooth muscle peristalsis. Kusakina (1963) attributes it to thermal denaturation of body cells and inactivation of cholinesterase. Pegel' and Remorov (1961) believed that, at high temperatures, the exceedingly high concentration of blood lactic acid results in changes in enzyme structure. Brett (1960) discussed the decrease in metabolic activity in the brain tissue of goldfish and largemouth bass, and Timet (1963) noted that very high temperatures caused injuries to living protoplasm and a decrease in oxygen consumption in marine fishes. It was believed by Agersborg (1930b) that death of fishes from high temperatures was caused by partial coagulation in the branchial capillaries followed by a rupture of the blood vessels of the respiratory mechanism. The failure of a number of biological systems above 30° C. has been discussed by Drost-Hansen (1965), who stressed that even though organisms may be alive at 33° or 34° C., their bodily mechanisms may be irreversibly evanescent. Mayer (1914) suggested that high temperatures produced death by causing asphyxiation, the oxygen being insufficient to sustain the increased metabolic activity of the animal. Orr (1955) believed that the killifishes Fundulus heteroclitus and F. majalis died at high temperatures primarily from a failure of some coordinating mechanism of the central nervous system, while Battle (1929a, 1929b) believed that heat shock in skates (Raja erinacea) was caused by a failure of the automatic mechanism of the heart.

Thermal death points have been tested by a number of workers. But before discussing temperature tolerances, it should be noted that the literature contains many references to the maximum temperature at which fish die. First, these experiments were largely designed to do just that—to tell us at what temperatures fishes die. They tell us nothing of the well-being of a fish before it approaches its thermal death point, or if a temperature well below the death point is sufficient to cause irreversible processes. For example, Cairns (1956b) points out that, in heat-survival experiments, despite fishes' surviving the initial shock, apparently healthy specimens may die after many days. Second, Davenport and Castle (1895) long ago indicated that thermal death points actually signify very little because they seldom take into consideration the thermal history of the organism and whether it has been acclimated or acclimatized to warmer temperatures during the course of the laboratory experiments. Acclimation is defined here as an experimental, more or less quick adaptation of the organism, while acclimatization is the adaptation which occurs slowly under natural conditions (Jones, 1964:156). Acclimatization or acclimation of fishes and the importance of their variance to survival times has been discussed by Loeb and Wasteneys (1912), Doudoroff (1942 et seq.), Keiz (1953), Brett (1956), Morris (1960), and Naylor (1965). Even a brief exposure to high temperature will produce a significant increase in temperature tolerance (Cairns, 1956b), but there is an upper limit beyond which a given species cannot survive for more than a few hours, regardless of acclimation. Cairns found this to be well below 40° C. for a series of freshwater fishes from the northeastern United States. Organisms cannot, however, be acclimatized to temperatures beyond the upper limit of the temperature at which they can survive normally, or to temperature changes greater than those encountered within the normal seasonal range of variation (Naylor, 1965). But the success of acclimation of fishes, as with their successful resistance to high temperatures, depends in part on the absolute temperature, the length of exposure to high temperatures, and, especially, upon the *rate* of temperature change (Gunter, 1957). The sculpin *Myoxocephalus octodecemspinosus,* which normally lives at about 14° C., was acclimated to 27.7° C. over a period of four days, an increase of over 13° (Britton, 1924). Hoak (1961) found that a Δt, of 5.6° C. in 10 minutes was the maximum change that could be tolerated by stream fishes in Pennsylvania. Kerr (1953) showed that 9° C. was the maximum Δt which could be tolerated by adult striped bass (*Roccus saxatilis*), a very hardy species. The duration of experimental exposure to high temperatures, in addition to Δt, is important in determining survival time. To produce heat death in *Fundulus heteroclitus* and *F. majalis* required 63 minutes at 34° C., 28 minutes at 36° C., 9 minutes at 37° C., and 2 minutes at 42° C. (Orr, 1955). Gradual exposure to high temperatures improves fishes' ability to survive subsequent high temperatures but lessens their ability to survive low temperatures (Fisher, 1958) because the lower lethal temperature limit rises (Tat'yankin, 1966). This factor would have considerable significance where heated effluents may cause acclimatization of local fish populations, which would be susceptible, during sudden cold snaps in winter, to sudden cooling of the water because heat adaptation is gained rapidly but lost slowly (Doudoroff, 1942).

There is evidence (Brett, 1956) that freshwater fishes under natural conditions can survive reasonably high temperatures through acclimatization, and that laboratory-maintained fresh- and saltwater fishes have their thermal death thresholds elevated remarkably. Binet and Marin (1934) increased the duration of resistance to high temperatures in *Gobius lota* from 50 to 100 percent through acclimation. But this seems to be true largely for fishes of polar, subpolar, and temperate regions, with tropical and subtropical species already living so close to their thermal death points (Mayer, 1914) that even a few degrees, may affect their well-being.

Thus, in our interpretation of maximum lethal temperatures which fish can withstand, we must consider acclimation time, the length of time the species was exposed to high temperature, and the rate of temperature change. Added to this is the common omission in the literature of essential factors such as the size of the experimental animal used, its condition or well-being, (i.e., if the fish was moribund when it was tested), the salinity and the dissolved oxygen concentration, and occurrence of ions which might contribute synergistically or agonistically to the effects of increased temperature (Doudoroff, 1957). And even though fishes died or were dying at these temperatures, the experiments do not disclose if sublethal, irreversible physiological reactions had occurred well below the so-called upper lethal limit. As Timet (1963) stressed, in studying lethal limits, the interval between complete survival and death is always extremely narrow.

Bearing these factors in mind, data are assembled from literature (Table 1) which report the upper lethal limits obtained for laboratory specimens of larval, juvenile, and adult marine fishes. Summarizing these data (Table 2), we see that most marine fishes do not survive above 35° C. If the adults are divided roughly into tropical, temperate, and arctic species, it may be noted that arctic species experimentally survive in water considerably warmer (20° to 25° C.) than that in which they normally live, with 29° C. being the upper limit, except for the single exception of *Gasterosteus aculeatus,* a species extremely tolerant to temperature and salinity extremes. The median value of upper lethal limits is 26° C. for arctic species. Temperate species reveal approximately the same range of upper

TABLE 1. Upper temperature tolerances reported in laboratory experiments for marine and estuarine fishes. Original data do not generally disclose if test animals were initially acclimated.

Species	Acclimation Temp., ° C.	Tolerance Limit Temp., ° C.	Duration Hours
Alosa pseudoharengus	15	23	90
Aspidophoroides monopterygius		24.4–25	
Atherinops affinis	20	31	24
Calotomus japonicus		28	
Caranx mate, prolarva and postlarva		30	
Clinocottus globiceps		26	
Clupea harengus, larva	15.5	23.0	24
Cyclopterus lumpus		25.5–26.9	
Cynoglossus lingua			
prolarva and postlarva		30	
juvenile		23	
Dussumieria acuta		31	
Enchelyopus cimbrius		27.2	
Fundulus heteroclitus		40	2
Fundulus heteroclitus	28	37	
F. parvipinnis	30	37	24
Gadus morhua		19.8–24.4	
embryo		10	
Gasterosteus aculeatus, adult		31.7–33	
larva		37.1	
Girella nigricans	20–28	31	72
Hemitripterus americanus		28	
Hippoglossoides platessoides		22.1–24.5	
Hypomesus olidus		10	
Limanda ferruginea		24	
Liopsetta putnami		31.6–32.8	
Macrozoarces americanus		26.6–29	
Megalaspis cordyla			
prolarva and postlarva		31	
Melanogrammus aeglefinus		18.5–22.9	
Microgadus tomcod		29	
2 cm		19–20.9	
14–15 cm		23.5–26.1	
22–29 cm		25.8–26.1	
Mugil cephalus			
prolarva and postlarva		32	
Myoxocephalus aeneus		26.3–27	
M. groenlandicus		25	
M. octodecemspinosus		28	
Oncorhynchus gorbuscha			
juvenile	20	23.9	168
O. keta, juvenile	20	23.7	168
O. kisutch, juvenile	20	25	168

Table 1 **(Continued)**

Species	Acclimation Temp., ° C.	Tolerance Limit Temp., ° C.	Duration Hours
O. masou, embryo		13	
O. nerka, juvenile	20	24.8	168
O. tshawytscha	20	25.1	168
Osmerus mordax		21.5–28.5	
Pagrosomus major		21	
Petromyzon marinus			
prolarva and postlarva	20	34.0	1.5
Plecoglossus altivelis		22	
Pleuronectes platessa, embryo		14	
Pollachius virens		28	
Polynemus indicus			
prolarva and postlarva		31	
yearling	9	23	40
Pseudopleuronectes americanus		27.9–30.6	
Salmo salar			
prolarva and postlarva		28	1
S. trutta trutta			
prolarva and postlarva		28	1
alevin	20	26	7
Saurida tumbil			
prolarva and postlarva		31	
Scomber scombrus, embryo		21	
Solea elongata			
prolarva and postlarva		32	
juvenile		23	
Tautogolabrus adspersus		29	
Triacanthus brevirostris			
prolarva and postlarva		30	
Ulvaria subbifurcata		27–29	
Urophycis chuss		27.3–28	
U. tenuis		24.5–25.2	
Raja erinacea		29.1–29.5	
juvenile		30.2	
R. ocellata		28	24
R. radiata		26.5–26.9	
Squalus acanthias		28.5–29.1	
Hippocampus sp.	30	Davenport and Castle, 1895	
Clupea harengus			
juvenile-adult	20.8–24.7	Huntsman and Sparks, 1924	
Fundulus heteroclitus			
juvenile-adult	40.5–42[1]	Huntsman and Sparks, 1924	

Source: Data based on Altman and Dittmer, 1966, and supplemental references. In the following additional references (p. 238) not listed by Altman and Dittmer (1966), it generally cannot be determined if the test animals were first acclimated.

Table 1 **(Continued)**

Species	Tolerance Limit Temp., ° C.	Reference
Roccus saxatilis, adult	32	Kerr, 1953
Gambusia nicaraguensis	"35–40"	Scholander et al., 1953
Abudefduf saxatilis	"	"
Lutjanus apodus	"	"
Haemulon bonariense	"	"
Scarus croicensis	"	"
Alosa pseudoharengus	26.7–32.2	Trembley, 1960
Clupea harengus, larva	18	Blaxter, 1956
Clupea harengus, larva	22–24	Blaxter, 1960
Clupea harengus, adult	19.5–21.2	Brawn, 1960
Fundulus parvipinnis larva	16.6–28.5	Hubbs, 1965
Atherinops affinis	12.8–26.8	"
Leuresthes tenuis	14.8–26.8	"
Hypsoblennius sp.	12.0–26.8	"
Alosa pseudoharengus	31.4	Huntsman, 1946
Alosa pseudoharengus	26.7–32.2	Trembley, 1960
Sardinella longiceps prolarva and postlarva	31	Kuthalingam, 1959
Fundulus sp.	35[2]	Loeb and Wasteneys, 1912
Pseudopleuronectes americanus adult	27	McCracken, 1963
Centronotus gunnellus larva	15–20	Qasim, 1959
Blennis pholis, larva	30	"
Box salpa	31	Timet, 1963
Mullus surmuletus	31	"
Mullus barbatus	32	"
Gobius paganellus	32	"
Scorpaena porcus	32	"
Sargus vulgaris	33	"
Crenilabrus ocellatus	33	Timet, 1963
Pleuronectes platessa	28–31	Waede, 1954
P. flesus	31–34	Waede, 1954
Girella nigricans	31.4	Brett, 1956
Menidia menidia, adult	22.5–32.5	Hoff and Westman, 1966
Pseudopleuronectes americanus juvenile	22–29	"
Sphaeroides maculatus juvenile	28.2–33.0	"

1. Recovered with temperature decrease
2. Acclimated

temperature limits, but the median upper lethal limit is higher (30° C.). Temperate species do not normally survive above 34° C., with the notable exception of several species of *Fundulus,* which can apparently recover from exposures of 40° to 42° C. (Huntsman and Sparks, 1924; Altman and Dittmer, 1966).

Insufficient data are available on tropical marine fishes. Six species of littoral Adriatic fishes, which inhabit seawater having a range of 22° to 25° C., were subjected by Timet (1963) to high temperatures. Adriatic fishes are zoogeographically intermediate between temperate and tropical species in their affinities, but are included under temperate forms in Table 2. Most species could withstand temperatures of up to 32° C. without any visible effects on their behavior, but at 33° C. all died. Even though the fish lived for three hours at 32° C., Timet pointed out that there was no indication that there would not have been bad effects. Scholander *et al.* (1953) reported on an experiment which included four tropical marine reef species from Atlantic Panamá in which "some were killed at 35° C., and all of them were dead at 40° C." (see Table 1). These species are accustomed to living in an annual range of 25.6° to 20.0° C., and are thus already living within a few degrees of their thermal death points (Mayer, 1914; Cairns, 1956a; Timet, 1963). Mayer (1914) observed in the Tortugas that the rate of activity of several phyla of marine invertebrates increased gradually to a point and suddenly declined at about 33° to 34° C. A similar cessation of biophysical activity has been noted in widely unrelated physical, chemical, and biological processes in nature by Drost-Hansen (1965). He believes that there are biophysical thermal boundaries outside of which the organism cannot function, and states that "it is likely that an upper thermal limit for the survival of many organisms will be found at 30° to 33° C., and this temperature range therefore represents a critical domain which cannot be transgressed."

Generally, we may conclude that the adults of arctic fishes can sometimes be slowly acclimated to temperatures far in excess of their normal environment, but that they are generally stenothermal; temperate species reveal a wide range of experimental lethal temperature limits; and tropical species are narrowly stenothermal, living close to their upper lethal ranges which are not far above those reported for temperate species. Experimentally acclimated fishes may survive in and even prefer higher temperatures than those in which they normally occur (Doudoroff, 1938; Mantel'man, 1958), depending upon the adaption temperatures. Brett

TABLE 2. Frequency of upper temperature tolerances in laboratory experiments on marine and estuarine fishes.

Temp., °C.	Arctic Larvae	Arctic Juveniles and Adults	Temperate Larvae	Temperate Juveniles and Adults	Tropical Larvae	Tropical Juveniles and Adults
10	1	1				
11						
12						
13	1					
14	1					
15						
16						
17						
18	1**1					
19						
20	1			1		
21		1	1	1		
22		1	1	1		
23	1			1		2²
24	1	4		1		
25		4		1		
26		2**	4**	1		
27		2		1		
28		5	1	4		
29		3		3		
30			1	4	2	
31				7**	4**	1
32				6	2	
33		1		3		
34			1	1		
35				1		5³
36						
37	1⁴			2		
38						
39						
40				1		
41						
42				1⁵		

Source: Data based on Table 1.

1. **: Median value
2. Juveniles
3. "Died between 35°–40° C." (Scholander et al., 1953)
4. Gasterosteus aculeatus (survived)
5. Fundulus heteroclitus (survived)

(1956) attributes this as possibly due to the phenomenon of summation of acclimation, in which acclimation is to the maximum rather than the mean temperature. In some instances, upper lethal temperature thresholds may thus be elevated 5° to 10° C. if the Δt is gradual. We may conclude that adults of most marine fishes will not survive temperatures above 33° C., but that, because there are so many inherent variables which must be considered—such as size of fish, length and temperature of acclimation, and genetic history of the fish—we must assume that these upper lethal temperatures probably represent maximum survival temperatures for hardy experimental fishes in the laboratory, and that these lethal limits are probably lower under conditions encountered in the environment. In considering establishment of upper limits for industrial effluents, a safety factor should be included so that temperatures will be 1° or 2° C. below this upper lethal limit.

Temperature Tolerance of Young Stages of Marine Fishes

It has been generally thought that the young stages of fishes—particularly marine forms—are more exacting than the adults (Brett, 1956; Gunter, 1957; Kinne, 1963; Naylor, 1965) in their environmental requirements. Morris (1962) found that the recently hatched larvae of a freshwater cichlid (*Aequidens*) showed no ability to compensate metabolically for either high or low temperatures, and thus newly hatched larvae may be quite susceptible to sudden changes within the environment. Data on upper lethal temperatures for young fishes are difficult to assess because the problem is partially semantic in defining the terms "early" or "young" stages which appear in the literature. Size of fish and, especially, its thermal history, are extremely important in determining its survival. Eggs and larvae are extremely exacting in their temperature requirements, while subjuveniles and juveniles appear to tolerate eurythermal conditions, and adults tend to be broadly stenothermal.

Huntsman and Sparks (1924) believed that young flounder, *Pseudopleuronectes,* from the Canadian Atlantic were more tolerant of high experimental temperatures than were the adults during a mortality from a natural water-temperature increase to 31.4° C. in Canadian maritime streams. Huntsman (1946) observed that young river herring (*Alosa* spp.), including the young of several freshwater fishes and salmonids (Huntsman, 1942), were more heat-tolerant than the adults. But for the few examples given in the literature (Table 1 and Altman and Dittmer, 1966) in which upper lethal temperature was determined for both young and adults of the same species, the adults do seem to be able to tolerate experimentally higher temperatures than the younger stages. These include the cod (*Gadus morhua*), tomcod (*Microgadus tomcod*), sculpin (*Myoxocephalus octodecemspinosus*), salmons, (*Oncorhynchus* spp.), flounder (*Pseudopleuronectes*), and herring (*Clupea*). Adult barracuda (*Sphyraena barracuda*) seldom if ever enter water warmer than 30° C., but subjuveniles (4–6 cm.) will enter water of 31° C. and occasionally 32° or 33° C. in shallow mangrove areas of the Bahamas and southern Florida; yet they become sluggish at temperatures above 30° C. (de Sylva, 1963). Notable exceptions are the stickleback (*Gasterosteus*), the larva of which can withstand 37.1° C., and the juvenile of the skate *Raja erinacea* which could tolerate 0.7° C. higher than the adult (Battle, 1929b), although this difference may not be significant. Another interesting exception is presented by Kuthalingam (1959) who found that the larvae of the flatfishes *Cynoglossus lingua* and *Solea elongata* had upper lethal temperatures of 30° to 32° C., respectively, while the juveniles perished above 23° C. This is probably because, following metamorphosis, the juveniles sink to deeper, cooler strata. Bishai (1960) believed that the larvae of migratory races of species of salmonids seemed to have less resistance to high temperatures than was indicated by lake- or stream-dwelling forms.

The exceedingly small tolerance of the larval stages of tropical marine fishes is witnessed in the data of Kuthalingam (1959) which disclose that the pelagic larvae of 10 tropical species from off Madras survived in the laboratory only within the lower ranges of 27° to 29° C. and the upper ranges of 30° to 32° C.

Based upon the few data on upper lethal temperatures reported from the literature (Table 1 and Altman and Dittmer, 1966), it is seen generally that larvae of arctic and temperate marine fishes have lower upper-lethal limits than do the adults (Table 2). The experimentally derived median upper temperature for larvae of arctic species is 18° C. and for the adults, 26° C.; that for the larvae of temperate species is 26° C. and for the adults is 30° C. The few data on tropical larvae show a median of 31° C. While the upper limits for larvae and adults of arctic and temperate species differ, the absolute *ranges* of temperatures tolerated are approximately identical. However, these data should be regarded as conservative estimates of upper lethal limits, because, as McCauley (1963) has stressed, lethal temperatures quoted in the literature usually have been determined for individuals of the more hardy stages of postembryonic development.

As with the adults, generalizations based upon laboratory data are difficult to make. For example, Qasim (1959) showed that acclimatized larvae of the mediterranean-boreal blenny *Blennius pholis* survived 25° C., but died at 30° C. At the lower temperature, there was initially no sign of shock, and the larvae began to feed, but suddenly the larvae began to die *en masse* following apparent survival, their maximum life span being four days. Thus, the problem of long-term sublethal temperature effects occurring through subtle biophysical changes must be considered in any such experiments.

Similarly, degree and length of acclimation temperatures are as important in determining survival and upper lethal limits of the larvae as they are in the adults (Bishai, 1960; Brett, 1956; Blaxter, 1956). Blaxter (1960) was able to increase the TL_m of herring larvae (6–8 mm.) to 24° C. by acclimating them from 7.5° to 15.5° C.

In nature, it is apparent that there would be little acclimatization of larval stages suddenly exposed to high temperatures of heated effluents. Even though larvae have functional thermal receptors on the first day after hatching (Evropeyzeva, 1944) and can detect slight temperature changes (Bull, 1937) and select them (Mantel'man, 1958), they may not be able to escape. Alabaster (1963; 1964) and Alabaster and Downing (1966) found that, in field studies of British freshwater fishes (salmonids, roach [*Rutilus*], and tench [*Tinca*]), even though the young stages (juveniles?) could withstand somewhat higher temperatures than the adults, they were in effect trapped because they could not migrate from areas receiving heated effluents. As we will discuss in a subsequent section, other factors associated with heated effluents may further reduce the ability of fishes, especially the young stages, to escape thermal shock.

Survival of the pelagic larvae of marine fishes is directly and indirectly related to temperature (Marak and Colton, 1961; Hermann and Hansen, 1965; Hermann et al., 1965), appropriate temperature being required for minimal or optimal growth of the larvae. Hermann (1953) and Hela and Laevastu (1962) have related strength of year-classes to temperature, with warm temperatures resulting in good growth of cod and herring, presumably as a result of the availability of an increased food supply. Similarly, low temperatures result in poorer year classes. That temperature may indirectly govern a year class of lemon sole (*Parophrys vetulus*) was shown by Ketchen (1952). Small annual differences in water temperature produced marked differences in the duration of the pelagic stage. Below average temperatures resulted in larvae being carried by currents for longer periods so that more larvae reach and are deposited on the nursery grounds. It is conceivable, conversely, that high temperatures accelerating development might result in fishes metamorphosing prematurely in depauperate areas prior to reaching the appropriate food-rich nursery grounds. That occasionally larvae may be carried adventitiously into unfavorable temperature conditions was discussed by Strasburg (1958). He collected numerous dead larvae of the frigate mackerel (*Auxis*) off Hawaii, where a temperature discontinuity of as much as 1.5° C. in about 1 mile was encountered. He hypothesized that the mass death was due to the larvae traversing an area having marked discontinuities in the surface-water temperatures. A similar mortality of larvae of yellowtail flounder (*Limanda ferruginea*) and hake (*Merluccius bilinearis*) was observed by Colton (1959), who encountered quantities of larvae in an area off the southern edge of Georges Bank which were believed to have been swept from an area of 6.7° C. to a region of 17.8° C. in a 24-hour period and over a distance of about 10 miles. While such instances are seldom reported in the literature, they do tend to confirm the laboratory findings of Kuthalingam (1959) that pelagic marine fish larvae have narrowly stenothermal requirements.

Effects of Temperature on Eggs of Marine Fishes

Perhaps the most serious implications of the effects of heated effluents on marine fish life would be upon the eggs themselves, which clearly are incapable of escaping unfavorable conditions. The early blastodermic stages of the rockling (*Enchelyopus cimbrius*) were found to be most susceptible to heat death (Battle, 1926) while Bonnet (1939) found greater mortality of cod eggs at higher temperatures. The eggs of the sea lamprey (*Petromyzon*) do not hatch above 25° C. (McCauley, 1963), and the thermal requirements for its embryonic stages were found to be more demanding than those of the postembryonic stages. He found that embryos could tolerate temperature extremes if they occurred late in development, and suggested that fishes only developed those mechanisms permitting them to resist high temperatures soon before or after hatching. Senō *et al.* (1926) observed that the eggs of the porgy *Calotomus japonicus* would not hatch above 31° C. Hatching ranges of many marine fishes occur essentially between 15° and 30° C. (Altman and Dittmer, 1966), adding credence to the theory of Drost-Hansen (1965) that these temperatures represent critical anomalies in the biophysical processes occurring in membrane phenomena. Temperatures above 16.5° C. prevented hatching of chinook salmon (Olson and Foster, 1955), and even a temperature of

16.5° C. produced a heavy mortality of larvae, which had hatched at this temperature, shortly after they began to feed. The authors suggest that this was due to early embryological damage that was not manifested until a later stage of development. Bishai (1960) reported a similar mass mortality for salmonid larvae which appeared healthy. Thermal lethalities resulting from eggs exposed to short durations of lethal temperatures were believed to be associated with failures at gastrulation, initiation of circulation, hatching, and perhaps melanophore formation. This critical yet thin line of temperature between life and death was also pointed out for adults of Adriatic marine fishes by Timet (1963).

Bergan (1960) reported that mitosis was prevented at high temperatures in the freshwater anabantoid fish *Trichogaster*. A critical tolerance level in the eggs of the cottid fish *Clinocottus analis* was shown by Hubbs (1966). He also pointed out the eggs taken from cold-water collecting stations were more cold-adapted than were warm-adapted eggs from warm-water stations in which it was found that, as the temperature was increased from 22° to 24° C., the percentage of gastrulated eggs dropped from about 70 percent to less than 5 percent, and that, as temperature rose from 21° to 24° C., the percentage of hatched embryos dropped from 75 to 0. Studying the eggs of four marine fishes from southern California, Hubbs (1965) observed that the summer breeders had warm-tolerant eggs and winter-breeders had cold-tolerant eggs. Although hatching occurred at a wide range of temperatures, the upper hatching temperature was 26.8° C. for three species and 28.5° C. for the fourth, indicating that these temperatures are sharply demarcated. In earlier experiments with the freshwater darter (*Etheostoma*), Hubbs (1964) found increased resistance of eggs and larvae to thermal shock as they were exposed to increased temperature fluctuations, although these studies dealt mainly with lower lethal temperatures.

The necessity of some marine organisms to be subjected to oscillations of warmer and colder water to initiate spawning has been discussed by Kinne (1963) and Naylor (1965), and this concept may be applied to the development of the eggs of fishes which require low temperatures to develop (Hynes, 1963). Thus, the continued addition of low-density heated effluents to salt water where fish eggs might be developing, in addition to the possibility of direct effects due to high lethal temperatures, might prevent eggs which require higher densities of salt water, characterized by the porgy *Calotomus japonicus* (cf. Senō *et al.*, 1926) and other species having psychrophilic eggs, from developing.

An indirect effect of temperature on egg survival could result from the addition of heated effluents lowering the density of ambient water (Jones, 1964). Such temperature changes, with salinity, affect the prevailing water density (Sverdrup et al., 1942) and therefore the buoyancy of the eggs (Hela and Laevastu, 1960). Thus, the addition of effluent water heated sufficiently to reduce density of the ambient water might cause fertilized pelagic eggs, which float at or near the surface and which appear to be impermeable to osmotic changes (Shelbourne, 1956), to sink toward or to the bottom. Conceivably, low oxygen, bottom silt, hydrogen sulfide, bacteria, or absence of adequate light could retard development or kill the egg.

Data obtained from Altman and Dittmer (1966) indicate the higher temperatures at which marine fish eggs will hatch (Table 3) and the frequency of occurrence of these hatching times (Table 4). From these data, however, one cannot deduce if these were the maximum temperatures at which the experiments were conducted, and thus upper temperatures lethal to eggs cannot be determined from these data. However, they do suggest that the incubating eggs of temperate marine fishes probably have an upper limit of well-being of about 28° C. No published data are available to me on tropical marine fishes.

Developmental Effects of Temperature on Fish Eggs

That temperature is an important factor in the rate of development of fish eggs is well known (Dannevig, 1894; Brett, 1956). A more rapid development of the eggs with increasing water temperature has been shown for the following marine species: herring, Clupea harengus (Meyer, 1878; Williamson, 1909, 1911; Blaxter, 1956); haddock, Melanogrammus aeglefinus (Williamson, 1909); plaice, Pleuronectes (Reibisch, 1902; Williamson, 1909; Johansen and Krogh, 1914); cod, Gadus morhua (Dannevig, 1894; Reibisch, 1902); ayu, Plecoglossus altivelis (Higurashi, 1925; Higurashi and Tauti, 1925); smelt, Hypomesus olidus (Higurashi and Tauti, 1925); stickleback, Gasterosteus aculeatus (Leiner, 1932); porgy, Pagrosomas major (Kajiyama, 1929); pilchard, Sardinops caerulea (Ahlstrom, 1943; Lasker, 1963, 1964); jack mackerel, Trachurus symmetricus (Farris, 1961); and anchovy, Engraulis anchoita (de Ciechomski, 1965). Most workers have learned that this increased rate of development is a very critical one, with the Q_{10} being linear only within certain limits (Blaxter, 1956), and with the Van't Hoff and Arhennius relationships not applicable for the temperature range tolerated (Kinne and Kinne, 1962). An abrupt retardation or cessation of egg development at upper temperature limits with consequent mortality has been noted

TABLE 3. Maximum temperatures reported at which eggs of marine fishes will hatch in laboratory experiments.

Species	Temp ° C	Hatching Time Days
Achirus fasciatus	23.3–24.4	1.5
Alosa aestivalis	22	2
Anchoa hepsetus	19–21	2
A. mitchilli	27.2–27.8	1
Anguilla rostrata	24–28	7
Apeltes quadracus	22	6
Archosargus probatocephalus	24.7	1.7
Atherinops affinis	26.8	8
Bairdiella chrysura	27.2–27.8	0.75
Calotomus japonicus	27.6	1.0
Chaetodipterus faber	26.7	1
Chasmodes bosquianus	24.5–27.0	11
Clupea harengus harengus	5.5	20–34
Cynoscion regalis	20.0–21.1	1.5–1.7
Fundulus heteroclitus	25	12
Fundulus parvipinnis	28.5	14
Gadus sp.	12	8.5
G. merlangus	14	5.8
G. morhua	14	8.5
Gasterosteus aculeatus	27	4.3
Gobionellus boleosoma	20	0.75
Gobiosoma bosci	26–28	4
Hypleurochilus geminatus	26–28	6–8
Hypomesus olidus	18.5	8.7
Hypsoblennius sp.	26.8	6
Hypsoblennius hentzi	24.5–27.0	10–12
Leuresthes tenuis	26.8	9
Melanogrammus aeglefinus	14	8.8
Menidia beryllina	26–28	8–10
M. menidia notata	22	8–9
Menticirrhus saxatilis	20.0–21.1	2
Merluccius bilinearis	22	2
Oncorhynchus masou	16.1	29
Pagrosomus major	21.8	1.4
Petromyzon marinus	25	6
Plecoglossus altivelis	24.0	8.5
Prionotus carolinus	22	2.5
Pseudopleuronectes americanus	20.6	15
Roccus saxatilis	17.9	2
Salmo salar	10	50
Scomber scombrus	21	2
Stenotomus chrysops	22	1.7
Tautoga onitis	22	1.7
Tautogolabrus adspersus	22	1.7
Urophycis chuss	15.6	4

SOURCE: Data in part from Altman and Dittmer, 1966.

TABLE 4. Frequency of maximum temperatures at which the eggs of marine fishes will hatch in laboratory experiments

Temp., ° C.	Arctic Species	Temperate Species
5	1	
10	1	
11		
12	1	
13		
14	3	
15		1
16	1	
17		1
18		1
19		
20		2
21		5
22		8
23		
24		3
25		2
26		4
27		5
28		5

SOURCE: Data from Table 3 and from Altman and Dittmer, 1966. (No data available for tropical marine fishes.)

(Higurashi and Tauti, 1925; Battle, 1929c; Kajiyama, 1929). Kinne and Kinne (1962) noted that extreme conditions of temperature, as well as other factors, produced developmental arrest in the killifish *Cyrinodon macularius,* which may cause irreversible damage. Working with brown trout, *Salmo trutta,* Orska (1956) found that eggs subjected to rapid and rather intense temperature increases of short duration at different stages of development caused considerable deviation in the mean number of vertebrae; and when this occurred in earlier stages, serious pathological changes occurred, including irregularly formed vertebrae or fused centra, or fused elements in the arches. Kawajiri (1927) observed deformed freshwater cyprinids which were reared above 19° C. And prolarvae of the sea lamprey, *Petromyzon marius,* reared at 25° C. were malformed and eventually died (McCauley, 1963). High temperatures are known to result in a decreased number of myomeres or vertebrae in marine fishes (Dannevig, 1894; Hempel and Blaxter, 1961), although its significance to survival of eggs and larvae subjected to heated effluents is not readily apparent. High temperatures were also responsible for abnormal development in the rockling, *Enchelyopus cimbrius* (Battle, 1929*c*), herring (Blaxter, 1956), and the ayu, *Plecoglossus altivelis* (Higurashi, 1925). Experimentally increased temperatures may exert sublethal effects, as in the instance found by Gray (1928*a*, 1928*b*) for the embryos of brown trout reared at high temperatures. Eggs incubated at lower temperatures (2.8° C.) yield significantly larger embryos than those hatched at higher temperatures (13° C.) because a higher proportion of the yolk is required for metabolism of embryonic tissue at higher temperatures. Thus, while the growth rate is increased, at the end of larval life, the full-term embryo is smaller.

Effects of High Temperatures on the Physiology and Metabolism of Fishes

Sublethal effects of pollution are increasingly important to pollution researchers, pollution-control officials, and legislative bodies, largely because the effects of man's additives are gradual and subtle, extremely difficult to assess, and seemingly impossible to control. Pollution-control ordinances virtually require that a fish must float before evidence of pollution can be sustained. But the complex results and side effects of long-term, sublethal exposure to a potentially harmful additive is seldom readily apparent. The effects of high temperatures on a biological system roughly increase the rate of a biochemical reaction, and each biochemical reaction within the system, from one to six times for each 10° C. increase (Gunter, 1957), although this rate does not necessarily hold for extreme temperatures. Thus, even a slight temperature increase may have far-reaching effects because a number of metabolic functions will be accelerated with a temperature increase even though the fish may not be killed outright (Kinne, 1963). What should be apparent to pollution-control legislative bodies— but apparently is not—is succinctly stated by Brett (1960), who stressed that thermal requirements for fishes should "permit survival at a level which allows for continuity of the species."

Although sublethal temperatures may not kill outright, they may produce, through heat stress, the loss of swimming ability (Cairns, 1956*b*). Loss of such metabolic functions could render the fish unable to perceive or capture its food, unable to escape more hardy predators, or unable to tolerate physico-chemical environmental changes, either naturally occurring or man-made, because it could lose its ability to react normally to its environment. Clearly, there is a need to study these complex effects under natural conditions (Brett, 1960; Alabaster, 1963; Kinne, 1963; Cadwallader, 1964; Jones, 1964; Naylor, 1965) and at different stages of the life cycle because the capacity of organisms to adapt to a Δt may be different at different ontogenetic stages (Kinne, 1964b).

Excellent discussions on the effect of high temperatures on the physiology and metabolism of fishes are presented by Stroganov (1956), Brett (1956, 1960), Fry (1957), Johnson (1957), and Paloheimo and Dickie (1966). Generally, the standard metabolic rate increases continuously with increasing temperature up to the lethal temperature in fishes allowed sufficient time for adjustment and not subject to other environmental stress, but the active rate of metabolism may level off or even decline at a temperature well below the upper lethal limit (Brown, 1946; Fry, 1957; Kinne, 1963). In tropical organisms, the rate of activity increases gradually to a point and suddenly declines at about 33°–34° C. (Mayer, 1914), while some Adriatic (approximately subtropical) marine fishes have a heat resistance of about 30° C. (Timet, 1963).

Laboratory experiments to determine the physiological effect of temperature extremes have been the classical theme of many studies; but, unfortunately, we can tell little of the actual effects in nature because the experimental animals were confined and possibly subjected either to many or to no additional stresses. And the determination of their survival does not take into account that the fish may have been moribund even though they may have appeared healthy. With these limitations in mind, we may refer to a number of laboratory experiments which suggest, at least, that the problem of heated effluents on fishes is a complex one. The heartbeat of the marine fishes *Fundulus* and *Menidia* and their hybrids was found to increase with temperature to a point, then suddenly decrease, depending on acclimation temperature (Loeb and Ewald, 1913). A reduced temperature caused a decrease in the frequency of heartbeat in three species of freshwater fishes from India (Hasan and Qasim, 1960), with younger individuals being more susceptible to a Δt. Britton (1924) found that the heartbeat of 11 species of marine fishes varied directly with the temperature of the surrounding water, while Glaser (1929) showed that the heart rate of embryos of *Fundulus heteroclitus* increased with temperature. In contrast, Bělehrádek (1931) observed that the rate of embryonic heartbeat in the shark *Scylliorhinus canicula* decreased with increased temperature. However, generally, we see that, to a point which approximates the thermal death point (roughly comparable to data presented in Tables 1 and 2), temperature increases heart rate.

Increased temperatures were found to cause inactivation of cholinesterase in freshwater stenothermal fishes from Lake Baikal (Kusakina,

1963). Baslow and Nigrelli (1964), working with a eurythermal killifish, *Fundulus heteroclitus,* found seasonal fluctuations in the quantity of brain cholinesterase, which they attributed to the ability of poikilotherms to adjust to the large seasonal variations in temperature which occur in the environment to which they have become adjusted, which was also suggested by Wells (1935*a*) for the ability of *Fundulus parvipinnis* to compensate in its respiratory metabolism. Pegel' and Remorov (1961) believed that during temperature adaptation the freshwater dace (*Leuciscus*) could actually regulate its body temperature within narrow limits, which was believed to be connected with a temperature adaptation of the enzymes. An increase of 40 to 50 percent in the blood lactic acid could be effected with a temperature increase. A water-temperature increase was also found to increase the activity of succinodehydrogenase and catalase in eels (*Anguilla*) by Precht (1961). Working with the Arctic sculpin (*Myoxocephalus quadricornis*), Musacchia and Clark (1957) found that, as the experimental temperature was slowly raised from ambient of 2°–5° to 15°–20° C., a decreased spontaneous activity in fish maintained at this high temperature was correlated with a reduced quantity of phospholipids.

The inactivation of digestive enzymes at higher temperatures is discussed subsequently under the section on "Feeding Rate."

Although Olivereau (1955) could not confirm that thyroid activity in eels, mullet, or sharks changed with increased temperature, Fortune (1958) correlated increased secretion of TSH (thyrotrophic hormone), largely in freshwater cyprinids, with elevated temperatures which was found to be independent of acclimation temperatures. Higher temperatures were shown by Lovern (1938) to increase the amount of saturated fat in eels (*Anguilla*).

A critical study of the effect of increased temperatures on herring (*Sardinops*) larvae was made by Lasker (1963, 1964). In his earlier paper, he discussed the drain on energy resources resulting from swimming, which can result in an oxygen consumption of up to 3.5 times the basal uptake. Should swimming be continuous, rather than punctuated with resting periods, and no food be located, the increase in its food requirements would further aggravate the energy deficiency already experienced. From this we may assume that a temperature increase (1) increases the oxygen demand of the larva because it must swim actively to avoid higher (i.e., undesirable) temperatures, and (2) the elevated temperature decreases the amount of dissolved oxygen available to the larva.

Osmoregulation in fishes could be affected by a temperature increase, both directly and indirectly, from hypersalinity resulting from heated salt water. Osmoregulation in marine animals is discussed by Pearse and Gunter (1957), and its extensive implications treated by Kinne (1960 *et seq.*; Kinne and Kinne, 1962). The ability of closely related forms to osmoregulate at different temperatures was shown by Waede (1954), who found that the flounder *Pleuronectes flesus,* a brackish-water species, could tolerate higher temperatures for longer periods than could *P. platessa,* a marine species, even at higher salinities. The rate of chloride excretion in eels (*Anguilla*) can be correlated with temperature (Fontaine and Callamand, 1940). The lesser ability of saltwater, stenohaline species to osmoregulate at high temperatures compared with brackish-water, euryhaline species is seen in the survival of *Fundulus* (Tables 1 and 2) and *Gasterosteus* (Huntsman and Sparks, 1924) at a wide range of temperatures and salinities. Even a relatively small temperature rise resulted in partial failure of osmoregulation in Pacific cottid fishes (Morris, 1960). He suggested that the increased temperatures affected one or more mechanisms involved in osmoregulation because of a sudden increase in the demand of one or more metabolic processes. The question of whether temperature increases are sufficient to cause hypersalinity *in situ,* and thus potential problems for marine teleosts, is rhetorical because the problem has not been investigated in the field. Yet it is not inconceivable, especially when we must consider that sources of heated effluents such as nuclear power-generating stations will most likely be associated in the future with desalination plants which discharge quantities of very hot, concentrated brine solutions. Such situations will require a precise knowledge of effluent characteristics of the discharging plants, together with an adequate knowledge of water characteristics and circulation in the receiving waters. These data must, clearly, be coupled with information on the physiological requirements, at all stages of their life histories, of the organisms to be affected at all stages of their life history.

Survival and development of fish eggs and larvae depend upon salinity, which itself may be a function of temperature. The theoretical ramifications of hypersalinity due to high temperatures are discussed subsequently under "Effects of Temperatures on the Physical Environment."

Effects of Temperature on Respiration

Fry (1957) discussed the general effects of experimentally increased temperatures on the respiration of fishes, largely freshwater species. He concluded that "the rate of standard metabolism increases continuously with increasing temperature up to the lethal temperature in animals allowed sufficient time for adjustment and not subject to other environmental stress." He also noted that "the active rate of metabolism may, however, level off or even decline at a temperature well below the upper lethal limit." For an extensive discussion of the problem, see Ivlev (1938).

Elevated temperature increases the rate of oxygen consumption of the eel *Anguilla* (Precht, 1961), and the rate of opercular movement of killifish, *Fundulus diaphanus*, increases as temperature rises (Kropp, 1947). The breathing rhythm of *Fundulus heteroclitus* increases steadily to about 24° C., above which it reaches a limiting value (Sizer, 1935). Even a 2° C. increase will cause a substantial increase in oxygen consumption of the freshwater Siberian dace, *Leuciscus leuciscus* (Pegel' and Remorov, 1961). Naylor (1965) concluded the effluent high temperatures also increased the rate of oxygen uptake by marine and freshwater organisms because solubility of oxygen is thus increased and, further, because the rates of oxygen utilization by bacteria are augmented at higher temperatures. Respiratory arrest in gobies, *Gobius lota*, subjected to varying concentrations of toxic materials, was reported by Binet and Marin (1934). The implications of other additives to the environment of fishes at higher temperatures is discussed subsequently under "Combined and Synergistic Effects."

An increase in the oxygen consumption of tautog, *Taulogolabrus*, occurs with increased temperature until a harmful relation is realized, resulting in a decrease in metabolism (Haugaard and Irving, 1943). They found respiration to be slightly higher in winter than in summer, and found winter adaptation, if any, to be very small. Freshwater killifish, *Crenichthys baileyi*, living in warm springs (35°–37° C.) have a higher rate of oxygen consumption than the same species living in a cool spring (21° C.), and the consumption by smaller fishes more affected by Δt than for larger individuals (Sumner and Lanham, 1942). Wells (1935a) also found a more pronounced difference in respiratory metabolism in small individuals of the Pacific killifish, *Fundulus parvipinnis*. He found an almost identical relationship between temperature and metabolism. Wells (1935a, 1935b) discussed the importance of acclimation to metabolic rate of the killifish. Fishes kept in cold water had a higher rate of metabolism than those maintained in warm water, but this depended upon the temperature of the water to which they were acclimated. Pegel' and Remorov (1961), working with the freshwater Siberian dace, *Leuciscus leuciscus*, found that oxygen consumption was reduced in unadapted fish, and that as the temperature exceeded the area of adaptation of the fish, the body temperature began to lag behind the changing ambient temperature. They suggested that, during temperature adaptation, the fish acquired the capacity to regulate its body temperature within narrow limits. Scholander *et al.* (1953), referring to arctic and tropical marine and freshwater fishes, stated that it could not be shown that organisms are adapted to seasonal or other fluctuations in temperature by having a low respiratory Q_{10}, i.e., by being metabolically insensitive to temperature changes. Blood from goldfish (*Carassius*) acclimated to temperatures near the extremes of their thermal range showed no difference from normal blood in its oxygen capacity (Anthony, 1961). Elevated temperatures increase respiration of marine animals; yet, some can acclimatize partially to extreme temperatures by reducing their metabolism (Nicol, 1967). The goby *Gillichthys mirabilis* showed reduced metabolic rates when exposed to high environmental temperatures for several weeks (Brett, 1956). Theoretically, the ability of an organism to shift its temperature coefficient (Q_{10}) would be advantageous in offsetting the effects of temperature changes, but the evidence for such a compensatory shift is still equivocal (Scholander *et al.,* 1953; Nicol, 1967).

Effects of Temperature upon Behavior

That temperature may act as a potent directive factor upon fishes was demonstrated by Bull (1937), who found that nineteen species of marine fishes could be conditioned to temperature changes as low as 0.03° C. That this critical perception occurs in the field was demonstrated by Breder (1951), who observed schools of the silverside *Jenkinsia lamprotaenia* which, even though it could withstand 35° C., consistently avoided 30° C., although it moved freely in a temperature range of 29.0° to 29.5° C. This inherent ability to perceive fine gradients of temperature apparently is exercised when an internal drive or environmental stress is elicited (Brett, 1956, 1960), and may account for the distinction between these fine gradients and the broad temperature preferenda of some fishes under natural conditions. Species in nature are distributed as a result of selection response to temperature gradients (Sullivan, 1954), which are usually in good agreement with the final preferenda determined in the laboratory for these species (McCracken, 1963).

Temperature perception occurs early in fishes. Evropeyzeva (1944) gave evidence that thermal receptors were functional in burbot (*Lota*) larvae on the first day after hatching. Mantel'man (1958) showed that young salmon (*Oncorhynchus*) and rainbow trout (*Salmo gairdneri*) in the early stages of postembryonic development reacted most quickly to temperature changes, and that there is a temperature zone characteristic for a given species of fish, this selected temperature depending upon the adaptation temperature according to the thermal requirements of a population. Adults of the opaleye, *Girella nigricans,* frequently selected temperatures (26°–27° C.) which were not those of the normal habitat (20.8° C. mean surface temperature), which Brett (1956) suggested as being due to the phenomenon of summation in acclimation (i.e., acclimation to maximum rather than mean temperature). The sensitivity of the thermal receptors of fishes to other factors was observed by Ogilvie and Anderson (1965), who found that DDT interfered with the normal ability of salmon, *Salmo salar,* to acclimate to thermal changes. It has been shown that adult salmonids, roach, and tench occurring in the heated effluent of power stations in England were able to detect temperature changes and escape, but small fish were mostly affected because their swimming ability was less (Alabaster, 1963, 1964; Alabaster and Downing, 1966).

Under certain environmental conditions, marine fishes react to slight temperature changes which result in a behavioral change. Temperature may affect fish behavior directly by increasing activity (Sullivan, 1954; Hela and Laevastu, 1960, 1962) or indirectly by affecting the behavior, availability, or distribution of the food organisms of fish (Fleming, 1956; Hela and Laevastu, 1960, 1962; Hynes, 1963; Kinne, 1963; Naylor, 1965; Mihursky and Kennedy, 1967).

Temperature may affect the swimming speed of fishes at sublethal levels, most studies on which have been performed on freshwater species (Brett, 1956; Fry, 1957). Acclimated goldfish (*Carassius*) revealed a steeply increased swimming speed at a range of 5° to 20° C., after which the speed remained fairly constant to 30° C., then dropped rapidly (Fry and Hart, 1948). Goldfish, bluegills (*Lepomis*), and bullheads (*Ictalurus*) which were adapted to low temperatures lost their ability to swim at temperatures 10° higher than the experimentally adaptive temperature (Roots and Prosser, 1962). Working with fresh- and saltwater fishes, Cairns (1956b) noted that at some sublethal temperatures, fishes appeared to have difficulty in swimming and in maintaining equilibrium, with an increase in many cases of mucous production on the body.

In marine fishes, Brett *et al.* (1958) showed that a sustained swimming speed of the sockeye salmon, *Oncorhynchus nerka,* was optimum at 15°

C., decreased past 20° C., and rapidly approached the lethal limit at 25° C. As in most metabolic experiments, the authors found that the speeds attained depended upon laboratory acclimation temperatures. Temperature increases the swimming rate of larval Pacific sardines (*Sardinops caerulea*) by increasing the metabolic rate (Lasker, 1963); yet, the temperature increase reduces the available dissolved oxygen concentration. The oxygen demand of the larvae is increased because it must swim actively to avoid higher temperatures. Because the larvae must alternate between periods of swimming and resting to conserve energy, even a slight temperature rise can be detrimental if it stimulates or increases swimming activity, for if swimming were to be continuous and if no food were found, the two-fold increase in its food requirements would further aggravate the energy deficit already experienced.

A temperature increase may affect the color and rate of color change in marine fishes. The darkening of some fishes when transferred from a light background to a dark one and the reverse is well known. Cole and Schaeffer (1937) observed that *Fundulus heteroclitus* could increase its rate of adaptation to light and dark backgrounds between 5° and 25° C., but at higher temperatures no adaptation occurred, the fish remaining dark and light, respectively, regardless of background. The higher critical temperature for darkening in sea water is near 25° C., and for blanching lies just above 29° C. (Cole, 1939). Pulsations of melanophores of *F. Heteroclitus* increased at between 12° and 27° C. with temperature increase, above which is a rapid decrease in the rate of pulsation and a cessation above 32° C. (Smith, 1931). Kruppert and Meijering (1963) reported that the duration of color change from sandy to black in *Pleuronectes* decreased at temperatures above 16° C., while Wells (1935*b*) also noted a loss of pigment in the Pacific killifish *Fundulus parvipinnis* at high temperatures. Ali (1964) found that the Atlantic salmon *Salmo salar* did not dark-adapt as well at high temperatures (20° C.) as at low temperatures (5° C.), and that the salmon's reactions slowed to a point where it could not react normally at warmer temperatures. The loss of compensatory color change according to environment would appear to affect the survival of a prey organism, and, for example, fishes living at high sublethal temperatures might not be able to avoid predators through light and dark color adaptations.

Temperature may affect the behavior of fishes by reacting with several factors in the environment. For example, Hynes (1963) observed that, in polluted fresh water, certain fishes tend to congregate about the outfalls where they feed upon the rich supply of worms and insects and on particles of sewage fungus. However, these organisms are often infauna succeeding what was a fauna characteristic of clear waters. Fishes ordinarily found in clear waters, therefore, and depending upon a clear-water food supply, might be unable to detect and capture replacement foods, also because of generally increased turbidity associated with heated effluents in combination with chemical pollutants. Thus, fishes could be replaced by more tolerant species, which often prove to be less desirable for sport and commercial purposes (de Sylva *et al.,* 1962). Further, such thermal increases result in enhanced algal productivity (Cairns, 1955, 1956*b*; Naylor, 1965; Warinner and Brehmer, 1966), which would also reduce the ability of fishes which feed by sight to perceive their prey. For example, Andrews (1946) reported that salmon reacted poorly to different light intensities at higher temperatures. Higher temperatures cause disorientation and cessation of directed activities of organisms. This critical thermal maximum (C.T.M.), discussed by Mihursky and Kennedy (1967), is the thermal point at which the locomotory activity becomes disorganized and the animal loses its ability to escape from conditions that will soon cause its death.

Effects of Temperature on Distribution, Migration, and Fisheries

Temperature affects so many biological and biophysical systems that one cannot state that any single factor is responsible for a marine organism being where it is at any given time. These many interlocked and complex theoretical factors have been reviewed critically by Sverdrup *et al.* (1942), Brett (1956), Hedgpeth (1957*a*, 1957*b*), Hela and Laevastu (1960, 1962), Kinne (1963), and Lauff (1967). From the practical aspect, however, the commercial and, occasionally, the sport fisherman have capitalized on a knowledge of temperature preferenda of marine fishes to capture them more efficiently and economically. A summary of the principles of fish concentration in relation to temperature is given by Fleming (1956) and by Hela and Laevastu (1960, 1962). A knowledge of the temperature optima for adults and the influence of temperature upon abundance, migrations, and shoaling of fish schools has been utilized by many countries for years, and commercial fishing vessels are often equipped with reversing thermometers, recording thermistors, or bathythermographs to determine suitable fishing areas such as are found along "fronts," zones of convergence or upwelling, and at the thermocline.

Perhaps the tunas (family Scombridae) reveal one of the closest associations between fish species and temperature (Kuroki, 1954; Koga, 1966; Nakagome, 1966). In some instances, the close correlation between catch rate and temperature permits an accurate prediction of the tuna catch (Matsudaira, 1965). In some instances, the correlation appears to be a direct one, while in others it is between temperature and the food supply, i.e., zooplankton, of tunas (Tawara and Tsuruta, 1966). Tuna concentrations may be concentrated within limits of only 2° or 3° C. Less spectacular but nonetheless potent correlations have been found between the distribution of warm temperatures and shad (Massmann and Pacheco, 1957), cod (McKenzie, 1934; Tremblay, 1942; Jean, 1964), and herring (Huntsman, 1933), plus a score of other examples cited by Hela and Laevastu (1962). Massmann and Pacheco (1957) showed no catches of shad below 8° C., and increasing catch rate to 15° C., and an abrupt slackening above this point. They pointed out that this temperature-catch relation was probably a combination of many factors. Most workers find a more or less constant, narrow range of temperature preferenda for pelagic and offshore demersal fishes, while estuarine and coastal species have somewhat wider preferenda for optimum catch rates.

Several factors are involved in stimulating and directing the migration of marine fishes (Chidester, 1924; Brett, 1956; Hela and Laevastu, 1962), including biological, chemical, and physical factors (Kinne, 1963). A correlation between temperature and migration time and pathways was shown for bonito, *Sarda sarda* (Acara, 1957); flounder *Pseudopleuronectes* (McCracken, 1963) during their riverward migration; eel fry, *Anguilla japonica* (Hiyama, 1952); and sockeye salmon (Naylor, 1965). The influence of naturally occurring increases in water temperature upon changes in fish faunas is discussed by Hubbs (1948), Edwards (1965), and Taylor *et al.* (1957), while gradual changes in invertebrate faunas from heated effluents from British power plants are discussed by Markowski (1962) and Naylor (1965). The implications of such artificial temperature changes in causing ecological replacements are wide-ranging.

Effects of Temperature on Feeding Rate

Feeding rate of fishes, as a metabolic rate, increases within limits as temperature rises (Brett, 1956; Paloheimo and Dickie, 1966). Freshwater fishes (family Centrarchidae) ate three times as much food per day at 20° C. than at 10° C. (Hathaway, 1927), while the rate of feeding was shown to increase with temperature in the eel, *Anguilla japonica* (Kubo, 1936). Markowski (cited in Naylor, 1965) reported a generally increased rate of food consumption in freshwater fishes held in ponds receiving heated effluents from power plants in England. One might suppose that this relation is linear for all temperatures, but inflection points in reduction of the feeding rate approximate sublethal limits of other metabolic activities for the same species. In Arctic waters, Komarova (1939) showed that the long rough dab, *Hippoglossoides platessoides,* had a maximum feeding rate at 1° C., which decreased rapidly at 4° C. to 10 percent of its maximum activity and at 7° C. to 5 percent of its peak rate.

Brown (1946) reported that brown trout, *Salmo trutta,* showed a constant increase in feeding rate and the quality of food eaten from 10° to 19° C., above which the rate declined abruptly. Largemouth bass, *Micropterus salmoides,* tested at 5° C. intervals between 5° and 25° C. showed a more rapid increase in the digestive rate at 5° to 10° C. than at high temperatures (Molnár and Tölg, 1962).

A temperature increase effects the motility of the alimentary canal in the killifish *Fundulus heteroclitus* (Nicholls, 1932) and the skates *Raja diaphanes* (=*eglanteria*) and *R. erinacea* (Nicholls, 1933). The upper limit for gastric contractions in the skates was about 24.5° C., while that for the killifish approached 30° C. Coupled with the reduction in gastric motility at elevated temperatures is the inactivation of digestive enzymes at higher temperatures (Mews, 1957) following their increased activity with increasing temperature. Although the rate of enzyme hydrolysis of food increases with a temperature rise, there is possibly enzyme inactivation at high temperatures in marine animals generally, and the higher the temperature at which digestion occurs, the more rapidly the enzyme is destroyed (Nicol, 1967).

Effects of Temperature on Growth

As a corollary to feeding rate and quantity of food consumed, the effect of temperature upon the growth of marine fishes is an important factor in considering the effects of heated effluents, but is one which has been studied essentially using freshwater fishes in the laboratory. The general relation between growth rate of fishes and temperature is discussed by Brett (1956), Fry (1957), and Paloheimo and Dickie (1966). Audigé (1921), working with several species of European freshwater fishes, found that growth was directly proportional to temperature and was controlled by seasonal changes. A maximum growth rate was found between 23° and 25° C. for eurythermal species and from 16° to 18° C. for stenothermal species. Brown (1946) showed that brown trout grew steadily to about 19° C., after which the growth rate became negative. A correlation between the production of rich year-classes of cod off West Greenland was found by Hermann (1953). This effect was ascribed by Hermann and Hansen (1965) as either directly or indirectly increasing cod activity and metabolism by increasing the abundance of available food. Markowski (cited in Naylor, 1965) obtained increased growth rates in freshwater fishes in England using warmed water from power plants, but these data tell us little of the long-term effects of heated effluents upon the metabolism and survival of natural populations. However, Trembley (1965) did not believe that increased growth rates of fishes actually occurred; rather, he assumed that the heated effluents merely attracted fishes during the cold months. Actually, we can only assume from any studies on tank-held organisms that, during the experiments, the surviving fish did not escape or die.

Growth rates proceed optimally only within genetically determined limits set for each species (Kinne, 1960). The upper limits for rapid growth are also often precariously close to thermal maxima for other physiological parameters such as respiration, heartbeat, and enzyme activity. That higher temperatures may not necessarily be beneficial in the long run is suggested by Kinne (1960). Working with the eurythermal fish *Cyprinodon,* Kinne obtained better initial growth rates at higher temperatures (to 30° C.), but these rates were not maintained later in life, and those fishes which were the slow-growing individuals at lower temperatures grew larger and lived longer. The limitations of short-term laboratory experiments, on which many projections are based, are seen in Kinne's findings that fast-growing fish adapted more rapidly to changes in temperature and salinity than slow-growing individuals. That these adapted organisms might not live long in nature appears significant. Initial rapid growth at high temperatures often occurs in marine organisms (Naylor, 1965), followed by a decrease of adult size and perhaps a shortening of the life cycle of some species. Cairns (1956*b*) cited experiments in which test fishes were subjected to "sublethal temperatures for extended periods of time . . . ," in which "the amount of food consumed was several times that of the control fish," yet, "despite this increased consumption of food, the fish became emaciated with little or no growth apparent."

Clearly, good growth rates obtained from short-term experiments to raise fishes for food or sport may be advantageous (Iles, 1963; Markowski, cited in Naylor, 1965); yet, they have little real significance in assessing effects of temperature on growth, life span, reproduction, and other vital functions necessary for the perpetuation of fishes in nature. Many long-term, carefully controlled and interpreted experiments are needed on typically stenothermal and stenohaline marine fishes before generalized conclusions can be drawn.

A problem ancillary to growth of fishes in heated effluents, requiring only superficial treatment at this stage of our knowledge, is the possible effect of growth of fishes in the effluent of nuclear power stations (Cronin, 1967). Radionuclides which find their way into the environment from pile effluent water or from natural diffusion through heating elements must be incorporated into the bodies of marine fishes, the organisms on which they feed, the sediments, and the water column. The question of whether these radioactive materials will affect growth of fishes, their reproductive capacity, or their compatibility with human digestive and life processes, may be purely academic. Presumably, because of the differential uptake by organisms of different nutrients at various temperatures, the resulting proportion of radionuclides accumulated and/or concentrated in an organism will be present differentially. Speculation must be followed by experimentation.

Effects of Temperature upon Reproduction

A temperature stimulus of some kind is the normal impulse for inducing sexual activity in marine animals (Orton, 1920). This threshold is usually quite critical, and may occur at a Δt of only $1°$ or $2°$ C. (Aleev, 1954; Brett, 1956). Brandhorst (1959) believed that spawning activity in herring was induced by the suddenness of the Δt rather than by the magnitude of the Δt *per se*. Generally, low temperatures during pre-spawning periods delay spawning and higher temperatures hasten it (Mánkowski, 1950; Brett, 1956). Mackerel and gadoids spawn within very narrow temperature limits, often with a range only of $1°$ C. (Orton, 1920). It appears that oceanic and neritic pelagic fishes are more critically stenothermal in their spawning requirements than are estuarine species (Hela and Laevastu, 1960, 1962; Naylor, 1965). For example, striped bass (*Roccus saxatilis*) spawn in California estuaries when the temperature is between $16.1°$ and $20.6°$ C. (Farley, 1966). The location of their spawning is also affected by temperature and a combination of water-quality conditions, such as turbidity, which, in turn, reflect the weather and the amount of spring runoff.

Warmer temperatures may also protract the spawning season of a species (Gunter, 1957; Naylor, 1965), but conceivably an increase in numbers in the population of one or more predator species could overgraze the food supply and negate the apparent benefits of an extended spawning season. Similarly, temperatures which are too high might put the development of larvae out of phase so that larval development occurs before the peak population of the proper plankton or other foods (Hela and Laevastu, 1960). Abnormally high temperatures also may cause a shift in spawning migrations of fishes and in their spawning and fishing grounds.

Temperature affects certain critical periods during the initial development and early maturation of the ova (Hodder, 1965). For example, Harrington (1959) showed that high temperature retarded early ovogenesis in the killifish *Fundulus confluentus*. Burger (1939) concluded that temperature was the cogent factor in the external environment required to induce spermatogenesis in *Fundulus heteroclitus*. Freshwater fishes are believed to cease reproducing above 36° C. (*Wastes Engineering*, 1961), and even tropical species which live in continually warm waters have temperature limits for breeding which may be relatively close to their thermal death points (Gunter, 1957; Kinne, 1963; Naylor, 1965).

While tropical forms appear to spawn during warmer months (Gunter, 1957), continued maintenance of high temperatures in aquatic organisms is presumed to be undesirable (Tarzwell and Gaufin, 1958). Species spawning at warmer temperatures require cooler water during the vegetative phases of gonad development (Chidester, 1924; Cairns, 1956*b*; Hynes, 1963), while tropical species which spawn in winter require these lowered temperatures to release the spawning mechanism. Kinne (1963) divided marine organisms into those having (1) negative eurythermy and reproductive polystenothermy, which require warm waters to reproduce, and those having (2) eurythermal requirements with reproductive oligostenothermy, which must migrate to colder areas for reproduction, or change the timing of their reproduction toward suitable temperatures. Nongenetic adaptations to such changes can be effected within narrow limits, but must occur during early stages of ontogenetic development to be successful adaptations (Kinne, 1964*b*). Merriman and Schedl (1941) believed that a combination of temperature plus a critical light intensity was required to stimulate gametogenesis in female sticklebacks (*Apeltes*). They stated that excessively high temperatures inhibited later phases of maturation of the oöcytes, and that reasonably low temperatures are essential to the completion of the normal process of oögenesis. At higher temperatures the necessary increase in yolk material cannot be effected successfully because the metabolic rate increases and the storage of food material is prevented. Temperatures which are higher than normal cause rapid completion of spermatogenesis, and much higher temperatures cause even greater spermatogenic activity and the discharge of sperm. It is thus seen that the constant addition of heat to the environment can cause complex effects on the reproductive cycle of organisms. Extensive narrowing of the seasonal range of temperature through the addition of heat would probably have considerable biological effects since a constant temperature of a particular value and fluctuating temperatures of that average value do not have the same biological effects (Brett, 1956; Kinne, 1963; Naylor, 1965).

Effects of Temperature upon Parasites and Diseases

Elevated water temperatures offer improved media for the growth, reproduction, and rate of infestation and infection of parasites and disease. Brett (1956) cited the near obliteration of the run of sockeye salmon, *Oncorhynchus nerka,* in the Columbia River in 1941, which was due to the combined effects of high temperature and bacterial infection. Stroud and Douglas (1968) reported that river warming has increased the incidence of columnaris disease in the Columbia River. Ordal and Pacha (1967) observed that "from the literature, it is evident that most fish diseases are favored by increased water temperature." Using hatchery-reared salmonids at the University of Washington, they found that higher water temperatures drastically increased the effect of kidney disease, furunculosis, vibrio disease, and columnaris in young fish. In the Delaware River estuary, de Sylva *et al.* (1962) found a high incidence of deformed fins in striped bass, *Roccus saxatilis,* which they believed to be associated with heated effluents from a nearby electric power station, but whether this was due to excessive temperatures during egg incubation *per se* or from a combination of high temperatures influencing the toxicity of industrial effluents on the eggs cannot be ascertained.

Fish diseases flourish with increased temperatures, within limits, in connection with sewage and other nutrient sources. Sewage fungus is usually associated with power-generating stations (Klein, 1957; Hynes, 1963), and thus the consequent rate of infestation can be expected to increase in those species which are suitable hosts. Where heated effluents are discharged into the marine environment where sewage is present, the medium may be conducive to growth and reproduction of such pathogens as fungi, bacteria, and viruses which live in the human body (37° C.), and which may be untouched by sewage treatment methods (Dr. Jack

W. Fell, personal communication). Whether these pathogens would live in and be transmitted from fishes back to the human consumer is problematical. In the effluent of a power-generating plant at Cutler Point, Biscayne Bay, Florida, gray snapper (*Lutjanus griseus*) frequently have been reported by anglers to have fungus-like lesions on the body, but scientific examination of these lesions has not been effected.

The complex life cycle of parasites requires continuity, but is regulated by weak links among intermediate and final hosts. Because low temperatures at a critical point in a parasite's life history may reduce its survival, sustained high temperatures in the environment from heated effluents would presumably favor an increased rate of host infection (Dr. Jesse White, personal communication).

Characteristics of the Marine Environment

Various parameters of the pelagic marine environment have been characterized by Sverdrup *et al.* (1942). There seems to be relatively little likelihood that steam-electric stations for power generation will ever materially affect the temperature regime of the open ocean, although various proposals to heat polar regions could result in long-term effects. However, we are concerned here with the possible effects of thermal effluents from power stations for electricity, most of which will be presumably located adjacent to the coastal environment. This coastal environment is a complex one, being acted upon by biological, chemical, physical, and geological factors of the open ocean, the neritic zone, the estuaries, the rivers, and the entire watershed which feeds it. These factors intrinsically and inextricably have attained an equilibrium within varying limits which have been reached over eons. This balance is now being changed by man's numerous activities, and it is increasingly difficult to interpret the dynamic forces at work in an "unspoiled" coastal environment.

The estuary offers perhaps the most widely fluctuating environment in its biological, chemical, and physical characteristics, while the coastal and neritic waters are subject to the smallest amount of fluctuation in their parameters. Organisms of tropical coastal environments are able to tolerate fluctuations least; polar forms are somewhat more plastic; and temperate organisms are most tolerant of varying conditions (Cairns, 1956a, 1956b; Kinne, 1963; Naylor, 1965). The impact of man's work in estuaries is summarized by Cronin (1967) and by Olson and Burgess (1967). For a discussion of recent work on estuaries, the reader is referred to Emery and Stevenson (1957), Hedgpeth (1957b), and Lauff (1967), while aspects of the ecology of coastal lagoons are treated by Smith *et al.* (1950), Voss and Voss (1955), and Hedgpeth (1957b). Each of the above-mentioned treatises reveals the intricacies and complexities of the coastal marine environment, and the dependency of this habitat upon a delicate balance among many dynamic factors.

Effects of Temperatures on the Physical Environment

The estuary and the coastal waters which receive the immediate effluent are unique in the ecosystem because of the complex mixing processes which tend to concentrate organisms and nutrients, possess peculiar flow patterns, and are the site of rich organic production and nursery areas (Hedgpeth, 1957b; Emery and Stevenson, 1965; Pritchard and Carter, 1965; Warinner and Brehmer, 1966). For example, Hockley (1963) and Alabaster and Downing (1966) showed that heated effluents in an estuary produced layering that held the main effluent stream of warm saline water below the surface. Naylor (1965) suggested the complexities which may ensue from altering the intricate coastal environment.

A temperature increase need not directly affect a fish in seawater. An increase may change physical properties of water such as salinity, temperature, density, solubility of dissolved gases, or turbidity from suspended particles. Or it may affect chemical processes such as dissolved solids, pH, and the effect of pollutants at higher temperatures. It may affect fishes by influencing their food, shelter, predators, or competitors. The factors are complex and interrelated (Jones, 1964; Hynes, 1963; Naylor, 1965). It is especially important to note the differences between chemical and physical properties of fresh and salt water (Sverdrup *et al.,* 1942). For this reason, we may expect exceedingly complex effects from elevated temperatures in salt water, and one should proceed with caution before deriving conclusions on the effects of heated effluents on salt water based on limnological evidence.

It is problematical if the present level of industrial discharges of heated water into the environment is sufficient to cause a salinity increase through evaporation *in situ* from high temperatures, or from the highly hypersaline effluents discharged from those generating stations which are associated with desalination plants. But it can be expected that with the growing shortage of fresh water, an increasing number of desalination plants will be located along the coast, especially in the tropics where fresh water may be at a premium, and it behooves us to investigate the theoretical effects of this hot, concentrated brine upon the coastal environment.

The effects of salinity on organisms have been discussed in detail by Pearse and Gunter (1957), Kinne (1964a), and Kinne and Kinne (1962). That salinity can act as a directive force in itself was shown by Bull (1957), who found that fishes could perceive differences as small as 0.06 °/oo. Osmoregulation in marine fishes is dependent upon salinity (Pearse and Gunter, 1957; Naylor, 1965), with estuarine organisms being able to osmoregulate better than stenohaline marine organisms (Naylor, 1965). The ability to survive of fishes which are accustomed to hypersaline waters is dependent upon "salinity history" of euryhaline marine fishes (Strawn and Dunn, 1967). They observed that even *some* salt in the water gives protection from heat shock in euryhaline fishes. The problem of osmoregulation in estuarine and marine fishes in varying salinities has already been alluded to under discussions of temperature effects on physiology and metabolism.

Salinity also affects developing fish eggs and larvae (Heuts, 1956). The controversy of whether marine fish eggs are impermeable to water and salt is discussed by Shelbourne (1956), but Kinne and Kinne (1962) believed that the variation in permeability reported in the literature probably is due to the osmotic gradient, time of exposure to osmotic stress, temperature, pH, and other factors influencing the different species, and the prevalent experimental conditions. They believed that the egg chorion of euryhaline species may actually be permeable. Senō *et al.* (1926) showed that salinity extremes reduced the percentage of hatching in the eggs of *Calotomus japonicus,* while Battle (1929c) found that the percentage of survival of the eggs of rockling, *Enchelyopus cimbrius,* dropped sharply above 50 °/oo. The eggs of herring (*Clupea*) are smaller when incubated at higher salinities (Holliday and Blaxter, 1960). They showed that eggs reared over a range of 6° to 50 °/oo yielded the greatest percentage of hatching at 24 °/oo, dropping to 35 percent at 50 °/oo, with the tolerance of newly hatched larvae being wider for the larva than for the adult. As with temperature, duration of exposure is important in survival, and herring eggs could withstand salinities of between 1.4 and 60.1 °/oo for 24 hours, but 2.5 to 52.5 °/oo for 168 hours. The cyprinodontid *Cyprinodon macularius* shows progressive retardation of development with an increase in salinity, even though the adult of this species withstands very high salinities (Kinne and Kinne, 1962). Salinity may accelerate or retard developmental rate in the stickleback (*Gasterosteus*), depending upon the genetic composition of the individuals. Within narrow limits, non-genetic adaptation to salinity ranges greater than those normally experienced can be effected, especially if they occur early in onto-genetic development (Kinne, 1964b).

It appears that the eggs and larvae of estuarine fishes require lowered salinities to survive. The eggs of striped bass (*Roccus*) survive at all salinities found under natural conditions, but low salinities enhance survival (Albrecht, 1964). Adult shad (*Alosa*) are sensitive to salinity changes and appear to follow them halotactically (Tagatz, 1961). During embryonic development of the Pacific cod, *Gadus macrocephalus,* low salinity causes the least osmotic stress (Forrester and Alderdice, 1966). And salmon smolts (*Salmo salar*) are less sensitive to an increase in temperature when exposed simultaneously to a transfer from salt water to fresh water (Ala-baster, 1967). Larval menhaden, *Brevoortia tyrannus,* require low salini-ties even though they transform into juveniles at salinities higher than those found in the nursery area (Lewis, 1966). He reported vertebral deformities in larvae reared above 15 °/oo, and some larvae were retarded in their growth rate. Increased salinity results in a higher number of myomeres in larval herring (Hempel and Blaxter, 1961), and this higher salinity would appear to have the same effect as low temperature in pro-longing larval life. The spawning environment of herring in the Kiel Canal was shown by Brandhorst (1959) to be limited to a range of 5° to 24 °/oo, with virtually no fertilization of eggs above 35 °/oo. He also found a dependency between the yield of the fishery and a critical salinity value. Thus, high salinities in or resulting from heated effluents could be detri-mental to those fishes requiring low salinities during development. Con-versely, at high salinities, eggs of stenohaline marine organisms survive better (Gunter, 1957), and Strawn and Dunn (1967) showed that some salt in the water gave protection to the euryhaline fish families Cyprinodon-tidae and Poeciliidae, although at very high temperatures as salinity in-creased (0° to 22 °/oo) mortality increased.

Salinity and temperature affect water density (Sverdrup et al., 1942), and heated effluents from power stations cause a change in water viscosity (Jones, 1964). Water density is important to the hatching of the marine fish *Calotomus japonicus* (Senō et al., 1926), the eggs of which will not hatch at specific gravities of less than 1.0212 g/cc. Salinity increases from heated effluents causing evaporation and consequent hypersalinity could increase the water density (Hela and Laevastu, 1960, 1962), causing eggs which are normally demersal, and which appear to be impermeable to osmotic changes (Shelbourne, 1956), to float to the surface, where they could be subjected to excessively high temperatures, light, or wave action to reduce survival. The resulting density would be balanced by the interplay of both increased temperature and salinity which, as has been cited by Hockley (1963) and Cronin (1967), may result in a lens of intermediate-density water floating beneath the surface. Thus, environmental physical factors affect marine fishes from egg to adult. Kinne (1956, 1964a) and Sweet and Kinne (1964) stated that the distribution of estuarine organisms is dependent upon the compensatory and complementary effects of both salinity and temperature, and that neither factor acts upon the organism independently.

Increased temperature reduces the solubility of oxygen in water, and increased salinity also decreases dissolved oxygen (Sverdrup et al., 1942). Thus, oxygen in or resulting from heated effluents might be expected to be reduced, other factors being equal. A reduced oxygen supply affects fishes in oblique ways. Hypothetically, heated effluents cause dissolved oxygen to diminish; yet, as in the example of the Pacific sardine (Lasker, 1963), the increased temperature would also heighten the metabolic rate of the fish. The resulting physiological stress from lowered oxygen may reduce survival (Haugaard and Irving, 1943). At temperatures of 26° to 30° C., the survival time for trout and roach is lower at low concentrations than at high ones (Alabaster and Welcomme, 1962). A temperature increase need not be great to affect a given organism in view of the other factors involved or the other factors present which compete for this oxygen. A Δt of 10° C. doubles the oxygen consumption of organisms (Klein, 1957). This may be a cogent factor in areas already depleted of oxygen by industrial and sewage wastes.

Low-oxygen tensions may retard growth rate of plaice larvae (Johansen and Krogh, 1914). Using several species of temperate freshwater fishes, Downing and Merkens (1957) showed that, at high temperatures which led to decreased oxygen saturation, at each temperature interval the period of survival decreased with a fall in oxygen tension. A rise of from 10° to 20° C. reduced the resistances to lack of oxygen in all species used. Henley (1952) found that dissolved oxygen affected survival of the larvae of cod, herring, and plaice by influencing the rate of gas disease in the embryos. Decreased oxygen may reduce food supply so that predator organisms such as fishes may be affected. Smith et al. (1950) reported that decreased resistance of plankton at high temperatures was associated with low dissolved-oxygen values. The problem of plankton productivity and energy interchange is a complex one in areas of heated effluents and needs to be investigated much further (Warinner and Brehmer, 1966).

Heated effluents also affect oxygen saturation by increasing the rate of oxygen uptake because rates of oxygen utilization by bacteria are increased as temperature rises, even in the face of reduced solubility (Naylor, 1965). Heated effluents stimulate the growth of sewage fungus, which results in a higher B.O.D. (Hynes, 1963). Organic waste loading associated with heated effluents may reduce dissolved oxygen, as occurred in the Grand River, Michigan, in which a 5.5° C. temperature rise was sufficient to produce a massive fish kill resulting from low oxygen (*Sewage Ind. Wastes,* 1956).

The equilibrium between dissolved oxygen and carbon dioxide is also affected by heated effluents (Hynes, 1963; Naylor, 1965), which in turn affect pH, which may affect fishes living under restricted situations (Klein, 1957; Hynes, 1963; Jones, 1964). The complex interdependency of temperature, oxygen, salinity, and other factors is readily apparent, though still little understood. The effects of temperature-salinity-oxygen combinations upon development of the killifish *Cyprinodon macularius* were studied by Kinne and Kinne (1962). In general, a salinity increase leads to progressive retardation of development, whereas an increase in oxygen content produces progressive acceleration. Both retardation and acceleration increased as the temperature rose. They concluded that the physiological effects of high salinity *per se* were rather limited, and that the problem had no simple answer. Sweet and Kinne (1964) observed that "changes in body dimensions of hatching [of *Cyprinodon*] fry may be related to the concomitant changes in the amounts of dissolved gases, especially oxygen, in the various temperature and salinity combinations employed." These environmental effects are realized during early ontogeny, and have paramount importance for the functional and structural properties of organisms. The need to study a combination of these effects using multivariable experimental techniques has been stressed by Alderdice (1963) and Mihursky and Kennedy (1967).

Effects of Temperature on Food and Habitat of Marine Fishes

In discussing the effects of heated effluents on marine fishes, one must consider the possible effects of a temperature increase upon other biotic factors such as food, predators, competitors, and shelter, the absence or presence of which may affect the occurrence and abundance of the more important species with which we may be especially concerned, or may affect the forage organisms on which they feed. The preceding discussions have been essentially limited to how temperature may affect the well-being of fishes or their physico-chemical environment, and these considerations must also be extended to food organisms on which marine fishes may depend at all stages of their life cycle (Hela and Laevastu, 1960, 1962; Naylor, 1965).

High temperatures may cause mortality or migration of food organisms to other areas not accessible to fishes, which could pose a serious problem for the less motile larvae feeding on plankton or benthos. Dr. Joseph P. Mihursky (personal communication) states that, in Chesapeake Bay, up to 95 percent of the plankton passing through certain power-plant condensers may perish. Hoak (1963) discussed the exceedingly complex effects of heated effluents and the food supply of fishes which must be considered in assessing effects, while Coutant (1965) discussed similar problems for estuarine invertebrates.

Tolerance of fishes to elevated temperatures may be greater than that of their food. For example, de Sylva et al. (1962) showed that juvenile striped bass (*Roccus saxatilis*) fed mainly on opossum shrimp (*Neomysis americana*) in the Delaware River estuary. The fish were found in temperatures up to 30° C., yet Mihursky and Kennedy (1967) found that unacclimated opossum shrimp in Chesapeake Bay did not survive above 28° C. Zooplankton live within a narrow temperature range (Smith et al., 1950) and many other forage planktonic and benthic organisms are relatively stenothermal in coastal and marine waters (Sverdrup et al., 1942; Gunter, 1957; Hedgpeth, 1957b). Larval marine fishes may be quite selective in their food habits, and this destruction of a major food supply for fishes on the nursery grounds could have serious effects if the species were unable to find another suitable food source.

Heated effluents increase the rate of photosynthesis and productivity. Detergents, which are often used to clean condensers of power stations, may interfere with these processes by decreasing the rate of re-oxygenation in water (Hynes, 1963). The equilibria among respiration, photosynthesis, oxygen production, and phytoplankton growth depend upon temperature (Alabaster, 1963; Hedgpeth, 1957b; Olson and Burgess, 1967; Lauff, 1967). Even slight temperature elevations could affect the rate of each and the combined interplay among the various dynamic processes. High temperatures increase the rate of oxygen uptake by organisms because solubility is reduced and the rates of oxygen utilization by bacteria are increased (Naylor, 1965). The effect on primary production and bottom invertebrates in the York River, an estuary of Chesapeake Bay, was studied by Warinner and Brehmer (1966). They found that production in winter increased with a Δt of 5° to 10° C., and at a Δt of 3° C., the production rate increased further. Above this, activity was depressed, and at natural temperatures of 25° C., thermal additions depressed activity. A sudden Δt greater than 5.5° C. similarly depressed production. The community structure and abundance of marine benthic invertebrates was affected 300–400 m from the thermal discharge, and they found evidence of heat stress on the benthic populations during months of high normal water temperature.

For a recent compendium on the effect of heated effluents on invertebrates and plankton, the reader is referred to the papers contained in the three-volume work on thermal pollution prepared for the U.S. Senate Subcommittee on Public Works (1968).

As temperature increases photosynthesis, phytoplankton populations increase. Dense blooms resulting from heated effluents could reduce the ability of those fishes depending on visual cues to locate and capture their prey. Hynes (1963) referred to the well-known observation that freshwater fishes may congregate about sewage outfalls, where they feed upon worms, insects, and particles of sewage fungus; and heated effluents, to a point, at least, would presumably increase this food supply. Whether this feeding actually benefits fishes or how it affects human consumers is unresolved.

The effects of high temperatures on competitors, predators, or symbionts is poorly understood. Cadwallader (1964) stressed the need to study freshwater species competing under natural conditions in areas of heated effluents. Hockley (1963) observed that heated effluents in temperate waters increased the occurrence of the gribble *Limnoria* and lengthened the breeding period in the mollusk *Teredo* in Southampton waters. Should these species compete with indigenous species for the same food and liebensraum, the most successful species would survive. In tropical waters, organisms are already living within a few degrees of their thermal death points (Mayer, 1914; Cairns, 1956a) and the balance of tropical communities is exceedingly complex (Voss and Voss, 1955). Mayer found that corals, sea urchins, and scyphozoans were killed at temperatures only a few degrees above those at which they normally live. Thus, the destruction of habitats such as coral reefs, turtle grass, mangroves, and eel grass could have far-reaching effects on marine fishes at various stages of their life histories because they are so closely dependent upon their environment.

Effect of Sources of Heated Effluents on Fishes

The previous discussion has largely centered on the theoretical aspects of high temperatures upon marine fishes, their life histories and metabolism, and their environment. Effects of heated effluents from power-generating stations and other sources have only been alluded to in the present paper but are the crux of the problem, from a practical standpoint. Thermal discharges and their effects upon aquatic organisms, mostly freshwater, are discussed by Cairns (1955, 1956a, 1956b), *Sewage Ind. Wastes* (1956) *Wastes Engineering* (1961), Klein (1957), Laberge (1959), Simpson and Garlow (1960), Trembley (1960, *et seq.*), Wurtz (1961), Hoak (1961, 1963), Arnold (1962), Markowski (1962), Hockley (1963), Hynes (1963), Alabaster (1963, 1964), Jones (1964), Cadwallader (1964), Dalton (1965a, 1965b), Edinger and Geyer (1965), Naylor (1965), Warinner and Brehmer (1966), Davidson and Bradshaw (1967), Douglas (1968), Stroud and Douglas (1968), and the U.S. Senate Committee on Public Works (1968).

Proponents of the thermal-enrichment school have attempted to point out the benefits of heated discharges to improve the environment, while opponents have proclaimed that any heated discharge is detrimental. The truth lies somewhere in between, depending upon the environmental conditions of a particular locality. Evaluation of the effects of heated effluents is difficult to accomplish because the effects are subtle, and only rarely are fishes killed outright from hot water, although this has indeed occurred (Alabaster, 1963, 1964; Alabaster and Downing, 1966). Dead fish are seldom seen in nature because they may sink to the bottom or, especially in the marine environment, are soon consumed by more resistant scavengers and predators, such as crabs and sharks. In tropical waters, for example, deaths due to heat have not been recorded in the literature because virtually no thermal plants have been established in areas of continually high temperature (Hoff and Westman, 1966). Thus, the presence or absence of dead fishes in heated effluents tells us very little. We must concern ourselves with sublethal, long-term effects, which are difficult to detect and to evaluate.

Theoretical effects are wide-ranging and are in most cases unstudied *in situ*. But in the few years that the effects of heated effluents have been studied, a number of interesting conclusions have been reached, all of which appear to stress that the problem is not a simple one of finding dead fish or finding no dead fish, but an extremely complex one requiring long-term, careful study (Arnold, 1962; Naylor, 1965). Alabaster (1963,

1964) has shown that heated effluents may result in stratification of the upper waters. Difficulties of predicting heat distribution are complicated by the complex interplay between temperature and salinity (Pritchard and Carter, 1965; Cronin, 1967), which pose prediction and evaluation problems not encountered in freshwater situations. For example, Hockley (1963) observed that a salinity gradient caused layering in the stream that held the main effluent stream of warm saline water below the surface.

Heated effluents may affect fishes in insidious and devious ways (Brett, 1960; Naylor, 1965). Temperature affects organisms not so much by the absolute value but by the Δt, and power plants often suddenly release large slugs of hot water which may not be tolerated. Discharges of hot, highly concentrated brine solutions from nuclear reactors have not yet been evaluated because desalination plants are presently few, but the potential effects should be studied.

Destruction of up to 95 percent of the plankton passing though condensing tubes in Chesapeake Bay has already been mentioned (Dr. Joseph P. Mihursky, personal communication), as has the possible effect of heated effluents increasing the rate of deformities and disease of fishes (cf. de Sylva et al., 1962). Excess heat may kill bacteria, small plants, and animals, turning them into dead organic matter which removes oxygen (Hynes, 1963), which, in turn, degrades the environment for fishes and their food. This greater rate of oxidation results in a higher production of carbon dioxide, which affects the quantity of dissolved oxygen available for adult and larval fishes (cf. Alabaster and Welcomme, 1962). Fish mortality in the Grand River (Michigan) was traced to organic waste loading aggravated by a 5.5° C. rise which reduced dissolved oxygen (Sewage Ind. Wastes, 1956).

Power-generating stations use various types of solutions to clean condensers, boilers, and pipelines (Cadwallader, 1964). Because of insufficient transfer of heat in thermal-power stations, corrosion of and leakage from cooling tubes occur, with a resultant mass growth of organisms on intake walls (Yamazaki, 1965). Chemicals used to remove these organisms or to treat cooling water are basically algicides, pH adjusters, corrosion and scale-corrosion inhibitors, and dispersing agents (Dalton, 1965b). Hynes (1963) observed that thermal plants may render the effluent water toxic either with residual chlorine or with compounds, such as cyanogen chloride, so that "the final effluent is therefore more polluting than was the water at the intake." Heated effluents containing detergents may reduce the rate of re-oxygenation of water (Hynes, 1963), or they may interfere with photosynthetic activity (Olson and Burgess, 1967). In addition to their inherent toxicity to fishes and their food organisms, these compounds may affect the palatability of fish by lending off-tastes to the fish flesh, thus possibly affecting the market value not only of the seafood taken in the immediate vicinity of the effluent but also the psychological market for seafood in the entire region.

Phytoplankton populations may increase, either due to heat or nutrients associated with sources of heated effluents, or both. The stimulation of phytoplankton growth may be beneficial, in that it provides food for zooplankton. However, such eutrophication increases turbidity, which may decrease the ability of visual-feeding fishes to locate their prey. Andrews (1946) showed that salmon reacted more poorly to different light intensities at higher temperatures. Thus, fishes living in turbid, heated effluents might be less able to locate food or to avoid more hardy predators. Turbidity from blooms may retard photosynthesis in benthic algae and vascular plants which might form an integral part of the nursery ground of larval or juvenile fishes.

Blue-green algae grow best at 30° to 40° C. (Cairns, 1955, 1956a), and grow well even a 46° C. (Trembley, 1965). This would inhibit competing green algae and permit the overgrowth of benthic and planktonic blue-green algae, especially since heated effluents may cause blooms of blue-green algae to occur earlier in the spring (Naylor, 1965) and thus compete with green algae. The larvae of many tropical marine fishes feed upon larval copepods, which themselves eat green algae (Mr. Charles A. Mayo III, personal communication), but whether larvae could successfully adapt to blue-green algae is not known. Blue-greens do not seem to be eaten by other organisms as much as green algae (Trembley, 1965). Blue-green algae are notoriously toxic to aquatic life or even to certain terrestrial mammals consuming them (de Sylva, 1963 and references contained therein; Trembley, 1965). They impart foul odors and fishy tastes when they decay, and may affect the taste of fish flesh.

Blue-green algae are believed to be responsible for the variety of fish poisoning known as ciguatera or ichthyosarcotoxism, peculiar to tropical reef fishes (Randall, 1958; de Sylva, 1963). Heated effluents which stimulate growth of blue-green algae around coral reefs might initiate the complex chain of events leading to the occurrence of toxic fish flesh. Periods of high air- and water-temperatures were believed by Ingle and de Sylva (1955) to be important in triggering the sudden growth of the toxic dinoflagellate *Gymnodinium breve,* responsible for the fish-killing "red tides" on the west coast of Florida, and thus heated effluents might stimulate growth of similar undesirable dinoflagellates.

Combined and Synergistic Effects from Sources of Heated Effluents

Heated effluents are not normally discharged into clean water, most being produced by industries and power stations situated in densely populated regions which are already polluted (Hynes, 1963). Chemicals used in the operation of power-generating stations, or other sources of heated effluents, or from nearby industrial sources, may affect fish life more acutely at higher temperatures (Binet and Marin, 1934; Sizer, 1936; Cairns, 1956b; Lloyd and Herbert, 1962; Alabaster, 1963). In some instances, this is because thermal stress itself lowers the resistance of fishes to acute toxins (Powers, 1920; Doudoroff, 1957), or because toxic materials in effluents increase the susceptibility of fishes to the relative lack of oxygen found in effluent waters (Lloyd and Herbert, 1962; Hynes, 1963; Naylor, 1965). Toxins in effluent discharges may act synergistically to complement the toxicity of one another (Wuhrmann and Woker, 1955; Laberge, 1959; Cottam and Tarzwell, 1960; Nikolsky, 1963), or may act agonistically to ameliorate the effects of one another (Doudoroff, 1957). But we should expect such reactions to occur much faster and for the toxicity to be greater at higher temperatures. Klein (1957) observed that a Δt of $10°$ C. doubles the toxicity of potassium cyanide, and that, because oxygen consumption of organisms doubles for each Δt of $10°$ C., we might expect that even a Δt of $1°$ C. taken in the aggregate might have markedly detrimental effects on fish life. Cairns and Scheier (1964) showed that the toxicity of zinc to freshwater sunfish (*Lepomis gibbosus*) was affected by a temperature increase, and although earlier (1957) they found that increased temperature did not speed up the process of zinc toxicity to bluegills (*Leopmis macrochirus*), the toxicity of zinc was greater in soft water. Increased toxicity of detergents to sunfish occurred at higher temperatures (Cairns and Schier, 1964). Eisler (1965) showed that the toxicity of detergents to several species of estuarine fishes was greater at higher salinities, which could presumably result from the hypersalinity of heated salt water or from discharged hot brine solutions from desalination-nuclear-power plants. Death of goldfish and guppies from insecticides occurs more quickly at higher temperatures (Adlung, 1957), the effect of cooler water being merely to delay the effect of the insecticide. Ogilvie and Anderson (1965) noted that Atlantic salmon (*Salmo salar*) which were acclimated to $8°$ C. and which had received 20–30 ppb of DDT and then were introduced into warmer water subsequently became extremely sensitive to cold water. They concluded that DDT may interfere with the normal acclimation mechanism. This study tends to confirm the importance of temperature upon the ability of fishes to resist toxins. The complexity of the entire environment in affecting the toxicity of poisons to fishes was stressed by Alderdice (1963), who suggested the use of multivariable experimental techniques to stimulate factors of the environment as accurately as possible (see also Cottam and Tarzwell, 1960; Arnold, 1962; Mihursky and Kennedy, 1967). The importance of different environmental factors in influencing the acute toxicity of poisons to fishes was studied by Wuhrmann and Woker (1955). They concluded that the problem of subacute concentrations of poisons at high temperatures was still unknown, and that there was an urgent need to study (a) factors directly influencing the physiological activity of a poison by changing the gill permeability, and (b) factors which attack the poison itself and change the concentration of its physico-chemical properties.

Effects of Heated Effluents on Faunistic Composition

That long-term shifts in fish faunas may occur from gradual, naturally increasing temperatures was shown for the Atlantic and Pacific coasts of the United States by Taylor *et al.* (1957) and Hubbs (1948), respectively. Agersborg (1930*a*) discussed the implications of high temperature in limiting distribution of cold-water species. Hela and Laevastu (1960, 1962) summarized the probable ways in which a long-term temperature rise can influence the distribution of fishes (in the Northern Hemisphere):

a) Spawning is diminished at its southern limit and increased at its northern limit.
b) An increase in bottom-water temperature can produce changes in spawning grounds.
c) New nursery and feeding grounds become available in the north
d) An increase in the amount of food is brought about by the rise in temperature in higher latitudes and changes occur in currents and in the amount of nutrient salts present.
e) The growth period is prolonged.
f) The limit at which larvae can survive is shifted farther north.

The substitution of replacement organisms which are pre-adapted to the peculiar conditions of a thermally polluted area is extensively discussed by Naylor (1965). Most examples given are invertebrates which have invaded temperate waters receiving heated effluents. Kinne (1956) believed, largely on the basis of estuarine invertebrates, that the distribution of organisms is dependent upon the compensatory and complementary effects of temperature *and* salinity, and there is evidence for this in fishes, as well (Pearse and Gunter, 1957; Kinne, 1964a).

Naylor (1965) presented numerous examples of invertebrate replacements. He cited an instance where a Mediterranean fish, the silverside *Atherina boyeri,* has become established and has spawned in South Wales. But it appears that some ecological replacements are not necessarily desirable. For example, stingrays and sharks from warmer southern waters have migrated northward and now occur in the heated effluents of a power plant at Los Alamitos, California (Dr. Joel W. Hedgpeth, personal communications). Hockley (1963) noted the increase in the population of the boring crustacean *Limnoria* and in the breeding period of the boring mollusk *Teredo* in Southampton waters. In the Delaware River estuary, de Sylva *et al.* (1962) discussed the gradual increase in skates, rays, small flounders, and burrowing invertebrates associated with waters increasingly polluted from many sources, including steam-electric stations. Mihursky and Kennedy (1967) showed that, of seven invertebrate species (six crustaceans and one coelenterate), the six crustaceans were less tolerant than the coelenterate, the sea nettle (*Chrysaora*). Thus, heated effluents could theoretically eradicate the crustaceans, which are valuable as fish food, while favoring growth and reproduction of the sea nettle which, incidentally, causes minor injury and is a major nuisance to swimmers. Economic losses from closed bathing beaches have stirred extensive financial support to eradicate or control sea nettles; it would be ironic if heated effluents stimulated their growth and extended their distribution while eliminating economically valuable species. Similarly, effluents from the power station at Los Alamitos, California, were presumed to warm the sea water for use by bathers and surfers. Yet, the beaches are reported to be periodically closed, and funds must be expended to seine out the stingrays and sharks attracted by the warm water.

Ecological replacements appear to be largely the more hardy species which may be undesirable to man's use of the sea. Scientifically, it could be argued that heated effluents, to be discharged from industrial plants in the future on a scale as yet inconceivable to us, will cause consternation to future ecologists, zoogeographers, and paleontologists trying to interpret the effects of thermal changes in faunal distributions which have occurred rather suddenly on the geological time scale.

Ecological replacements in polar waters have not been reported, because few power stations exist in such areas. One might expect the invasion of temperate species with heated effluents; yet, polar fishes are quite steno-thermal in their requirements, and the detrimental effects of heated efflu-ents on indigenous fish populations could be locally serious. Mid-latitudes, inhabited by typically eurythermal temperate fishes, are more likely to be amenable to receiving replacement fish faunas from tropical and subtropical waters. However, thermal effluents discharged into tropical waters pose a serious problem, especially where circulation is limited or the water is shallow. Tropical waters are already at 25°–30° C. in summer, and may exceed these values in isolated patches of tropical lagoons or tidepools. Tropical organisms live within 5° C. of their maximum activity, and within 10°–15° C. of their thermal death points (Mayer, 1914). Even Mediterranean fishes, which may be considered intermediate between temperate and subtropical fishes in their thermal requirements, are living only about 6°–8° C. below their upper lethal limit (Timet, 1963). Tropical marine fishes seldom live in waters above 31° or 32° C., and thus conventional power stations, whose effluent may be 6–9° C. above am-bient, and nuclear power stations, whose effluent may be up to 11° C. above ambient, could locally eradicate fishes in areas of limited circulation (de Sylva, 1968). In such areas, there could be no replacement fauna, as these temperature values represent approximately the upper limits for the existence of life (Davenport and Castle, 1895; Mayer, 1914; Drost-Hansen, 1965).

The Use of Heated Effluents to Improve the Environment

It has been wisely proposed that waste heat from industrial effluents be utilized for the good of mankind. Domestic heating, the rearing of aquatic organisms, warming of beach areas for year-round swimming, the amelioration of severely cold waters in aquatic habits, the maintenance of ice-free conditions for shipping, and the heating of areas to attract sport and commercial fishes have been suggested (Klein, 1957; Naylor, 1965). Iles (1963) noted that, in Great Britian, by 1970, it is expected that 20,000 million gallons of water will be heated daily to 7°–8° C., pointing out that this vast volume of heat is being wasted. Among his suggestions are the rearing of tropical aquarium fishes, farming of important sport fishes, and the rearing and farming of fish and shellfish. The rearing of disease-resistant and fast-growing hybrids grown in warmed waters could provide both scientific and practical information.

Elser (1965) found that angling in the Potomac River (Maryland) improved in areas receiving heated effluents during the coolest nine months of the year, but became poor in the summer, especially for channel catfish (*Ictalurus*). There was no difference in the size of the fish—mostly catfish, smallmouth bass, and sunfish—caught in heated waters and control areas. Most observers have noted that heated waters do not actually produce more fish or increase the growth rate, but merely concentrate the fish during the cold months (Trembley, 1965), so that they are available to anglers. During colder months, snappers, barracuda, needlefish, sharks, and mullet congregate in the warm effluent of power plants at Cutler and Turkey points on Biscayne Bay, Florida (de Sylva, 1963), where they are sought by anglers. However, reports reaching the Institute of Marine Sciences of the University of Miami that grey snappers caught in these effluents sometimes have body lesions should cause one to reflect on the possible side effects of power plants upon fishes. Hoak (1963) summarized the problem by stating that "the discharge of heated water has an effect that pleases fishermen but not aquatic biologists." It seems axiomatic that heated water should not be considered as a panacea to improve sport or commercial fisheries, but that each geographic area must be first investigated to determine the organisms present, their life cycles, and their environment, and what the long-term effects might be on the total environmental complex.

Summary

The ecological effects of heated effluents on marine fishes are many, complex, and difficult to study and evaluate. At present, we do not understand, nor can we predict with a great deal of confidence, these complex effects of heated water on marine fishes and their environment. The cohesive force joining the many factions interested in the problems of thermal pollution is the agreement that more research is needed. There is much need for work on sublethal, long-term, and side effects of heated effluents and the ancillary pollutants which accompany hot-water discharges. The public, politicians, administrators, and scientists must be convinced that a dead fish on the beach may be only the *end* result of pollution.

Kinne (1963) rightly concluded that "a detailed assessment of thermal effects necessitates long-term rearing and breeding experiments in the laboratory under a variety of conditions and parallel investigations in the field." Research must be carried out on fishes typical of the marine environment in question, and not on eurythermal species such as carp, goldfish, and killifish, which appear to defy death. Multivariate techniques must be used which approach those conditions of the natural environment (Alabaster, 1963, 1964; Alderdice, 1963; Naylor, 1965; Mihursky and Kennedy, 1967). Only in this way can we learn of the complexities of how heated water may affect aquatic life.

Prior to the establishment of sites for power plants or other potential sources of heated pollution, careful biological, chemical, and oceanographic surveys should be made at least two years in advance of the site selection, with a study to be effected of such factors as currents, winds, tides, runoff, bottom type, chemical and physical characteristics of the environment, and the type and number of organisms present. An excellent example of a carefully planned survey in California is reported by Kerr (1953). The original survey studied fish behavior, velocity studies determining fish-swimming ability, screen mesh size, water temperature tolerance, traveling screen operation, fish stratification and dispersion, electric fish screens, velocity barriers to fish, and development of a device for safely collecting fishes and returning them to the river unharmed. Naylor (1965) concluded that "the siting of heated effluents in marine and estuarine localities should therefore be based on a sound ecological approach, and future studies would clearly benefit by investigation of faunal and hydrographic characteristics of a region before a heated effluent is discharged." Finally, there is no substitute for adequate pollution controls by the industry discharging the heated effluents into the environment. The technology is available to industry for constructing the necessary cooling towers or closed-circuit cooling systems, and these should be included in the design of plants that will discharge hot water, with the installation and operation costs to be routinely included in the consumer's bill.

But there is an increasing urgency for electric power from fossil- and nuclear-power plants concomitant with a shortage of time, money, and qualified scientists to carry out the necessary surveys and laboratory experiments. The need for ecological studies is more urgent in the tropical and subtropical lower latitudes, where hydroelectric power is scarce because of low stream-gradient. It is here, also, that the least research has been done.

In the United States, scientists called upon to testify before congressional investigating committees, or called upon to make judgments on the effect of power stations and other industries, have been understandably loath to predict the effect of thermal loading upon marine fishes and their environment, because there have been so few actual investigations upon which they can base their recommendations. But the rate of application for permits and construction of power plants is proceeding at a rate much faster than adequate laboratory and field studies can be carried out. Surely the scientific community can make some *a priori* predictions based on principles of the life history, ecology, and physiology of marine fishes in general, and upon the marine environment in general. Faced with the impending construction of a power plant in an area which has not been studied, our tendency often is to state that on this site nothing has been done, we know nothing, and we can predict nothing. It seems gratifying that power companies sometimes retain professional aquatic scientists to guide them in their location and construction of power plants and other industries to minimize damage to the natural resource; surely this is better than guesswork in establishing sites.

The alternative is to say that we don't know. Until that day of nirvana when adequate personnel, funds, and time are available for ecological studies, we may be forced to play God on a limited budget and make some quick and dirty guesses about our environment. Otherwise, we may not have much environment left to play with.

REFERENCES

Acara, A. 1957. "Relation Between the Migration of *Sarda sarda* Bloch and Prevailing Temperature. *Proc. tech. Pap. gen. Fish. Coun. Mediterr.* (26):193–196.

Adlung, K. G. 1957. "Zur Kenntnis der Fisch-Toxizität von Insektiziden und ihrer Temperaturabhängigkeit." *Naturwissenschaften* 44:622–623.

Agersborg, H. P. K. 1930a. "Does High Temperature in a Frigid Country Limit Diversification of the Species?" *Trans. Ill. St. Acad. Sci.* 22:103–114.

———. 1930b. "The Influence of Temperature on Fish." *Ecology* 11:136–144.

Ahlstrom, Elbert H. 1943. "Influence of Temperature on the Rate of Development of Pilchard Eggs in Nature." In *Studies on the Pacific Pilchard or Sardine* (Sardinops caerulea). *Spec. scient. Rep. U.S. Fish. Wildl. Serv.* (23):26 pp.

Alabaster, John S. 1963. "The Effect of Heated Effluents on Fish." *Int. J. Air Wat. Pollut.* 7:541–563.

———. 1964. "The Effect of Heated Effluents on Fish." *Adv. Wat. Pollut. Res.* 1:261–292.

———. 1967. "The Survival of Salmon (*Salmo salar* L.) and Sea Trout (*S. trutta* L.) in Fresh and Saline Water at High Temperature." *Water Res.* 1(10):717–730.

Alabaster, John S., and A. L. Downing. 1966. "A Field and Laboratory Investigation of the Effect of Heated Effluents on Fish." *Fishery Invest., London, Ser. I* 6(4): 1–42.

Alabaster, John S., and R. L. Welcomme. 1962. "Effect of Concentration of Dissolved Oxygen on Survival of Trout and Roach in Lethal Temperatures." *Nature, Lond.,* 194:107.

Albrecht, Arnold B. 1964. "Some Observations on Factors Associated with Survival of Striped Bass Eggs and Larvae." *Calif. Fish Game,* 50(2):101–113.

Alderdice, D. F. 1963. "Some Effects of Simultaneous Variation in Salinity, Temperature, and Dissolved Oxygen on the Resistance of Young Coho Salmon to a Toxic Substance." *J. Fish. Res. Bd. Can.* 20(2):525–550.

Aleev, Yu. G. 1954. "The Role Played by Low Temperatures in the Stimulation of Trophoplasmatic Growth of Oöcytes in Fish." *Dokl. Akad. Nauk SSSR* 110(3):4– 493. (In Russian).

Ali, M. A. 1964. "Über den Einfluss der Temperatur auf die Geschwindigkeit der retinomotorischen Reaktionen des Lachses (*Salmo salar*)." *Naturwissenschaften* 51:471.

Altman, Philip L., and Dorothy S. Dittmer, editors. 1966. *Environmental Biology.* Bethesda, Md.: Fed. Amer. Soc. Exper. Biol.

Andrews, C. W. 1946. "Effect of Heat on the Light Behavior of Fish." *Trans. R. Soc. Can.* 40(5):27–31.

Anthony, E. H. 1961. "The Oxygen Capacity of Goldfish (*Carassius auratus* L.) Blood in Relation to Thermal Environment." *J. Exp. Biol.* 38(1):93–107.

Arnold, G. E. 1962. "Thermal Pollution of Surface Supplies." *J. Am. Wat. Wks. Ass.* 54:1332–1346.

Audigé, P. 1921. "Influence de la Température sur la Croissance des Poissons." *C. r. Séanc. Soc. Biol.* 84:67–69.

Baslow, M. H., and Ross F. Nigrelli. 1964. "The Effect of Thermal Acclimation on Brain Cholinesterase Activity of the Killifish, *Fundulus heteroclitus*." *Zoologica,* N. Y. 49:41–51.

Battle, Helen I. 1926. "Effects of Extreme Temperatures on Muscle and Nerve Tissue in Marine Fishes." *Trans. R. Soc. Can., Ser. 5.* 20:127–143.

———. 1929a. "A Note on Lethal Temperature in Connection with Skate Reflexes." *Contr. Can. Biol. Fish.* 4(31):495–500.

———. 1929b. "Temperature Coefficients for the Rate of Death of Muscle in *Raja erinacea* (Mitchill) at High Temperatures." *Contr. Can. Biol. Fish.* 4(32): 501–526.

———. 1929c. "Effects of Extreme Temperatures and Salinities on the Development of *Enchelyopus cimbrius* (L.)." *Contr. Can. Biol. Fish.* 5(6):107–192.

Bĕlehrádek, Jan. 1931. "Influence de la Température sur la Fréquence Cardiaque Chez les Embryons de la Roussette, *Scylliorhinus canicula*." *L. C. r. Sanc. Soc. Biol.* 107(20):727–729.

Bergan, P. 1960. "On the Blocking of Mitosis by Heat Shock Applied at Different Mitotic Stages in the Cleavage Divisions of *Trichogaster trichopterus* var, *sumatranus* (Teleostei: Anabantidae)." *Nytt Mag. Zool.* 9:37–121.

Binet, L., and A. Marin. 1934. "Action de la Chaleur sur les Poissons." *J. Physiol. Path. gén.* 32:372–379.

Bishai, H. M. 1960. "Upper Lethal Temperatures for Larval Salmonids." *J. Cons. perm. int. Explor. Mer.* 25(2):129–133.

Blaxter, John H. S. 1956. "Herring Rearing. II. The Effect of Temperature and Other Factors on Development." *Mar. Res.* No. 5:19 pp.

———. 1960. "The Effect of Extremes of Temperature on Herring Larvae." *J. Mar. biol. Ass. U. K.* 39:605–608.

Bonnet, D. E. 1939. "Mortality of the Cod Egg in Relation to Temperature." *Biol. Bull. Mar. Biol. Lab., Woods Hole* 76:428–441.

Brandhorst, Wilhelm. 1959. "Spawning Activity of Herrings and the Growth of their Larvae in Relation to Temperature and Salinity Conditions." *Int. Oceanogr. Congr., Preprints.* Edited by M. Sears, pp. 218–221. Washington, D.C.: Am. Ass'n. Adv. Sci.

Brawn, Vivien M. 1960. "Temperature Tolerance of Unacclimated Herring (*Clupea harengus* L.)." *J. Fish. Res. Bd. Can.* 17(5):721–723.

Breder, Charles M. 1951. "Studies on the Structure of the Fish School." *Bull. Am. Mus. Nat. Hist.* 98:5–27.

Brett, J. R. 1956. "Some Principles in the Thermal Requirements of Fishes." *Q. Rev. Biol.* 31(2):75–87.

———. 1960. "Thermal Requirements of Fish—Three Decades of Study, 1940–1970." *In: Biological Problems in Water Pollution.* 2d seminar. *Robert A. Taft Sanit. Engng. Cent. Tech. Rep., W60–3:* 110–117.

Brett, J. R., M. Hollands, and D. F. Alderdice. 1958. "The Effect of Temperature on the Cruising Speed of Young Sockeye and Coho Salmon." *J. Fish. Res. Bd. Can.* 15(4):587–605.

Britton, S. W. 1924. "The Effects of Extreme Temperatures on Fishes." *Am. J. Physiol.* 67(2):411–421.

Brown, Margaret E. 1946. "The Growth of Brown Trout (*Salmo trutta* Linn.). III. The Effect of Temperature on the Growth of Two-Year-Old Trout." *J. Exp. Biol.* 22:145–155.

Bull, H. O. 1937. "Studies on Conditioned Responses in Fishes. Part VII. Temperature Perception in Teleosts." *J. Mar. Biol. Ass. U. K.* 21(1):1–27.

———. 1957. "Behavior: Conditioned Responses." In *The Physiology of Fishes* by Margaret E. Brown. 2:221–228. New York: Academic Press.

Burger, J. W. 1939. "Some Experiments on the Relation of the External Environment to the Spermatogenetic Cycle of *Fundulus heteroclitus* (L.)" *Biol. Bull. Mar. Biol. Lab., Woods Hole* 77:96–103.

Cadwallader, L. W. 1964. "Thermal Pollution of Watercourses." *Proc. 19th Ind. Waste Conf. Purdue Univ., Engng. Ext. Ser.* (117):9–11.

Cairns, John, Jr. 1955. "The Effects of Increased Temperature upon Aquatic Organisms." *Proc. Tenth Purdue Ind. Waste Conf.* 346–354.

———. 1956a. "Effects of Increased Temperatures on Aquatic Organisms." *Ind. Wastes* 1(4):150–152.

———. 1956b. "Effects of Heat on Fish." *Ind. Wastes* 1(5):180–183.

Cairns, John, Jr., and Arthur Scheier. 1957. "The Effects of Temperature and Hardness of Water upon the Toxicity of Zinc to the Common Bluegill (*Lepomis macrochirus* Raf.)." *Notul. Nat.* (299):1–12.

———. 1964. "The Effects of Sublethal Levels of Zinc and of High Temperature upon the Toxicity of a Detergent to the Sunfish, *Lepomis gibbosus* (Linn.)." *Notul. Nat.* (367):1–3.

Chidester, F. E. 1924. "A Critical Examination of the Evidence for Physical and Chemical Influences of Fish Migration." *J. Exp. Biol.* 2:79–118.

de Ciechomski, Juana D. 1965. "Observaciones sobre la reproducción desarrollo embrionario y larval de la anchoíta argentina (*Engraulis anchoita*)." *Boln. Inst. Biol. Mar., Mar del Plata* (9):1–29.

Cole, William H. 1939. "The Effect of Temperature on the Color Change of *Fundulus* in Response to Black and to White Backgrounds in Fresh and in Sea Water." *J. Exp. Zool.* 80(2):167–172.

Cole, William H., and Kenneth F. Schaeffer. 1937. "The Effect of Temperature on the Adaption of *Fundulus* to Black and to White Backgrounds." *Bull. Mt. Desert Isl. Biol. Lab.* (1936):26–30.

Colton, J. B. 1959. "A Field Observation of Mortality of Marine Fish Larvae due to Warming." *Limnol. Oceanogr.* 4(2):219–222.

Cottam, Clarence, and Clarence M. Tarzwell. 1960. "Research for the Establish-
ment of Water Quality Criteria for Aquatic Life." In *Biological Problems in
Water Pollution*. Trans. 1959 Seminar, Robert A. Taft Engng. Cent., Publ.
Hlth. Serv., Tech. Rep., W60–3:226–232.

Coutant, C. C. 1965. "Effect of Thermal Pollution on Delaware River Invertebrates." In *Minutes, Third Annual Conf. on the Patuxent Estuary Studies, Nov.
13–14, 1964*. Chesapeake Biol. Lab., Ref. No. 65–23.

Cronin, L. Eugene. 1967. "The Role of Man in Estuarine Processes." In *Estuaries*,
edited by George K. Lauff, pp. 667–689. Am. Assoc. Adv. Sci., Publs., (83):757 pp.

Dalton, T. F. 1965*a*. "Cooling Water Treatment. Part 1. Some Principles to Aid in
Pollution Control." *Wat. Wks. Wastes Engng*. 2(11):61.

———. 1965*b*. "Cooling Water Treatment. Part 2. Problems . . . and Some
Treatment Practices." *Wat. Wks. Wastes Engng*. 2(12):53–56.

Dannevig, H. 1894. "The Influence of Temperature on the Development of the Eggs
of Fishes." *Rep. Fishery Bd., Scotl*. 13:147–153.

Davenport, C. B., and W. E. Castle. 1895. "Studies on Morphogenesis. III. On the
Acclimatization of Organisms to High Temperatures." *Arch. EntwMech. Org*.
2:227–249.

Davidson, Burton, and Robert W. Bradshaw. 1967. "Thermal Pollution of Water Systems." *Environ. Sci. Technol*. 1(8):618–630.

Doudoroff, Peter. 1938. "Reactions of Marine Fishes to Temperature Gradients."
Biol. Bull. Mar. Biol. Lab., Woods Hole 75(3):494–509.

———. 1942. "The Resistance and Acclimatization of Marine Fishes to Temperature
Changes. I. Experiments with *Girella nigricans* (Ayers)." *Biol. Bull. Mar. Biol.
Lab., Woods Hole* 83:219–244.

———. 1945. "The Resistance and Acclimatization of Marine Fishes to Temperature Changes. 2. Experiments with *Fundulus* and *Antherinops*." *Biol. Bull.
Mar. Biol. Lab., Woods Hole* 88:194–206.

———. 1957. "Water-Quality Requirements of Fishes and Effects of Toxic Substances. In *Physiology of Fishes*, vol. 2, edited by Margaret E. Brown, pp. 403–
430. New York: Academic Press.

Douglas, Philip A. 1968. "Heated Discharges and Aquatic Life." *Bull. Sport Fish.
Inst*. (198):1–5.

Downing, K. M., and C. J. Merkens. 1957. "The Influence of Temperatures on the
Survival of Several Species of Fish in Low Tensions of Dissolved Oxygen." *Ann.
Appl. Biol*. 45(2):261–267.

Drost-Hansen, Walter. 1965. "The Effects of Biologic Systems of Higher-Order Phase
Transitions in Water." *Ann. N. Y. Acad. Sci*. 125(2):471–501.

Edinger, J. E., and J. C. Geyer. 1965. "Heat Exchange in the Environment, a
Study of the Physical Principles Relating to Power Plant Condenser Cooling
Water Discharges." *Edison Elect. Inst. Publ. No. 65–902*:1–259.

Edwards, Robert L. 1965. "Relation of Temperature to Fish Abundance and Distribution in the Southern New England Area." *Spec. Publs. Int. Commn. NW
Atlant. Fish., No. 6*:95–110.

Eisler, Ronald. 1965. "Some Effects of a Synthetic Detergent on Estuarine Fishes."
Trans. Am. Fish. Soc. 94(1):26–31.

Elser, Harold J. 1965. "Effect of a Warmed-Water Discharge on Angling in the
Potomac River, Maryland, 1961–62." *Progve. Fish Cult*. 27(2):79–86.

Emery, K. O., and R. E. Stevenson. 1957. "Estuaries and Lagoons. I. Physical Characteristics." *Treatise on Marine Ecology and Paleoecology,* by Joel W. Hedgpeth, pp. 673–693. Geol. Soc. Amer., Mem., 67(1).

Evropeyzeva, N. V. 1944. "Preferred Temperature of Fish Larvae." *Dokl. Akad. Nauk SSSR* 423:138–142.

Farley, T. C. 1966. "Striped Bass, *Roccus saxatilis,* Spawning in the Sacramento-San Joaquin River Systems During 1963 and 1964." In *Ecological Studies of the Sacramento-San Joaquin Delta. Part II. Fishes of the Delta,* edited by J. L. Turner and D. W. Kelley. Bull. Dep. Fish. Game St. Calif. (136):28–42.

Farris, David A. 1961. "Abundance and Distribution of Eggs and Larvae and Survival of Larvae of Jack Mackerel (*Trachurus symmetricus*)." *Fishery Bull. Fish. Wildl. Serv.,* 61(187):247–279.

Fisher, Kenneth C. 1958. "An Approach to the Organ and Cellular Physiology of Adaptation to Temperature in Fish and Small Mammals." *Physiological Adaptation,* edited by C. L. Prosser, pp. 3–48. Washington, D. C.: Amer. Physiol. Soc.

Fleming, Richard H. 1956. "The Influence of Hydrographic Conditions on the Behavior of Fish." *Fish. Bull. F.A.O.* (9):181–196.

Fontaine, Maurice, and Odette Callamand. 1940. "Influence de la température sur l'élimination chlorée de l'Anguille." *C. R. hebd. Séanc. Acad. Sci.,* Paris, 211:488–489.

Forrester, C. R., and D. F. Alderdice. 1966. "Effects of Salinity and Temperature on Embryonic Development of the Pacific Cod (*Gadus macrocephalus*)." *J. Fish. Res. Bd. Can.* 23(3):319–340.

Fortune, P. Y. 1958. "The Effect of Temperature Changes on the Thyroid-Pituitary Relationship in Teleosts." *J. Exp. Biol.,* 35:824–831.

Fry, F. E. J. 1957. "The Aquatic Respiration of Fish." In *The Physiology of Fishes, 1,* edited by Margaret E. Brown, pp. 1–63. New York: Academic Press.

Fry, F. E. J., and J. S. Hart. 1948. "Cruising Speed of Goldfish in Relation to Water Temperature." *J. Fish. Res. Bd. Can.* 7(4):169–175.

Gameson, A. L. H., H. Hall, and W. S. Preddy. 1957. "Effects of Heated Discharges on the Temperature of the Thames Estuary." *Engineer,* Lond., 204:850–893.

Glaser, Otto. 1929. "Temperature and Heart Rate in *Fundulus* Embryos." *J. Exp. Biol.* 6:325–339.

Gray, J. 1928a. "The Growth of Fish. II. The Growth Rate of the Embryo of *Salmo fario.*" *J. Exp. Biol.* 6:110–124.

———. 1928b. "The Growth of Fish. III. The Effect of Temperature on the Development of the Eggs of *Salmo fario.*" *J. Exp. Biol.* 6:125–130.

Gunter, Gordon. 1957. "Temperature." In *Treatise on Marine Ecology and Paleoecology,* edited by Joel W. Hedgpeth, pp. 159–184. Geol. Soc. Amer. Mem., (67), 1.

Harrington, Robert W., Jr. 1959. "Effects of Four Combinations of Temperature and Day Length on the Ovogenetic Cycle of a Low-Latitude Fish, *Fundulus confluentus* Goode and Bean." *Zoologica,* N.Y., 44(11):149–168.

Hasan, R., and S. Z. Qasim. 1960. "Studies on Fish Metabolism. I. The Effect of Temperature on the Heart Rate." *Proc. Indian Acad. Sci., Sect. B.* 53(5):230–239.

Hathaway, E. S. 1927. "The Relation of Temperature to the Quantity of Food Consumed by Fishes." *Ecology,* 8(4):428–434.

Haugaard, Niels, and Laurence Irving. 1943. "The Influence of Temperature upon the Oxygen Consumption of the Cunner (*Tautogolabrus adspersus* Walbaum) in Summer and Winter." *J. Cell. Comp. Physiol.* 21(1):19–26.

Hedgpeth, Joel W., editor. 1957a. *Treatise on Marine Ecology and Paleoecology.* Geol. Soc. Amer., Mem., (67), 1:1296 pp.

————. 1957b. "Estuaries and lagoons. II. Biological aspects." In *Treatise on Marine Ecology and Palaeoecology,* edited by Joel W. Hedgpeth, pp. 693–729. Geol. Soc. Amer. Mem., (67), *1.*

Hela, Ilmo, and Taivo Laevastu. 1960. "Influence of Temperature on the Behavior of Fish." *Suomal. eläin- ja kasvit. Seur. van Tiedon. Pöytäk, 15*(1–2):83–103.

————. 1962. *Fisheries Hydrography.* London: Fishing News (Books) Ltd.

Hempel, Gotthilf, and John H. S. Blaxter. 1961. "Einfluss von Temperatur und Salzgehalt auf Myomerenzahl und Körperprösse von Heringslarven." *Z. Naturf.* 16b (3):227–228.

Henley, Eva. 1952. "The Influence of the Gas Content of Sea Water on Fish and Fish Larvae." *Rapp. P.-v. Réun. Cons. Perm. Int. Explor. Mer, 1931:*24–27.

Hermann, Frede. 1953. "Influence of Temperature on Strength of Cod Year-Classes." *Annls. Biol. Copenh.,* 9:31–32.

Hermann, Frede, and Paul M. Hansen. 1965. "Possible Influence of Water Temperature on the Growth of West Greenland Cod." *Spec. Publs. Int. Commn. NW Atlant. Fish., No.* 6:557–564.

Hermann, Frede, Paul M. Hansen, and Sv. Aa. Horsted. 1965. "The Effect of Temperature and Currents on the Distribution and Survival of Cod Larvae at West Greenland." *Spec. Publs. Int. Commn. NW Atlant. Fish., No.* 6:389–395.

Heuts, M. J. 1956. "Temperature Adaptation in *Gasterosteus aculeatus* L." *Pubbl. Staz. Zool. Napoli* 28:44–61.

Higurashi, Tadashi. 1925. "Optimum Water Temperatures for Hatching the Eggs of *Plecoglossus altivelis* T & S." *J. Imp. Fish. Inst., Tokyo* 20:12–14.

Higurashi, Tadashi, and Moriasburô Tauti. 1925. "On the Relation Between Temperature and the Rate of Development of Fish Eggs." *J. Imp. Fish. Inst., Tokyo* 21: 5–9.

Hiyama, Yoshio. 1952. "Thermotaxis of Eel Fry in Stage of Ascending River Mouth." *Jap. J. Ichthyol.* 2:23–30.

Hoak, R. D. 1961. "The Thermal-Pollution Problem." *J. Wat. Pollut. Control Fed.* 33(12):1267–1276.

————. 1963. "Thermal Loading of Streams." A.S.T.M. Special Publ. No. 337. In *Pap. Ind. Wat. Ind. Waste Wat.,* (1962):20–31.

Hockley, A. R. 1963. "Some Effects of Warm Water Effluents in Southampton Water." *Ann. Rep. Challenger Soc.* 3(15):37–38.

Hodder, V. M. 1965. "The Possible Effects of Temperature on the Fecundity of Grand Bank Haddock." *Spec. Publs. Int. Comm. NW Atlant. Fish.* No. 6:515–522.

Hoff, J. G., and James R. Westman. 1966. "The Temperature Tolerance of Three Species of Marine Fishes." *J. Mar. Res.* 24(2):131–140.

Holliday, F. G. T., and John H. S. Blaxter. 1960. "The Effects of Salinity on the Developing Eggs and Larvae of the Herring." *J. Mar. biol. Ass. U.K.* 39:591–603.

Hubbs, Carl L. 1948. "Changes in the Fish Fauna of Western North America Correlated with Changes in Ocean Temperature." *J. Mar. Res.* 7(3):459–482.

Hubbs, Clark. 1964. "Effects of Thermal Fluctuations on the Relative Survival of Greenthroat Darter Young from Stenothermal and Eurythermal Waters." *Ecology* 45(2):376–379.

————. 1965. "Developmental Temperature Tolerance and Rates of Four Southern California Fishes, *Fundulus parvipinnis, Atherinops affinis, Leuresthes tenuis* and *Hypsoblennius* sp." *Calif. Fish Game* 51(2):113–122.

————. 1966. "Fertilization, Initiation of Cleavage and Developmental Temperature Tolerance of the Cottid Fish, *Clinocottus analis.*" *Copeia,* 1966 (1):29–42.

Huntsman, A. G. 1933. "Heat and Cold Make Herring Seasons in Passamaquoddy." *Prog. Rep. Atl. Biol. Stn.* (7):3–6.

————. 1942. "Death of Salmon and Trout with High Temperature." *J. Fish. Res. Bd. Can.* 5(5):485–501.

————. 1946. "Heat Stroke in Canadian Maritime Stream Fishes." *J. Fish. Res. Bd. Can.* 6:476–482.

Huntsman, A. G., and M. I. Sparks. 1924. "Limiting Factors for Marine Animals. 3. Relative Resistance to High Temperatures." *Contr. Can. Biol. Fish., N.S.* 2(6): 97–114.

Hynes, H. B. N. 1963. "The Biology of Polluted Waters." In *Heat, Salts and Pollution of Lakes*, pp. 136–145. Liverpool: Liverpool Univ. Press.

Iles, R. B. 1963. "Cultivating Fish for Food and Sport in Power Station Water." *New Scient.* 17(324):227–229.

Ingle, Robert M., and Donald P. de Sylva. 1955. *The Red Tide.* Educ. Ser., Fla. St. Bd. Conserv., (1):31 pp. (revised).

Ivlev, V. S. 1938. ["The Influence of Temperature on the Respiration of Fishes."] *Russk. Zool. Zh.* 17(4):645–660. (In Russian.)

Jean, Y. 1964. "Seasonal Distribution of Cod (*Gladus morhua* L.) Along the Canadian Atlantic Coast in Relation to Water Temperature." *J. Fish. Res. Bd. Can.* 21(3):429–460.

Johansen, A. C., and August Krogh. 1914. "The Influence of Temperature and Certain Other Factors upon the rate of Development of the Eggs of Fishes." *Publs. Circonst. Cons. Perm. Int. Explor. Mer* (68):1–44.

Johnson, F. J., editor. 1957. *Influence of Temperature on Biological Systems.* Washington, D. C.: Am. Physiol. Soc.

Jones, J. R. E. 1964. "Thermal Pollution: The Effect of Heated Effluents." *Fish and River Pollution*, Chap. 13. London: Butterworths.

Kajiyama, Eiji. 1929. "On the Influence of Temperature upon the Development of Eggs of *Pagrosomas major* (Temminck & Schlegel)." *J. Imp. Fish. Inst., Tokyo* 24(5):109–113.

Kawajiri, Minoru. 1927. "The Influence of Variation of Temperature of Water on the Development of Fish Eggs." *J. Imp. Fish. Inst., Tokyo* 23:65–77.

Keiz, G. 1953. "Über die Beziehungen zwischen Temperatur-Akklimatisation und Hitzeresistenz bei eurythermen und stenothermen Fischarten (*Squalius cephalus* L. und *Trutta iridea* W. Gibb.)." *Naturwissenschaften* 40:249–250.

Kennedy, V. S., and Joseph A. Mihursky. 1967. *Bibliography on the Effects of Temperature in the Aquatic Environment.* Univ. Maryland, Nat. Resources Inst., Contr., No. 326.

Kerr, James E. 1953. "Studies on Fish Preservation at the Contra Costa Stream Plant of the Pacific Gas and Electric Company." *Fish. Bull. Calif.* (92):1–66.

Ketchen, K. S. 1952. "Factors Influencing the Survival of the Lemon Sole (*Parophrys vetulus* Girard) in Hecate Strait, British Columbia." *J. Fish. Res. Bd. Can.* 13(5): 647–694.

Kinne, Otto. 1956. "Über Temperatur und Salzgehalt und ihre physiologischbiologische Bedeutung." *Biol. Zbl.* 75(5/6):314–327.

————. 1960. "Growth, Food Uptake, and Food Conversion in a Euryplastic Fish Exposed to Different Temperatures and Salinities." *Physiol. Zoöl.* 33:288–317.

————. 1963. "The Effects of Temperature and Salinity on Marine and Brackish-Water Animals. I. Temperature." *Oceanogr. Mar. Biol. Ann. Rev.* 1:301–340.

————. 1964a. "The Effects of Temperature and Salinity on Marine and Brackish-

Water Animals. II. Salinity and Temperature Salinity Combinations." *Oceanogr. Mar. Biol. Ann. Rev.* 2:281–339.

———. 1964*b*. "Non-Genetic Adaptation to Temperature and Salinity." *Helgoländer wiss. Meeresunters.* 9:433–458.

Kinne, Otto, and Eva-Marie Kinne. 1962. "Rates of Development in Embryos of a Cyprinodont Fish Exposed to Different Temperature-Salinity-Oxygen Combinations." *Can. J. Zool.* 40:231–253.

Klein, L. 1957. *Aspects of River Pollution.* London: Butterworths.

Koga, S. 1966. ["Latitudinal Variation of Surface Water Temperature Optimum to the Fishing of the Albacore."] *J. Shimonoseki Coll. Fish.* 14(3):37–41. (In Japanese.)

Komarova, I. V. 1939. ["Feeding of the Long-Rough Dab in the Barents Sea in Connection with Food Resources."] *Trudý vses. nauchno-issled.* Inst. morsk. ryb. Khoz. Okeanogr., 4:298–320. (In Russian.)

Kropp, B. N. 1947. "The Effect of Temperature on the Rate and Variation of Opercular Movement in *Fundulus diaphanus diaphanus*." *Can. J. Res.* ser. D 25: 91–95.

Kruppert, H. H. B., and M. P. D. Meijering. 1963. "Die Beziehungen zwischen Temperatur und Hell-Dunkel-Farbwechsel bei der Scholle (*Pleuronectes platessa* L.)." *Zool. Anz.* 170:55–61.

Kubo, T. 1936. "Feeding Velocity of the Eel, *Anguilla japonica* (T. & S.), in Relation to Water Temperature and Other Environmental Conditions." *Bull. Jap. Soc. Scient. Fish.* 4:335–338.

Kruoki, T. 1954. "On the Relation Between Water Temperature and the Response for Stimuli. The Investigation to Decide the 'Optimum Temperature.' " *Mem. Fac. Fish. Kagoshima Univ.* 3(2):19–24.

Kusakina, A. A. 1963. "Relation of Muscle and Cholinesterase Thermostability in Certain Fishes to Specific Environmental Temperature Conditions." *Fedn. Proc. Fedn. Am. Socs. exp. Biol.* 22:T123–T126.

Kuthalingam. M. D. K. 1959. "Temperature Tolerance of the Larvae of Ten Species of Marine Fishes." *Curr. Sci.* 28:75–76.

Laberge, R. H. 1959. "Thermal Discharges." *Wat. Sewage Wks.* 106(12):536–540.

Lasker, Reuben W. 1963. "The Physiology of Pacific Sardine Embryos and Larvae." *Rep. Calif. Coop. Oceanic Fish. Invest.* 10:96–101.

———. 1964. "An Experimental Study of Effect of Temperature on the Incubation Time, Development, and Growth of Pacific Sardine Embryos and Larvae." *Copeia* 2:399–405.

Lauff, Geroge H., editor. 1967. *Estuaries.* Publs. Ass. Advmt. Sci., (83).

Leiner, M. 1932. "Die Entwicklungsdauer der Eier des dreistacheligen Stichlungs in ihrer Abhängigkeit von der Temperatur." *Z. vergl. Physiol.* 16:590–605.

Lewis, Robert M. 1966. "Effects of Salinity and Temperature on Survival and Development of Larval Atlantic Menhaden, *Brevoortia tyrannus*." *Trans. Am. Fish. Soc.* 95(4):423–426.

Lloyd, R., and D. W. M. Herbert. 1962. "The Effect of the Environment on the Toxicity of Poisons to Fish." *Instn. Publ. Hlth. Engrs. J.* 7:132–145.

Loeb, Jacques, and W. F. Ewald. 1913. "Die Frequenz der Herztätigkeit als eindeutige Funktion der Temperatur." *Biochem. Z.* 58:177–185.

Loeb, Jacques, and Hardolph Wasteneys. 1912. "On the Adaptation of Fish (*Fundulus*) to Higher Temperatures." *J. Exp. Zool.* 12:543–557.

Lovern, J. A. 1938. "Fat Metabolism in Fishes. XII. Factors Influencing the Comp-position of the Depot Fat of Fishes." *Biochem. J.* 32:1214–1224.

McCauley, R. W. 1963. "Lethal Temperatures of the Developmental Stages of the Sea Lamprey, *Petromyzon marinus.*" *J. Fish. Res. Bd. Can.* 20:483–490.

McCracken, F. D. 1963. "Seasonal Movements of the Winter Flounder, *Pseudo-pleuronectes americanus* (Walbaum) on the Atlantic Coast." *J. Fish Res. Bd. Can.* 20(2):551–586.

McKenzie, R. A. 1934. "Cod and Water Temperature." *Prog. Rep. Atl. Biol. Stn.* (12):3–6.

Mánkowski, W. 1950. ["The Influence of Thermal Conditions on the Spawning of Fish."] *Biul. Morsk. Inst. Ryb. Gdyni* 5:65–70. (In Polish with English and Russian summaries.)

Mantel'man, I. I. 1958. "Distribution of the Young of Certain Species of Fish in Temperature Gradients." *Izv. vses. nauchno-issled. Inst. ozern. rechn. ryb. Kohz.* 47(1):1–63. Fish. Res. Bd. Can., Transl. Ser. 257. Mimeographed.

Marak, Robert R., and John B. Colton, Jr. 1961. "Distribution of Fish Eggs and Larvae, Temperature, and Salinity in the Georges Bank-Gulf of Maine Area, 1953." Spec. Scient. Rep., U.S. Fish and Wildl. Serv.—Fish. (398)

Markowski, S. 1962. "Faunistic and Ecological Investiagtions in Cavendish Dock, Barrow-in-Furness." *J. Anim. Ecol.* 31:45–52.

Massmann, William H., and Anthony L. Pacheco. 1957. "Shad Catches and Water Temperatures in Virginia." *J. Wildl. Mgmt.* 21(3):351–352.

Matsudaira, Yasuo. 1965. ["An Opinion on the Forecast of Fishing Grounds and on the Fundamental Sea States from the Water Temperature Distribution in the Neighboring Sea of Japan."] *J. Fac. Fish. Anim. Husb. Hiroshima Univ.* 6(1):313 321. (In Japanese with English summary.)

Mayer, Alfred G. 1914. "The Effects of Temperature upon Tropical Marine Ani-mals." *Pap. Tortugas Lab.* 6(1):1–24.

Merriman, Daniel, and H. P. Schedl. 1941. "The Effects of Light and Temperature on Gametogenesis in the Four-Spined Stickleback, *Apeltes quadracus* (Mitchill)." *J. Exp. Zool.* 88(3):413–449.

Mews, H. H. 1957. "Über die Temperaturadaptation der Sekretion von Verdauungs-fermenten und deren Hitzresistenz." *Z. vergl. Physiol.* 40:345–355.

Meyer, Heinrich A. 1878. "Beobachtungen über das Wachsthum des Herings in westlichen Theile der Ostsee." *Wiss. Meeresunters.* 3:227–252.

Mihursky, Joseph A., and V. S. Kennedy. 1967. "Water Temperature Criteria to Pro-tect Aquatic Life." *Am. Fish. Soc., Spec. Publ. No. 4*:20–32.

Molnár, Gyula, and István Tölg. 1962. "Relation Between Water Temperature and Gastric Digestion of Largemouth Bass (*Micropterus salmoides* Lacépède)." *J. Fish. Res. Bd. Can.* 19(6):1005–1012.

Morris, Robert W. 1960. "Temperature, Salinity, and Southern Limits of Three Species of Pacific Cottid Fishes." *Limnol. Oceanogr.* 5(2):175–179.

———. 1962. "Body Size and Temperature Sensitivity in the Cichlid Fish, *Aequidens portalegrensis* (Hensel)." *Amer. Nat.* 96:35–50.

Musacchia, X. J., and M. R. Clark. 1957. "Effects of Elevated Temperatures on Tissue Chemistry of Arctic Sculpin, *Myoxocephalus quaricornis.*" *Physiol. Zoöl.* 30(1):12–17.

Nakagome, Jun. 1966. ["On the Causes of Annual Variation of Catch of Bigeyed Tuna in the Tropical Eastern Pacific Ocean. II. On the Monthly and Annual Variation of Surface Water Temperature in Fishing Ground and Relation Between these and the Monthly and Annual Variation of Hook Rate."] *Bull Jap. Soc. Scient. Fish.* 32(10):856–861. (In Japanese with English Summary.)

Naylor, E. 1965. "Effects of Heated Effluents upon Marine and Estuarine Organisms." *Adv. Mar. Biol.* 3:63–103.

Nicholls, John V. V. 1932. "The Influence of Temperature on Digestion in *Fundulus heteroclitus.*" *Contr. Can. Biol. Fish.* 7(5):45–55.

————. 1933. "The Effect of Temperature Variations and of Certain Drugs upon the Gastric Motility of Elasmobranch Fishes." *Contr. Can. Biol. Fish.* 7(36):447–463.

Nicol, J. A. Colin. 1967. *The Biology of Marine Animals.* 2d ed. New York: John Wiley & Sons, Inc.

Nikolsky, Georgi V. 1963. *The Ecology of Fishes.* Transl. from the Russian by L. Birkett. New York. Academic Press.

Ogilvie, D. M., and J. M. Anderson. 1965. "Effect of DDT on Temperature Selection by Young Atlantic Salmon, *Salmo salar.*" *J. Fish. Res. Bd. Can.* 22(2):503–512.

Olivereau, Madeleine. 1955. "Température et Fonctionnement Thyroïdien chez les Poissons." *J. Physiol.,* Paris 47(1):256–258.

Olson, P. A., and R. F. Foster. 1955. "Temperature Tolerance of Eggs and Young of Columbia River Chinook Salmon." *Trans. Am. Fish. Soc.* 85:203–207.

Olson, Theodore A., and Frederick H. Burgess, editors. 1967. *Pollution and Marine Ecology.* New York: Interscience Publ.

Ordal, Erling J., and Robert E. Pacha. 1967. "The Effects of Temperature on Disease in Fish." In *Water Temperature; Influences, Effects, and Control.* Proc. 12th Pacific Northwest Symposium on Water Pollution Research. Portland, Oregon: Fed. Water Pollut. Contr. Admin.

Orr, Paul R. 1955. "Heat Death. I. Time-Temperature Relationships in Marine Animals." *Physiol. Zoöl.* 28(4):290–294.

Orska, Janina. 1956. "The influence of Temperature on the Development of the Skeleton in Teleosts." *Zoologica Polon.* 7(3):272–325.

Orton, J. H. 1920. "Sea-Temperature, Breeding and Distribution in Marine Animals." *J. Mar. Biol. Ass. U. K.* 12:339–366.

Paloheimo, J. E., and L. M. Dickie. 1966. "Food and Growth of Fishes. II. Effects of Food and Temperature on the Relation between Metabolism and Body Weight." *Fish. Res. Bd. Can.* 23(6):869–908.

Pearse, A. S., and Gordon Gunter. 1957. "Salinity." In *Treatise on Marine Ecology and Palaeoecology,* Mem. Geol. Soc. Am. by Joel W. Hedgpeth. 67, 1:129–158.

Pegel', V. A., and V. A. Remorov. 1961. ["Effect of Heating and Cooling of Water on the Gas Exchange and Blood Lactic Acid Concentration of Fish Adapted to a Certain Temperature."] *Nauch. Doki. výssh. shk., Biol. Nauki,* 1:58–61. Translation by Off. Tech. Serv., U.S. Dept. Comm., Washington, D.C.

Picton, Walter L. 1960. "Water Use in the United States, 1900–1980." Bus. Def. Serv. Adm., U.S. Dep. Comm., March 1960.

Powers, Edwin B. 1920. "Influence of Temperature and Concentration on the Toxicity of Salts to Fishes." *Ecology* 1(2):95–112.

Precht. H. 1961. "Beiträge zur Temperaturadaptation des Aales (*Anguilla vulgaris* L.)." *Z. vergl. Physiol.* 44:451–462.

Pritchard, Donald W., and H. H. Carter. 1965. "On the Prediction of the Distribution of Excess Temperatures from a Heated Discharge in an Estuary." Tech. Rep. Chesapeake Bay Inst. (33).

Qasim, S. Z. 1959. "Laboratory Experiments on some Factors Affecting the Survival of Marine Teleost Larvae." *J. Mar. Biol. Ass. India* 1(1):13–25.

Randall, J. E. 1958. "A Review of Ciguatera, Tropical Fish Poisoning, with a Tentative Explanation of its Cause." *Bull. Mar. Sci. Gulf Caribb.* 8(3):236–267.

Raney, Edward C., and Bruce W. Menzel. 1967. "A Bibliography—Heated Effluents and Effects on Aquatic Life with Emphasis in Fishes." *Philad. Elec. Co. and Ichthyol. Assoc. Bull.* (1).

Reibisch, J. 1902. "Ueber den Einfluss der Temperatur auf die Entwicklung von Fisch-eiern." *Wiss. Meeresunters., (N.F.), Abt. Kiel* 6:215–231.

Roots, Betty I., and C. Ladd Prosser. 1962. "Temperature Acclimation and the Nervous Systems in Fish." *J. Exp. Biol.* 39:617–629.

Scholander, P. F., W. Flagg, V. Walters, and L. Irving. 1953. "Climatic Adaptations in Arctic and Tropical Poikilotherms" *Physiol. Zoöl.* 26:67–92.

Senō, Hidemi, Ken-ichi Ebina, and Takuo Okada. 1926. "Effects of Temperature and Salinity on the Development of the Ova of a Marine Fish *Calotomus japonicus* (C. & V.)" *J. Imp. Fish. Inst., Tokyo,* 21(4):44–48.

Sewage Industrial Wastes. 1956. "Heat—A New Pollutant." 28:705.

Shelbourne, J. E. 1956. "The Effect of Water Conservation on the Structure of Marine Fish Embryos and Larvae." *J. Mar. Biol. Ass. U.K.* 35:275–286.

Simpson, R. W., and W. A. Garlow. 1960. "Urban Redevelopment and Thermal Discharge." *Wat. Sewage Wks.,* Nov. 1960 107(2):449.

Sizer, Irwin W. 1936. "Stimulation of *Fundulus* by Oxalic and Malonic Acids and Breathing Rhythm as Functions of Temperature." *J. Gen. Physiol.* 19(5):693–714.

Smith, Dietrich C. 1931. "The Effect of Temperature Changes upon the Pulsations of Isolated Scale Melanophores of *Fundulus heteroclitus.*" *Biol. Bull. Mar. Biol. Lab., Woods Hole* 60(3):269–287.

Smith, F. G. Walton, Robert H. Williams, and Charles C. Davis. 1950. "An Ecological Survey of the Subtropical Waters Adjacent to Miami." *Ecology* 31:119–146.

Strasburg, Donald W. 1958. "An Instance of Natural Mass Mortality of Larval Frigate Mackerel in the Hawaiian Islands." *J. Cons. Perm. Int. Explor. Mer* 24:255–263.

Strawn, Kirk, and James E. Dunn. 1967. "Resistance of Texas Salt- and Freshwater Marsh Fishes to Death at Various Salinities." *Tex. J. Sci.* 19(1):57–76.

Stroganov, N. S. 1956. "Physiological Adaptation of Fish to Environmental Temperature Conditions." *Moskva,* Izdatel'stvo Akad. Nauk SSSR. Translated by M. Roublev for the National Science Foundation, Washington, D.C., 1962.

Stroud, Richard H., and Philip A. Douglas. 1968. "Thermal Pollution of Water." *Bull. Sport Fish. Inst.* (191):1–8.

Sullivan, Charlotte M. 1954. "Temperature Reception and Responses in Fish." *J. Fish. Res. Bd. Can.* 11(2):153–170.

Sumner, F. B., and Urless N. Lanham. 1942. "Studies of the Respiratory Metabolism of Warm and Cool Spring Fishes." *Biol. Bull. Mar. Biol. Lab.. Woods Hole* 82(2):313–327.

Sverdrup, Harald U., Martin W. Johnson, and Richard H. Fleming. 1942. *The Oceans, their Physics, Chemistry, and General Biology.* New York: Prentice-Hall, Inc.

Sweet, John G., and Otto Kinne. 1964. "The Effects of Various Temperature-Salinity Combinations on the Body Form of Newly Hatched *Cyprinodon macularius* (Teleostei)." *Helgoländer wiss. Meeresunters.* 11(2):49–69.

de Sylva, Donald P. 1963. "Systematics and Life History of the Great Barracuda, *Sphyraena barracuda* (Walbaum)." *Stud. Trop. Oceanogr., Miami* 1:1–179.

————. 1968. Statement on thermal pollution before the Subcommittee on Air and Water Pollution of the Senate Committee on Public Works, Miami, Florida, April 19, 1968. *Thermal Pollution: 1968(Part 2)*: 768–775.

de Sylva, Donald P., Frederick A. Kalber, and Carl N. Shuster, Jr. 1962. "Fishes in the Shore Zone and Other Areas of the Delaware River Estuary." *Inf. Ser. Univ. Delaware, Publ.* (5):1–164.

Tagatz, Marlin E. 1961. "Tolerance of Striped Bass and American Shad to Changes of Temperature and Salinity." Spec. Scient. Rep., U.S. Fish Wildl. Serv.—Fish, (388).

Talbot, Gerald B. 1966. "Estuarine Environmental Requirements and Limiting Factors for Striped Bass." Am. Fish. Soc., Spec. Publ. No. 3:37–49.

Tarzwell, Clarence M., and A. R. Gaufin. 1958. "Some Important Biological Effects of Pollution often Disregarded in Stream Surveys." *Proc. 8th Purdue Ind. Waste Conf.*

Tat'yankin, Yu. V. 1966. ["Upper Temperature Threshold of Cod and Pollack Fry and its Dependence on the Adaptation Temperature."] *Dokl. Akad. Nauk SSSR* 167(5):1159–1161. (In Russian.)

Tawara, S., and A. Tsuruta. 1966. ["Water Temperature Relevant to Distribution of Zooplankton Biomass in the Tuna Fishing Ground in the Eastern Tropical Pacific."] *J. Shimonoseki Univ. Fish.* 14(3):43–57. (In Japanese with English summary.)

Taylor, Clyde C., Henry B. Bigelow, and Herbert W. Graham. 1957. "Climatic Trends and the Distribution of Marine Animals in New England." Fish. Wildl. Serv. U.S., *Fishery Bull.* 57:293–345.

Timet, Dubravko. 1963. "Studies on Heat Resistance in Marine Fishes. I.Upper Lethal Limits in Different Species of the Adriatic Littoral." *Thalassia jugosl.* 2(3):5–21.

Tremblay, J.-L. 1942. "La Pêche de la Morue et la Température d'Eau de Mer." *Prog. Rep. Atl. Biol. Stn.* (33):7–9.

Trembley, F. J. 1960. "Research Project on Effects of Condenser Discharge Water on Aquatic Life." Inst. Research, Lehigh Univ., *Prog. Rep., 1956–1959.*

————. 1961. "Research Project on Effects of Condenser Discharge Water on Aquatic Life." Inst. Research, Lehigh Univ., *Prog. Rep.*

————. 1965. "Effects of Cooling Water from Steam-Electric Power Plants on Stream Biota." *Biological Problems in Water Pollution,* Edited by C. M. Tarzwell. *Publ. Hlth. Serv. Publ. No. 999-WP-25:334–345.*

U.S. Senate Committee on Public Works. 1968. "Thermal Pollution—1968." Hearings before the Subcommittee on Air and Water Pollution of the Committee on Public Works, Pts. 1 and 2. United States Senate, Nineteenth Congress. Washington, D.C.: U.S. Government Printing Office.

Voss, Gilbert L., and Nancy A. Voss. 1955. "An Ecological Survey of Soldier Key, Biscayne Bay, Florida. *Bull. Mar. Sci., Gulf Caribb.* 5:203–229.

Waede, M. 1954. "Beobachtungen zur osmotischen, chemischen und thermischen Resistenz der Scholle (*Pleuronectes platessa*) und Flunder (*Pleuronectes flesus*)." *Kieler Meeresforsch.* 10:58–67.

Warinner, J. E., and M. L. Brehmer. 1966. "Effects of Thermal Effluents on Marine Organisms." *Proc. 19th Industr. Waste Conf. Purdue Univ., Engng. Ext. Serv.* (117):479–492.

Wastes Engineering. 1961. "Is Heat Pollution a Threat to Fish Life?" 32(7):348.

Wells, Nelson A. 1935a. "The Influence of Temperature upon the Respiratory Metabolism of the Pacific Killifish, *Fundulus parvipinnus*." Physiol. Zoöl. 8:196–227.

――――. 1935*b*. "Variations in the Respiratory Metabolism of the Pacific Killifish, *Fundulus parvipinncs*, due to Size, Season, and Continued Constant Temperature." *Physiol. Zoöl.* 8:318–336.

Williamson, H. C. 1909. "Experiments to Show the Influence of Cold in Retarding the Development of the Eggs of the Herring (*Clupea harengus* L.), the Plaice (*Pleuronectes platessa* L.) and the Haddock (*Gadus aeglefinus* L.)." *Rep. Fishery Bd. Scotl.* 27:100–128.

――――. 1911. "Experiments in Retarding the Development of Eggs of the Herring." *Rep. Fishery Bd. Scotl.* 28(3):16–23.

Wuhrmann, K., and H. Woker. 1955. "Influence of Temperature and Oxygen Tension on the Toxicity of Poisons to Fish." *Verh. int. Verein. theor. angew. Limnol.* 12:795–801.

Wurtz, C. B. 1961. "Is Heat a New Pollution Threat?" *Wastes Engng.* 32(12): 684–686; 706.

Yamazaki, Masao. 1965. "Problems in Using Sea Water for Condenser Cooling in Thermal Power Stations." *Adv. Wat. Pollut. Res.* 3:117–132.

Chapter **10**   R. E. Nakatani

EFFECTS OF HEATED DISCHARGES
ON ANADROMOUS FISHES

THE physical and biological aspects of Columbia River water temperatures have been studied at Hanford for more than 20 years. Research in the first decade centered on the consequences of radioactive materials in reactor effluent and the relationships between reactor-production efficiency and river temperatures (Davis, Watson, and Palmiter, 1956). The emphasis in early years on the radiological aspects, rather than the heat aspects, was necessitated by the paucity of radiological information. However, the effects of elevated temperatures, particularly on salmonids (Nakatani and Foster, 1966), were always studied along with the radioactivity and chemical-toxicity features of the effluent.

While information on water-temperature problems has accrued at Hanford over the years, only relatively recently has national attention been focused on the biological effects of heated discharges into aquatic systems. The growth of the U.S. population, concomitant with demands for more electric power, the advent of nuclear power plants, the passage of the Federal Water-Quality Act of 1965 are all important factors in our national concern about water quality and preservation of aquatic environments.

Investigations on the effects of water temperature on aquatic organisms,

This paper is based on work performed under U.S. Atomic Energy Commission Contract AT(45–1)1830. The work of over ten principal investigators associated with aquatic ecology at Hanford is gratefully acknowledged. Their findings have been freely used in this review. The fine co-operation of the Washington and Oregon State Fishery Agencies and the U.S. Fish and Wildlife Service made many of these studies possible. J. H. Johnson, Bureau of Commercial Fisheries, was especially helpful on the sonic-tagging experiments.—R.E.N.

294

including anadromous fishes, have generated much interest and research activity. Recently, V. S. Kennedy and J. A. Mihursky (1967) published a bibliography listing 1,220 references on the effects of temperature in the aquatic environment. E. C. Raney and B. W. Menzel (1967) provided a further list. H. Harty, *et al.* (1967), reported on nuclear power-plant-siting problems in the Pacific Northwest in a voluminous report for the Bonneville Power Administration. They evaluated possible sites from geological, hydrological, seismological, meteorological, socio-economical, and ecological standpoints. J. R. Brett (1956, 1960) has excellent review articles on fish thermal requirements, particularly for salmonids. R. E. Burrow (1963) reported on optimal temperatures for maximum productivity of salmon. Very likely, no other single environmental parameter in the aquatic environment has received so much attention by aquatic biologists.

Despite this wealth of biological information on temperature effects, a salmon-biologist faced with the pressing pragmatic problem of predicting or assessing the biological costs or benefits to a fishery resource of adding some small heat increment to a specific river system encounters an extremely complex and difficult task. He finds gaps in our knowledge which do not, in a practical sense, permit him to integrate and apply laboratory and field findings to his particular problem. Part of the difficulties arise because many excellent studies were not specifically designed to help resolve complex field problems in a specific river ecosystem. Further, the extrapolation of laboratory findings to field situations always poses problems. The biologist may find field data related to thermal plants located in another river system helpful (e.g., a steam-electric plant on the Delaware River), but not representative of a salmon-river. When he turns to fundamental biologists for basic biochemical, physiological, behavorial, and ecological information, he finds insufficient information available. The general effects of increased water temperatures on fish are well known. Increases in metabolic rates and oxygen requirements, reduction in stamina, added sensitivity to toxic materials and fish diseases are but some of the effects (usually inferred from laboratory experiments) associated with increased temperatures; but these generalizations in our present state of knowledge are no substitute for extensive on-site studies and evaluations. Thus, the practical investigator is likely to use the direct observational approach of investigating the effects of elevated temperatures on salmon or other desired species "on site", using local fish and on-site water. Essentially, this approach has been used at Hanford.

The purpose of this paper is to review some of the highlights of biological research at Hanford on the effects of elevated water temperatures, especially on salmonids. About 30 species of freshwater fishes have been identified in the Hanford reach. They are all exposed to reactor effluent in varying degrees, but our research has emphasized the Pacific salmon, *Oncorhynchus,* and rainbow trout, *Salmo gairdneri,* because of their sensitivity to warm waters and their well-known dominant economic importance to commercial and sport fisheries.

Hanford Reactors and the Biological Problem

One of the primary reasons Hanford reactors were located on the Columbia River was the availability of large volumes of cold river water for reactor-coolant. Potential hazards to fishery resources caused by discharging heated effluents into the Columbia River were recognized before the reactors were built (Hines, 1962). As early as 1945, young salmonids were reared, from "green" eggs to fingerling size in different concentrations of reactor effluent, to observe their growth and mortality. Three characteristics of reactor effluent are of major biological importance. These are heat, radioactivity, and certain chemicals (mostly sodium dichromate, used routinely as a corrosion-inhibitor in reactor-cooling tubes). A review of the research on the radioactivity and chemical toxicity features of effluent is beyond the scope of this paper, but it is noteworthy that biological reasons were largely responsible for modifying reactor operations to reduce the amount of phosphorous-32 and sodium dichromate discharged in effluent water.

Figure 1 shows the Hanford Reservation with the reactor areas along the right bank of the river near the northern boundary of the 600-square-mile reservation. Our main experimental hatchery and wet laboratory is located at 100-F area, downstream from all reactors, and uses untreated Columbia-River water; hence it functions as a "built-in" continuous monitoring station. Thus, our hatchery fish are exposed to effluent-contaminated river water, and the stress of any major release would presumably wipe out our stock. For intensive studies of effluent, we have another smaller hatchery located at a reactor farther upstream. In this facility, various dilutions of effluent are used for rearing salmon and trout.

Detailed information about Hanford's reactor effluent or the actual heat contribution is classified, and the public and the fishery agencies receive only the strong impression of "thermal pollution" from observing

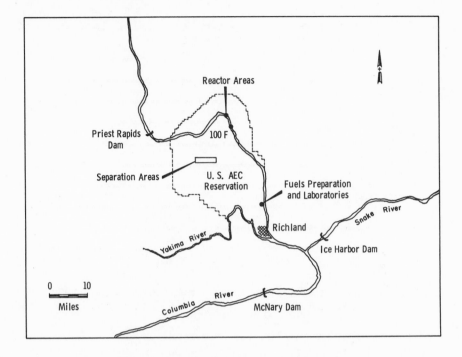

Fig. 1—.U.S. AEC: Richland operations.

Photograph by Pacific Northwest Laboratory, operated by Battelle Memorial Institute for the U.S. Atomic Energy Commission

the towering columns of steam which rise several hundred feet above each retention basin in the Hanford reactor areas. However, a study of the heat contributed by Hanford reactors to the Columbia shows immediately that Hanford adds only a relatively small heat increment to the widely-variable seasonal river temperature. The history of the operation of Hanford reactors (summarized in Figure 2) shows a total of six reactors operating from 1944 to 1955, eight from 1955 to 1964, and the maximum of nine during 1964. At present, there are only four reactors operating.

The biological problem is to evaluate the effect of Hanford's heat increment on fishes, especially on the valuable salmon populations. This problem defies a direct, definitive answer, and much of the evidence on Hanford effect or lack of effect is circumstantial or inferred from limited field studies, laboratory experiments, and the literature. Although Hanford temperature effects are emphasized because of the theme of this meeting, it must be remembered that many other complex factors control the health

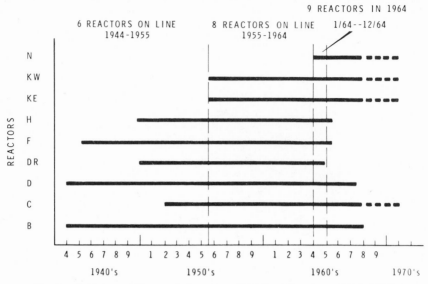

Fig. 2——.Initial operation and closures of Hanford Reactors, 1944–1968.

Photograph by Pacific Northwest Laboratory, operated by Battelle Memorial Institute for the U.S. Atomic Energy Commission

of the salmon populations. Hanford temperature effects need to be kept in context with the major factors which contribute to the mortality of the salmon throughout its life cycle and recent history.

River Temperatures

Before we discuss the temperature effects of reactor effluents, we should examine water use upstream from Hanford. On the main stem of the Columbia River, upstream from Hanford, there are seven dams: Priest Rapids, Wanapum, Rock Island, Rocky Reach, Wells, Chief Joseph, and, finally, Grand Coulee. Obviously, variation in the methods of operation of these dams, especially the pattern of daily and seasonal discharges, greatly influences the base river temperatures experienced at Hanford and also the temperature regimes below Hanford.

The notion that dam construction invariably raises the water temperature of the Columbia is questionable, if not erroneous. R. T. Jaske and J. B. Goebel (1967) have shown that the erection of low-head reservoirs on the main stem of the Columbia River has not produced significant changes in the average temperature of the river, and that dams and reservoirs actually decrease the expected seasonal range of water temperature. Their review of the history of Columbia River temperatures shows that, with the advent of dams and multiple water use, there has been a supression of the maximum temperatures during the summer months, an elevation of the minimum temperatures during the winter months, and little change in the annual mean temperature. Further, they demonstrated that construction of impoundments has created a shift in the timing of the seasonal temperature cycle towards the later months by lengthening the time of water passage. The maximum temperatures formerly observed in early August are shifted about 30 days and are now seen in early September. This shift has caused a gradual increase in the fall temperatures of the main-stem river, and a corresponding decrease in spring temperatures.

Temperature Effects on Adults—Transients and Residents

Both salmon and steelhead trout (*Salmo gairdneri gairdneri*) spawn in fresh water, and the progeny spent varying periods in fresh water before migrating to sea, where they obtain most of their growth, returning as mature adults to their home stream to start the cycle again. Dams have inundated most of the spawning grounds in the main stem of the Columbia, but a small spawning area still remains in the only free-flowing stretch of about 42 miles, from Priest Rapids Dam to Ringold (15 miles upstream from Richland). Some 33 miles of this reach are within the confines of the Hanford Reservation. Virtually all of the salmon spawning now occurs in streams tributary to the Columbia, and thus most adult fish entering the Hanford reach are transients bound for areas upstream from Hanford. Anadromous species in the Hanford reach are chinooks (*O. tshawytscha*), sockeye (*O. nerka*), silver or coho (*O. kisutch*), steelhead (*Salmo gairdneri gairdneri*), shad (*Alosa sapidissima*), white sturgeon (*Acipenser transmontanus*), and Pacific lamprey (*Lampetra tridentata*).

In a particular year, 50,000 to 100, 000 adult sockeye salmon migrating to spawning areas in Osoyoos Lake and Lake Wenatchee may pass the Hanford reach during July. A study by the Fish Commission of Oregon and the Washington Department of Fisheries (1967) of the sockeye counts at different dam sites as the fish progress upstream does not indicate delays in the Hanford reach. Most of the sockeye counted over McNary Dam (below Hanford) are accounted for by summation of subsequent counts at Ice Harbor Dam (Snake River) and Priest Rapids Dam (above Hanford). On the other hand, counts of summer-run steelhead at these dam sites show significantly high and, as yet, unaccountable losses above McNary Dam. The welfare of this summer steelhead population is of deep concern to fishery agencies.

A direct approach to determine the migration behavior of adult salmon

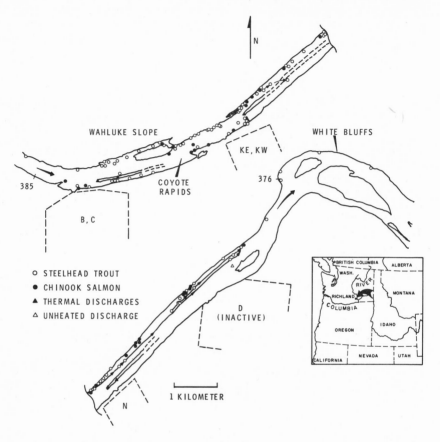

Fig. 3—.Observed locations of sonic tagged steelhead trout and chinook salmon, August-September 1967

Photograph by Pacific Northwest Laboratory, operated by Battelle Memorial Institute for the U.S. Atomic Energy Commission

and steelhead trout near reactor plumes in the Hanford reach was started in the fall of 1967 in cooperation with the Bureau of Commercial Fisheries (Coutant, 1968). A sonic fish tag developed by the Bureau was used to track fish movements. The tag emits a high-frequency signal which can be monitored with hydrophones mounted on a boat. Some 70 adult salmon and steelhead trout were tracked in the Hanford reach. The data showed no temperature block of fish, but most fish migrated upstream in relatively shallow waters (4 to 9-foot depth) along the opposite bank from reactor sites (Figures 3 and 4).

At a recent meeting of the Ben Franklin Technical Committee of

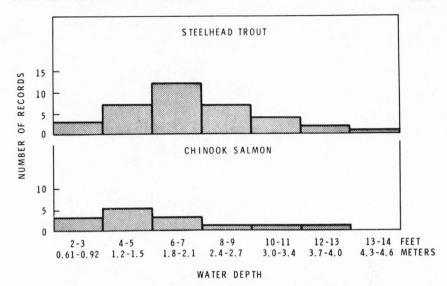

Fig. 4——.Frequency of water depths recorded for upstream-migrating chinook salmon and steelhead trout, August-September 1967.
(Depth data not taken for all fish.)

Photograph by Pacific Northwest Laboratory, operated by Battelle Memorial Institute for the U.S. Atomic Energy Commission

FWPCA, engineers discussed the presence of a cooler "wedge" of water along the shore opposite the reactors and speculated that this cooler "wedge" provides the route for migrating salmon. However, our data show that fall chinook salmon are not limited to the cool wedge. Even when the number of reactors discharging effluent into the river was much greater than it is at present, spawning occurred year after year near the reactor side of the river in waters in which complete mixing of the effluent had not taken place. The implication that the routes of travel of migrating salmon is influenced solely, or nearly so, by water temperatures in the Hanford Reservation is not supported by field observations. Since it is known that salmon are extremely responsive to variance in water velocity, an effort will be made next season to obtain detailed velocity measurements on the migration route along the banks.

An aerial census of fall chinook spawning in the Hanford reach has been made annually since 1947 (Watson, 1968). These records of the number of salmon nests observed each year for over 20 years are perhaps the best circumstantial evidence of the continued viability of the fall chinook population that spawns within the Hanford Reservation. Of the total escapement of fall chinook salmon above Bonneville Dam, about 6 to 13 percent

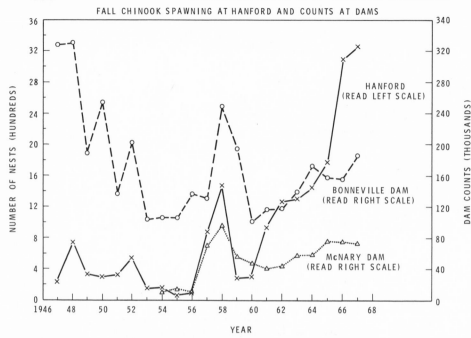

Fig. 5——. Fall chinook spawning at Hanford and counts at dams.

Photograph by Pacific Northwest Laboratory, operated by Battelle Memorial Institute for the U.S. Atomic Energy Commission

spawn here. This race of chinook salmon enters the Columbia and migrates to its spawning grounds during the late summer, when the river temperature is near its annual peak. The eggs and young are certainly much more subject to any effects of the Hanford Plant than races that spawn elsewhere. There is no evidence, however, that the local race has been adversely affected. On the contrary, the number of salmon nests observed has increased from about 300 annually, prior to 1950, to a record number of over 1,700 in 1965, 3,100, in 1966, and to a 21-year record high of 3,300 in 1967 (Fig. 5). This increase in spawners is most likely related in part to spawners displaced when Priest Rapids Dam was completed in 1959. It is interesting that we have seen a continual increase for the last eight years. Five of these years coincide with maximum reactor operation. This increase in local spawners has occurred during a period of years when the runs of fall chinooks to other parts of the river system, e.g., the Snake River, have declined appreciably.

Not only do we have a census of spawners, but we also have annual records on the distribution of the nests in the Hanford reach. Figure 6 sum-

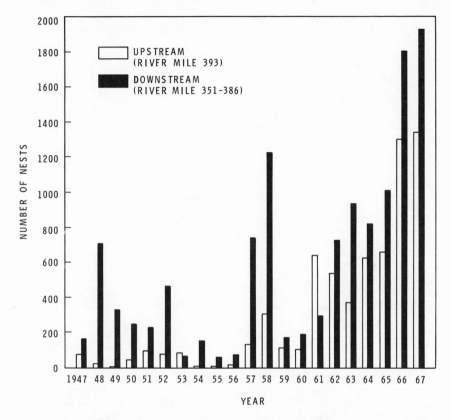

Fig. 6—.Fall chinook salmon spawning, upstream and downstream from reactors.

Photograph by Pacific Northwest Laboratory, operated by Battelle Memorial Institute for the U.S. Atomic Energy Commission

marizes the number observed in areas above and below the reactors. Before 1959, the area above the reactors had an average of 17 percent of the total count; but after 1959, this increased to 42 percent. The extent and distance of "fall-back" from the dam is unknown, but it could also be responsible for increased spawning below the reactors. In other words, the greater utilization of the Hanford reach for spawning may be related to unfavorable spawning areas upstream or downstream from Hanford. Nevertheless, for over 20 years, the Hanford reach has been an acceptable area for salmon spawning.

It is well known that a female adult salmon is normally highly sensitive to the environmental requirements necessary for successful nest-building, spawning, and incubation of eggs. Much work has been done to under-

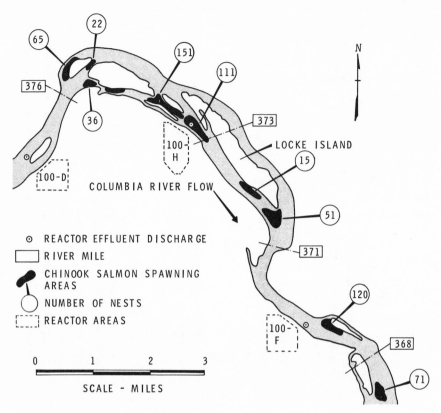

Fig. 7——.Distribution of chinook spawning in relation to reactor effluent outfalls, 1962.

Photograph by Pacific Northwest Laboratory, operated by Battelle Memorial Institute for the U.S. Atomic Energy Commission

stand these requirements, pertaining to gravel size, water flow, and water temperatures, especially in the interests of construction of artifical spawning channels (Chambers, 1964; Allen and Moser, 1967). Since spawning normally occurs only on a limited range of sites, we have considered the possibility of displacement of our local spawners in the Hanford reach due to reactor plumes. No data on the distribution of spawning areas prior to the Hanford operations are available, but a comparison between the distribution of the spawning areas during operations of eight reactors in 1962 and four reactors in 1967 should indicate some measure of the influence of reactor plumes on spawning behavior.

Figure 7 shows the distribution of salmon nests of fall chinooks counted by aerial survey in 1962. A total of 712 nests (shown in Figure 6) are clearly in Hanford effluent zones. The main river channel between the 100-H outfall (river-mile 371 to 373) and the end of Locke Island has attracted salmon spawners from 1947 to the present. During much of this time, these fish were in a zone of incompletely mixed effluent from 100-H reactor. Similar observations have been made in the river immediately below 100-D and 100-F, during periods when these reactors were in operation. The distribution shown is typical of many years of observation. It is of interest to note that last fall, when 100-D, 100-H, and 100-F were no longer operating, the distribution in this reach showed little change. The reactor plumes apparently are not a significant factor in the distribution of the spawning areas in this reach. The spawning areas appear to be determined more by gravel size, current velocity, and water depth.

Temperature Effects on Eggs and Young Salmon Resident in the Hanford Reach

Our success in rearing young salmonids in reactor effluent is reported in over 20 A.E.C. reports issued in years since 1946. This work provides estimates of critical concentrations and gave some assurance that the average concentrations of reactor effluent found in the river will not induce growth depression or cause excessive mortality. These laboratory tests have been continued through the years because of variations in upstream water use and technical changes in reactor operations. Usually effluent concentrations greater than 4 percent resulted in excessive mortality, but acceleration of growth occurred in all concentrations up to about 6 percent because the effluent raised water temperatures. When concentrations exceeded about 7 percent effluent, growth depression occurred due to toxicity of hexavalent chromium in the effluent. No radiation damage could be demonstrated at these levels of effluent concentrations.

Because the eggs and young of salmon are especially sensitive to warm waters, considerable research has been done at Hanford on testing the biological response of young salmon and trout reared under various heat increments above base river temperatures (Olson and Foster, 1955; Olson and Nakatani, 1968). For example, during the 1966–67 season, young chinook salmon were reared for over five months under seven temperature regimes. Each treatment followed the Columbia River seasonal pattern, with an increment of 2° F. over the next lower temperature, giving an over-all range of 12° F. The effects on growth and mortality are summarized in Table 1. Temperature increments exceeding 4° F. above the base river temperature caused excessive mortality, i.e., greater than 20 percent, in fish hatched from spawn collected in October. The Decem-

TABLE 1. Mortality and Growth of Young Chinook Salmon Reared Under Elevated River Temperatures

Date Spawned		10/30/66		11/14/66		11/23/66		12/8/66		
Date Terminated		4/24/67		5/9/67		5/18/67		6/2/67		
Initial Base Temp. °F		56		54		53		48		
Lot	Temperature Condition	% Mortality	Mean wt, g	% Mortality	Mean wt, g	% Mortality	Mean wt, g	% Mortality	Mean wt, g	
1	Control[a]	4.6	0.95	10	0.63	5.5	0.64	4.8	0.63	
2	+2 °F	3.6	1.40	11	0.96	9.3	0.97	5.5	0.97	
3	+4 °F	11	1.92	16	1.53	12	1.37	7.5	1.42	
4	+6 °F	28	2.72	19	2.55	9.4	1.96	7.3	1.79	
5	+8 °F	60	4.28	17	4.17	6.9	2.61	17	2.31	
6	+10 °F	97	—	43	5.52	10	3.82	14	3.18	
7	+12 °F	100	—	93	5.65	37	5.28	12	4.24	

[a]Similar to temperature at Priest Rapids Dam

ber spawn tolerated an incremental addition of as much as 12° F. without significant mortality. In the river, however, most of the spawning is over by mid-November, so there are few December spawners. The particular heat increments tested in the laboratory are not important in themselves, but serve as a demonstration that more heat can be added during late fall and winter seasons without significant detrimental effects. Furthermore, they show that warmer waters favor the survival of young salmon from the later spawners of fall-run chinook salmon.

Increased water temperatures clearly accelerated fish growth in all lots. Fish in the warmest lots were as much as eight times heavier than those in the coldest lots at the conclusion of the test. Do larger-sized migrants have better survival? There have been several extensive studies which attempt to answer this question (Wallis, 1968). The general conclusion has been that there is some evidence that larger downstream migrants do in fact have higher survival rates if other conditions are equal.

Most of the local fall chinook spawn in water temperatures of from 50° to 59° F., but the late spawners in a cold year may deposit their eggs in temperatures of about 41° F., well below the optimum. The importance of the time of deposition of eggs or spawning of all chinook salmon was indicated by the mortalities of spawn taken in December 1965. Eggs incubated in river water in 1965 showed higher mortalities than did those incubated in water warmed by addition of reactor effluent. Therefore, warming the colder waters of early winter might actually be beneficial for egg incubation.

From November 1954 to May 1955, temperatures were measured in the river-bottom gravel of a simulated salmon nest approximately ½ mile downstream from 100-F reactor effluent outfall (Watson, 1968). This location was chosen to obtain information on the thermal contribution of incompletely mixed reactor effluent. Temperatures were taken at positions 6 inches above the river bottom and at 6 inches, 15 inches, 21 inches, and 30 inches below the surface.

During the period of salmon incubation (November-February), there was seldom more than 1° C. (1.8° F.) difference between the river gravel and the water above. There was generally a lag (and this was substantiated the following year with continuous temperature records) in temperature change between the water below and above the river bottom; and temperatures ranged from 1° to 4° C. (1.8° to 7.2° F.) higher in the simulated nest than in the river immediately upstream from the effluent outfall. It should be borne in mind that these measurements were made in an area of incompletely mixed effluent and are not necessarily typical of salmon spawning areas near Hanford.

Temperature Effects on Seaward Migrants

Information on the distribution (in time and space) of the young seaward migrants passing through the Hanford reach is necessary for assessment of effects of Hanford discharges. A question is often raised about the possibilities of seaward-migrant kills in reactor plumes. To determine whether this is a source of mortality, we need to know the distribution and behavior of the fish in relation to the distribution of water velocity and temperature of the plumes.

The most thorough work to date on distribution and movement of seaward migrants passing Hanford is reported by E. M. Mains and J. M. Smith (1964) of the Washington State Dept. of Fisheries. The work was done with large fyke nets rigged to five different river barges located on a transect across the Columbia River at Byer's Landing (downstream about 25 miles from 100-F). They found the seaward migration of chinook salmon occurred between March and July, during the high flows. A peak in April was mostly fry or 0-age chinooks (mean length, 39 mm), while peaks in June and July were largely yearlings (mean length, 72 mm in June).

Although many fish were taken at all stations, 50 to 60 percent of the catch per unit effort occurred near the banks. The seaward migrants also exhibited a preference for the surface zone. About 44 percent of the catch occurred within 30 inches' depth from the surface.

Currently, we are testing the possibilities of kills in reactor plumes by drifting juvenile chinook salmon, 0-age) in live-boxes through the plumes and warmed shoreline areas. Water temperature is recorded continuously. Immediate and latent mortalities are scored. An inclined plant scoop-trap is also being used in the river downstream from a reactor outfall, to sample the natural run of seaward migrants. In both the live-box drifts and the trap collections, no mortalities attributable to heat have been observed this spring.

This lack of termal mortalities is not surprising, because high water temperatures were not experienced by the fish. The water temperatures observed during the spring migration of 1968 at the fish-trap anchored about 400 meters downstream in a center of a reactor discharge plume showed a range of 51° to 60° F.

Lethal temperature ranges for the local stock of young chinook salmon are being studied in the laboratory to define the temperatures and exposures which may cause mortalities (Dean and Coutant, 1968). Some of the questions posed are as follows:

1. How does the resistance to thermal shock of local chinook compare with published data?
2. How does the developmental temperature regime affect the response to thermal shock?
3. Is fish size a significant factor?
4. What are the relationships between initial loss of equilibrium and times to physiological death under thermal stress?

Our preliminary laboratory work suggests that local chinooks may be slightly more sensitive to heat than those studied by Brett (1956, 1960). The thermal history of rearing does not seem to change thermal resistance significantly, at least for the ranges studied. Within the range of weights (1.6 g to 10.8 g) studied, size had little effect. Average loss of equilibrium occurs at a predictable time prior to death; however, studies on individual fish were inconclusive. Loss of equilibrium may be a more valid and useful ecological death point than the commonly used physiological death.

Our laboratory work on lethal temperatures is directed toward providing empirical data for developing a model to predict the likelihood of fish mortalities under fluctuating thermal conditions encountered in a heated-discharge plume. Initial attempts to construct a computer model simulating fish death in fluctuating temperatures have been encouraging, but much more work needs to be done on the model (Coutant and Cole, 1968).

Apparently, Hanford temperatures have little effect on the downstream Chinook migrants due to the relationship between time of migration past Hanford and prevailing river temperatures. Two waves of downstream migration occur between March and late July, during a period of rising river temperatures from about 43° to 60° F. (6° to 15.5° C.) Several factors help to temper any Hanford thermal effect in the spring:

1. Bulk of the migration occurs during the spring months during cool waters.
2. Some fish migrate later in the season during warmer but still satisfactory temperatures.
3. The yearly freshet, which peaks during late May or early June, helps to minimize any Hanford thermal effects at this season. The critical tests must be carried out in late August and September, during the warm-water period.

Columnaris: A Warm-Water Fish Disease

Our columnaris research, initiated in 1959, deals with the possible aggravation of columnaris disease by the introduction of reactor effluent into the Columbia River. Columnaris has been endemic in the river for many years prior to Hanford operations, and the etiological agent, *Chondrococcus columnaris,* was positively identified in the river system in 1942 (Fish and Rucker, 1943). Because reactor effluent is thermally hot and radioactive, some biologists have speculated that Hanford might be a serious focal point for the columnaris disease (Ordal and Pacha, 1963). Some of the speculations ran as follows:

1. Radioactivity and warm waters might seriously aggravate the columnaris problem and induce higher mutation rates which could lead to more virulent strains of columnaris.
2. Initial infection of fish and development of the disease could occur in the warm water sloughs of the Hanford Reservation.
3. Columnaris may be a serious threat to the salmon populations and could be responsible for the declining runs.

Early work in our laboratory showed that columnaris exposed to radiation levels many orders of magnitude higher than the radiation levels in undiluted reactor effluent did not change mutation rates or increase in virulence (Fujihara, Olson, and Foster, 1960). Mutants of columnaris were induced readily by a 10kR X-ray dose; however, the virulence of these mutants was no greater than the parent strain.

Monthly field sampling of coarse fish, mostly large-scale sucker (*Catostomus macrocheilus*), for an entire year at sites upstream and downstream from Hanford provided evidence that coarse fish serve as reservoirs for columnaris, especially during the winter months, when the disease is apparently dormant. We found the number of "incidences" (positive, but fish not in a diseased state) and diseased fish at Hanford differed little from other sites on the main stem of the river. Incidence ranged from 6 to 9 percent and diseased fish ranged from 1.9 to 2.2 percent for all sites in 1965.

Although high temperatures are often associated with columnaris, experience with our hatchery trout (maintained throughout the year in Columbia River waters) indicates the further complexity of the problem. Columnaris disease is well-established in the hatchery when the water temperatures reach 63° to 64° F. (17° to 18° C.) in early July. Fingerlings suffer mortalities for a few weeks, but losses then taper off sharply, and few deaths occur, despite the continual rise in the river-water temperature to about 70° F. (21° C.) in late August. Serious problems may arise, however, when salmon are confined and crowded in warm waters, as exemplified by the considerable loss suffered at Oxbow Dam in the Snake in 1961. Laboratory work has also demonstrated that severe crowding and high temperatures can cause serious columnaris mortalities.

Our recent studies suggest those fish-ladders at Columbia River dam sites which have a heavy fish population may be major focal points of columnaris exposure and infection. The resident coarse fish in ladders showed 87 percent "incidence" and 24 percent infection. These ladders present a potentially serious columnaris disease problem for anadromous salmon as they pass through ladders heavily saturated with the pathogen. Monthly water samples taken near the top and bottom of fish-ladders at four sites—Bonneville, McNary, Priest Rapids, and Rocky Reach Dams—served to measure the incidence of the pathogen in the water before, during, and after the warm-water season. Figure 8 summarizes the results.

A columnaris population becomes evident in the ladder when water temperature rises above 50° F. (10° C.) and starts to decline when the temperature begins to decrease. The counts represent organisms released

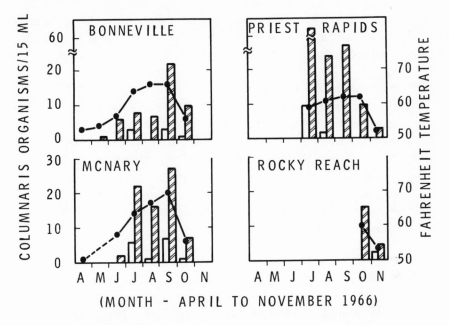

(MONTH - APRIL TO NOVEMBER 1966)

1. ☐ SAMPLES NEAR TOP OF LADDER

2. ▨ SAMPLES NEAR BOTTOM OF LADDER

Fig. 8—.Columnaris population in fish ladders.

to the water by infected fish in the ladder, rather than organisms entering the ladder from the river. This is demonstrated by the large increase of organisms in the lower section of the ladder when compared to the upper section.

Our basic laboratory studies on antibody production and immune response of rainbow trout to *C. columnaris* provided the basis for the development of a technique of measuring the agglutinating antibodies in fish-blood sera against columnaris (Fujihara, Olson, and Nakatani, 1965). Measurements of agglutinating titers in blood sera of over 900 suckers sampled at four sites—Hanford, McNary Dam, Wenatchee, and Bonneville Dam—showed a single cyclic pattern during a year. Residual antibodies are carried by the fish during the cooler waters of winter and spring, but a sharp increase in the average agglutinating titers occurs during September to November, the warm-water period. Last year, both the downstream- and upstream-migrant sockeye salmon were sampled along their migration routes. The downstream migrants showed little change in titers during their migration in water temperatures of 45° to 57° F. during April to June; however, the adults migrating in warmer waters of 51° to 66° F. during the summer showed progressively higher titers with upstream migration (Fujihara, 1968).

These columnaris studies illustrate the probable critical significance of fish-ladders in the etiology of the columnaris disease in the Columbia River. Maximum concentration of columnaris organisms occurs in ladders during periods of warm-water temperature when concentration of resident fish and passage of anadromous salmon through the ladders are at a peak. These salmonids are not only repeatedly subjected to columnaris exposure at each dam site, but may constitute a major vector for spread of the disease throughout the river system.

Discussion and Summary

In discussing the effects of Hanford's heated discharges on salmonids, it is convenient to think of the Hanford reach as two areas, namely the area of the reactor plumes and the area of "mixed" zones farther downstream. While the temperature of the undiluted reactor effluent is obviously lethal to fish, it is not so obvious whether any significant number of fish actually experience such extreme exposures because of the swimming behavior of fish and the nature of the hydraulic characteristics at the outfall. Observations from a boat over the reactor outfalls show periodic and turbulent upwellings from the discharge pipe on the river bottom. Hence, it is likely many of the seaward migrants will be swept away into cooler waters. The outfalls are in fast waters of about five fps and the small migrant—of, say, 38 mm length—can barely swim against one fps velocity.

Estimates of the average temperatures in the mixed zones downstream from reactor plumes are essential data; however, what the fish actually experiences may vary considerably from the averages. Adults have the swimming ability needed to avoid unfavorable temperatures and thus may experience temperatures significantly cooler than the average. On the other hand, a tendency of young fish to hug shoreline waters may expose them to higher temperatures. Although we can discuss the average temperatures of the river or selected extreme temperatures in an effluent plume, the actual temperature experience of a salmon migrating past Hanford is mostly not yet established.

The laboratory and field findings presented here need to be analyzed in light of the seasonal temperature regime of the Columbia River near Han-

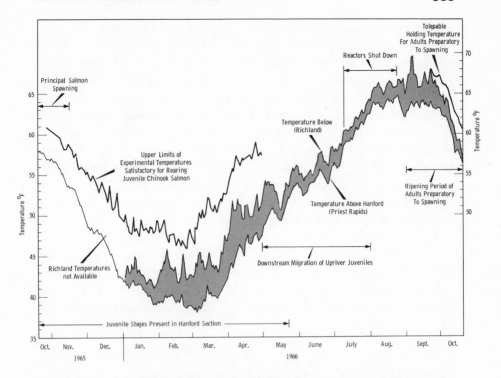

Fig. 9——.Relationship between chinook salmon and Columbia River temperatures near Hanford, 1965–1966.

Photograph by Pacific Northwest Laboratory, operated by Battelle Memorial Institute for the U.S. Atomic Energy Commission

ford. Figure 9 shows the river-temperature profile above and below Hanford in 1965–66. The difference between Priest Rapids and Richland temperatures is the result of not only Hanford reactors, but also of heat gains or losses from other sources. In fact, for over 40 days (starting in early July) of 1965, the reactors were shut down because of a strike. Hence, the increment of about 3° F. seen during this period is the result of natural heat gains during the summer as the water flows from Priest Rapids to Richland. Clearly, the river temperatures near Hanford increase during the summer, even when the reactors are not operating.

Extrapolation of laboratory data to field situations must be treated cautiously, but must be attempted, due to gaps in the field data. Figure 9 shows the history of temperatures experienced in our laboratory which yielded a total mortality no greater than 20 percent from egg to migrant size. The curve shows that more heat can be added during the cooler periods of late fall, winter, and spring, without a threat to young salmon.

The migration of seaward chinooks based on the work of E. M. Mains and J. M. Smith in 1955 (1964) occurred before the warm-water season of August and September; however, more recent work of Bureau of Commercial Fisheries' sampling of seaward migrants from turbine-intake gate-wells at Priest Rapids Dam showed sockeye, coho, yearling chinook, and steelhead trout migrated in May and June, whereas 0-age chinook migrated primarily in July and August (Raymond, 1967). This shift in the migration time of 0-age chinooks from early spring to summer months exposes the fish to warmer waters. The reason for this shift is not yet clear, but there are some indications that reservoirs of upstream low-head dams delay rates of movement (Raymond, 1967).

Probably the most critical temperature is that at spawning. This temperature not only affects the adults but also the survival of the egg, especially in the early stages of development (Burrows, 1963). Tolerable holding temperatures of adults preparatory to spawning are not very well known for our local chinooks. However, those fish spawning earlier (in October) face a greater threat from heat than later spawners (in November).

Salmon eggs and young are not so vulnerable to heat from the Hanford effluent today as they are to any significant shifts in the seasonal temperature cycles of the river. A marked shift in the temperature regime of the river, so that temperatures now experienced in September occur in October, could affect the survival of those salmon eggs deposited during the early portion of the spawning period. Most of the local chinooks spawn from about mid-October through mid-November, when the temperature is rapidly declining from about 57° to 48° F. (13.9° to 8.9° C.). Incubation of the eggs and development of the young occurs from October through April when the Columbia River is sufficiently cold that heat increments added by the Hanford reactors cannot raise the river temperature above an acceptable range. Temperatures above 60° F. (15.5° C.) appear to be too high for initial incubation of chinook eggs. The U.S. Fish and Wildlife Service (1967) recommended the following generalized "optimum" temperature ranges for the following activities:

Migration: 45° to 60° F.

Spawning: 45° to 55° F.

Rearing: 50° to 60° F.

These limits may be "optimal", but they are not very realistic for the main stem of the Columbia. During the warm summer months, the Columbia River has always been a marginal habitat for salmonids. Recorded temperatures in the main stem, either before the period of accelerated dam construction or even the data of more recent years, show 60° F. was always exceeded. The Columbia River in its pristine state must have exceeded 60° F.

A review of the laboratory and field studies concerning the biological effects of Hanford heat on salmonids shows no demonstrable evidence that the present level of heat burden introduced by Hanford reactors creates damage to our valuable salmon resources. However, the problem of evaluating and assessing the biological effect of the relatively small heat increments added by the Hanford reactors to the Columbia River or added heat from any source on the health of salmon populations defies a direct, definitive answer at this time because of large gaps in our knowledge about those complex factors which determine survival.

We believe the best practical method to investigate the effects of elevated temperatures on salmon or other desirable species in the Columbia River is the direct approach of working on-site, using local fish and Columbia River water. We have the only hatchery that successfully uses Columbia River water. Ideally, we hope to develop basic principles from our findings for application on a broader basis. At present, however, we hope the biological data obtained from our experiments will help to determine the quantity of heat which can be added to the river without significant damage to salmon.

REFERENCES

Allen, R. L., and A. C. Moser. 1967. "Rocky Reach Fall Chinook Salmon Spawning Channel." Mimeographed. Ann. Rept., 1965–66 Season. Olympia, Washington: Washington State Dept. of Fisheries.

Brett, J. R. 1960. "Thermal Requirements of Fish—Three Decades of Study, 1940–1970." In *Biological Problems in Water Pollution*, Trans. of 2nd Seminar, U.S. Public Health Service, Cincinnati, 1959, pp. 111–117. [Washington, D.C.: U.S. Gov't. Printing Office.]

Brett, J. R. 1956. "Some Principles in the Thermal Requirements of Fishes." *Quart. Rev. Biol.* 31:75–87.

Burrows, R. E. 1963. "Water Temperature Requirements of Maximum Productivity of Salmon." In *Water Temperature, Influences, Effects, and Control*. Proceedings of the Twelfth Pacific Northwest Symposium on Water Pollution Research November 7, 1963. U.S. Department of Health, Education, and Welfare, U.S. Public Health Service Publication, pp. 29–38. [Washington, D.C.: U.S. Gov't. Printing Office.]

Chambers, J. S. 1964. "Propagation of Fall Chinook Salmon in McNary Dam Experimental Spawning Channel." Rept., Research Div., Washington Department of Fisheries.

Coutant, C. C. 1968. "Behavior of Adult Salmon and Steelhead Trout Migrating

Past Hanford Thermal Discharges." In *Pacific Northwest Laboratory Annual Report, 1968, to U.S. AEC Div. of Biol. and Med.*, vol. 1, edited by R. C. Thompson, P. Teal, and E. G. Swezea, pp. 9–10. Richland, Washington: Biol. Sci. BNWL–714.

Coutant, C. C., and C. R. Cole. 1968. "Modeling of Aquatic Systems." In *Pacific Northwest Laboratory Annual Report, 1967, to U.S. A.E.C. Div. of Biol. and Med.*, vol. 1, edited by R. C. Thompson, P. Teal, and E. G. Sewzea, pp. 9–30. Richland, Washington: Biol. Sci. BNWL–714.

Davis, J. J., D. G. Watson, and C. C. Palmiter. 1956. "Radiobiological Studies of the Columbia River Through December 1955." Richland, Washington: U.S. AEC Doc. HW–36074.

Dean, J. M., and C. C. Coutant. 1968. "Lethal Temperature Relations of Juvenile Columbia River Chinook Salmon." In *Pacific Northwest Laboratory Annual Report, 1967, to U.S. AEC, Div. of Biol. and Med.*, vol. 1, edited by R. C. Thompson, P. Teal, and E. G. Swezea, pp. 9–30. Richland, Washington: Biol. Sci. BNSL–714.

Fish Commission of Oregon and Washington Department of Fisheries. 1967. "The 1966 Status Report of the Columbia River Commercial Fisheries." Portland, Oregon.

Fish, F. F., and R. R. Rucker. 1943. "Columnaris as a Disease of Cold-Water Fishes." *Trans. Am. Fish. Soc.* 73:32–36.

Fujihara, M. P. 1968. "Columnaris Exposure and Antibody Production in Seaward and Upstream Migrant Sockeye Salmon." In *Pacific Northwest Laboratory Annual Report, 1967, to U.S. AEC, Div. of Biol. and Med.*, vol. 1, edited by R. C. Thompson, P. Teal, and E. G. Swezea, pp. 9–16. Richland, Washington: Biol. Sci. BNWL–714.

Fujihara, M. P., P. A. Olson, and R. E. Nakatani. 1965. "Antibody Production and Immune Response of Fish to C. columnaris." In *Pacific Northwest Laboratory Annual Report, 1964, to U.S. AEC, Div. of Biol. and Med.*, vol. 1, edited by R. C. Thompson, P. Teal, and E. G. Swezea, pp. 194–196. Richland, Washington: Biol. Sci. BNWL–714.

Fujihara, M. P., P. A. Olson, and R. F. Roster. 1960. "Mutation and Temperature Effects in C. columnaris." In *Biol. Res. Ann. Rept. for 1959.* U.S. AEC Doc HW–65500, p. 188. Richland, Washington.

Harty, J., R. F. Corlett, R. E. Brown, C. C. Coutant, J. F. Fletcher, H. E. Hanthorn, R. T. Jaske, C. L. Simpson, W. L. Templeton, W. A. Watts, M. A. Wolf, J. B. Burnham, J. G. Rake, and G. L. Wilfert. 1967. "Final Report on Nuclear Power Plant Siting in the Pacific Northwest." Richland, Washington: Pacific Northwest Laboratories.

Hines, N. O. 1962. *Proving Ground—An Account of the Radiological Studies in the Pacific, 1946–1961.* Seattle, Wash.: University of Washington Press.

Jaske, J. T., and J. B. Geobel. 1967. "Effects of Dam Construction on Temperatures of Columbia River." *J. Am. Water Works Assoc.* 59:935–942.

Kennedy, V. S., and J. A. Mihursky. 1967. "Bibliography on the Effects of Temperature in the Aquatic Environment." Univ. of Maryland, Natural Resources Institute, Contr. No. 326.

Mains, E. M., and J. M. Smith. 1964. "The Distribution, Size, Time, and Current Preferences of Seaward Migrant Chinook Salmon in the Columbia and Snake Rivers." *Fisheries Res. Papers, Wash. Dept. of Fisheries* 2:5–43.

Nakatani, R. E., and R. F. Roster. 1966. "Hanford Temperature Effects on Columbia River Fishes." U.S. AEC Doc. BNWL–CC–591, Richland, Wash.

Olson, P. A., and R. F. Roster. 1955. "Temperature Tolerance of Eggs and Young of Columbia River Chinook Salmon." *Trans. Amer. Fish. Soc.* 85:203–207.

Olson, P. A., and R. E. Nakatani. 1968. "Effect of Elevated Temperatures on Mortality and Growth of Young Chinook Salmon." In *Pacific Northwest Laboratories Annual Report, 1967, to U.S. AEC, Div. of Biol. and Med.*, vol. 1, edited by R. C. Thompson, P. Teal, and E. G. Swezea, p. 9.3–9.10. Richland, Washington: Biol. Sci. BNWL–714.

Ordal, Erling, and R. E. Pacha. 1963. "The Effects of Temperature on Disease in Fish." In *Proceedings of the 12th Pacific Northwest Symposia on Water Pollution Research,* Corvallis, Oregon. U.S. Department of Health, Education, and Welfare, Public Health Service. [Washington, D.C.: U.S. Gov't. Printing Office.]

Raney, Edward C., and Bruce W. Menzel. 1967. "A Bibliography. Heated Effluents and Effects on Aquatic Life with Emphasis on Fishes." *Philadelphia Electric Company and Ichthyological Associates, Bull. No. 1.*

Raymond, H. L. 1967. "A Summary of the 1966 Outmigration of Juvenile Salmonids in the Columbia Basin." Mimeographed. Seattle, Washington: U.S. Bureau of Commercial Fisheries, 1967.

U.S. Department of the Interior, Bureau of Commercial Fisheries. 1967. "Nuclear Thermal Power Plants and Salmonid Fish." Mimeographed. Seattle, Washington: U.S. Bureau of Commercial Fisheries, 1967.

Wallis, Joe. 1968. *Recommended Time, Size, and Age for Release of Hatchery Reared Salmon and Steelhead Trout.* Fish Comm. of Oregon, Res. Div., Clackamas, Oregon.

Watson, D. G. 1968. "Chinook Salmon Spawning Near Hanford—1967. In *Pacific Northwest Laboratories Annual Report, 1967, to U.S. AEC, Div. of Biol. and Med.*, vol. 1, edited by R. C. Thompson, P. Teal and E. G. Swezea, pp. 9–14. Richland, Washington: Biol. Sci., BNWL–714.

Watson, D. G. 1968. Personal communication on unpublished data. Richland, Washington.

DISCUSSION/ George R. Snyder

I WOULD like to thank the program chairmen for giving me the opportunity to review Dr. Nakatani's paper. I feel that the subject material presented in his paper and the research objectives being pursued finally by the competent staff at Hanford are of concern to all individuals who place a high value on the populations of cold-water anadromous fish found in the Pacific Northwest.

My comments on Dr. Nakatani's paper will be specifically limited to the problems of thermal pollution. The problem of thermal pollution at Hanford first became critical in 1958, when river temperatures in July were averaging 4° to 5° C. above the 1957 figures. This occurred during a time when eight production reactors were on the line.

Considerable time, money, and personnel were assigned to an examination of this problem, since it was apparent that production losses at Hanford were imminent. A plan was proposed and implemented to use cold storage-waters and increased flows from Grand Coulee Dam for relief. Temperatures at Hanford were held from 1° to 3° C. below the forecasted highs during 1958 as a result of this program, as will be shown subsequently.

The problem still exists today, even though the number of reactors in operation has been reduced to four, as has been shown in Dr. Nakatani's paper. A review of weekly water-temperature records above and below Hanford that are assembled and distributed by the U.S. Geological Survey will reveal the current magnitude of this problem.

I do not feel it is necessary to dwell on any specific aspects of Dr. Nakatani's paper. It is well-written and thoroughly documented. In essence, I agree with the approach and findings reported in his paper, and I will summarize them as follows.

Dr. Nakatani's present staff and their predecessors have spent over twenty years on the investigation of problems of thermal pollution caused

318

by the Hanford reactors. The basic objectives of these investigations were to relate the effects of thermal pollution to each stage of the life cycle of selected salmonids for proper evaluation of the magnitude of the Hanford problem. We find that adult chinook salmon that move through the reactor areas seem to avoid the plumes during upstream travel but apparently build redds near these plumes, to spawn, as river temperatures cool. We also find that more fall chinook spawn above the reactors than below them.

We have been shown that, if water temperatures are increased by 4° F. in October (and October appears to be the time fish spawn, in this area), excessive mortalities will occur (this is listed by Dr. Nakatani as over 20 percent), and that, although spawn tolerated a 12-degree F. increase in December, most of the spawning was over by mid-November. Thus, fish that spawn below the reactors are faced with loss of viability of spawn if unseasonal conditions occur; and fish that spawn above the reactors have viable spawn, but as the fry develop from this spawn and emerge from the gravel, they are swept downstream, where they could be subjected to the reactor discharge plumes.

We have been further told that, while the temperature of undiluted reactor effluent is obviously lethal to fish—and I will quote that one, because I didn't hear Dr. Nakatani say that: he wrote that, while the temperature of the undiluted reactor effluent is "obviously lethal to fish"—and I will qualify it—"it is not so obvious whether any significant number of fish actually experience such extreme exposures because of the swimming behavior of fish and the nature of the hydraulic characteristics near the outfall."

Although I agree with Dr. Nakatani's approach and his research findings, I would like to comment on his conclusions. Dr. Nakatani's conclusions are, and I quote,

A review of the laboratory and field studies concerning the biological effects of Hanford heat on salmonids showed no demonstrable evidence that the present level of heat burden introduced by Hanford reactors creates damage to our valuable salmon resources.

It is apparent, then, that although conditions at Hanford could produce mortalities, there is little evidence that these mortalities are occurring. One wonders how much effort has been devoted to this task; but, more important, one realizes that dead fish 50 to 100 mm in length are very difficult to find in the Columbia, under any circumstances.

I am reminded of the study, "An Investigation of Adult Chinook Salmon

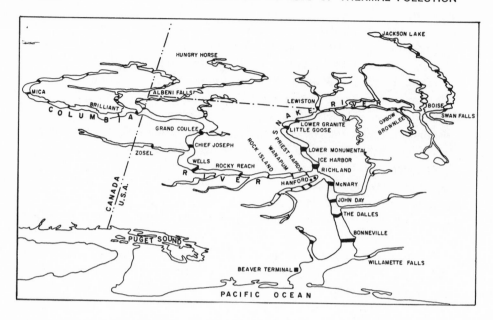

Fig. 1—.General location of dams on the Columbia River and its principal tributaries.

Mortality in the Vicinity of Bonneville Dam", carried out during 1954 and 1955 on the Columbia River, where adult salmon carcasses were released purposefully to evaluate the study objectives. During 1955, 1169 adult chinook salmon were released below Bonneville Dam, yet only 32 were recovered. Crews assigned to this study carried out daily searches for these dead fish, and they were aided by large numbers of fishermen on the banks and in boats.

What are the chances that dead fish 50 to 100 mm long will be found in a river as large as the Columbia, in a restricted area where only classified boat-travel is allowed, and just as important, on weekends when these problems occur? I feel that more tangible evidence is needed on the investigation of predator stomach contents or results from traps that are being carried out now by Battelle to establish the magnitude of mortalities that are caused by the problems of thermal pollution in the area.

Thus, the problem of thermal pollution at Hanford has been adequately presented by Dr. Nakatani. What I would like to present next is how the problems of thermal pollution at Hanford fit into the general problem of thermal pollution of the Columbia River.

TABLE 1. Changes in water temperature in the Columbia River from Grand Coulee Dam to Hanford, Washington, as a result of the "Artificial Cooling Program." 1958–1964.

	DEGREES PER DAY SAVED AT H.A.P.O.						
	1958 °F.	1959 °F.	1960 °F.	1961 °F.	1962 °F.	1963 °F.	1964 °F.
AUG 1	2.0		1.8	7.9			
AUG 7	2.3	1.8	3.1	6.7			
AUG 15	1.3	2.9	4.3	4.7		1.4[1]	1.4[2]
SEPT 1	.9	1.6	1.6	5.8		2.7	
SEPT 7		.9	1.4	2.5			
SEPT 15		3.1	.7	2.0			
SEPT 30		.7	2.2	.5			
MAXIMUM COOLING	2.7	3.1	4.9	9.2		2.7	1.4
6 HR PEAK	6.3						
AVERAGE FOR YEARS[3] PROGRAM					2.7		

H.A.P.O. Hanford Atomic Products Operation

[1] 2.7° F. on August 25

[2] 1.4° F. on August 25

[3] Temperature reduction averaged 2.7° F. for controlled-cooling period of 69 days—no other details forwarded

* NOTE   Years 1958–1962 from Kramer (1959–63).
Years 1963–1964 from Jaske (1966).

Figure 1 shows the Columbia Basin hydroelectric development and is indicative of the extensive hydroelectric development that has occurred in the Columbia Basin. The only free-running stretch of the Columbia River left in the United States, above tidal influence, is the area contained in the Hanford Reservation. The problems of thermal pollution in the Columbia River include (1) hydroelectric impoundments, (2) the Hanford production reactor discharges, (3) proposed thermal electric plants, (4) river-flow reductions, and (5) nitrogen gas saturation.

In 1958, the Columbia River cooling-water study was implemented by the General Electric Company at Grand Coulee Dam, and the program continued until 1965. Benefits in water-temperature reduction at the Hanford Reservation have been summarized as shown in Table 1. Maximum benefits were realized in 1961, when temperatures were reduced by 9.2° F.

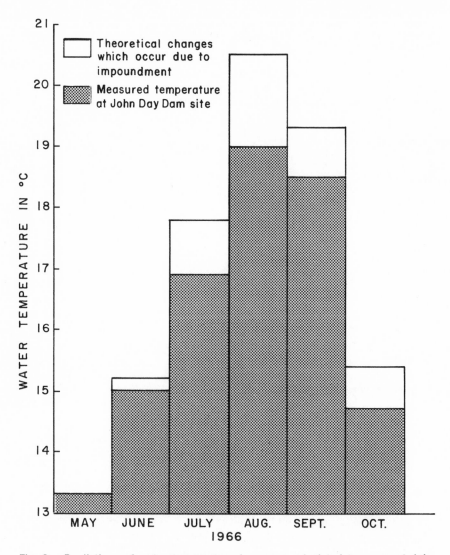

Fig. 2—.Predictions of water temperature increases calculated to occur at John
Day Dam compared with actual measure temperatures.
(Novotny and Clark 1968)

Predictions of water-temperature increases calculated to occur at John
Day Dam, a Columbia River run-of-the-river dam, indicate the magnitude
of increases typical of these impoundments. Minor increases occur during
June, July, and August, extending the temperature peak, and a shift in the
cycle will occur during September and October. Notice that the major in-

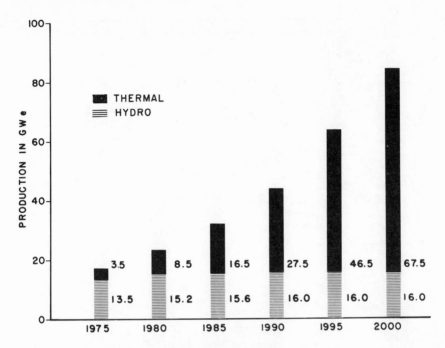

Fig. 3—.Assumed schedule for thermal generating plant additions in the Pacific
Northwest, 1975–2000.
(Battelle Northwest 1966)

creases occur during the time when fall chinook salmon spawn, as Dr.
Nakatani has illustrated and as indicated on Figure 2.

Figure 3 shows a need-histogram of electrical power in the Pacific
Northwest. A change will occur in Federal Power Commission Region #7,
from a predominately hydroelectric network to one that will be predominately thermal-electric. By 1985, thermal power production will exceed
hydro production in this region.

Preferred thermal-electric plant sites in the Pacific Northwest are shown
in Figure 4. The electrical production that has been announced for these
plants on the Columbia River is given. So far, only three plants have
formally announced intentions to construct, i.e., Bellinghans, Trojan, and
Kalama. Sixteen plants will be needed by 1980, and 22 plants by 1985.

Figure 5 shows Canadian-U.S. treaty dams. The Columbia River-
Canadian Treaty Program included provisions for the installation of four
dams and a total storage of over 25 million acre-feet of water. The storage
and subsequent release of this quantity of water may contribute to thermal

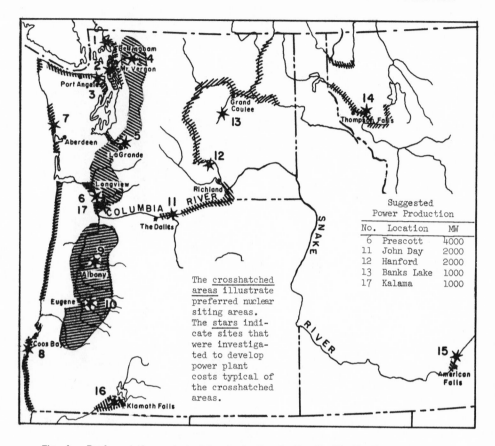

Fig. 4——.Preferred thermal electric plant sites in the Pacific Northwest.
(B.P.A. 1967)

pollution in the lower river by altering flow characteristics (velocity and volume).

Water released from completed Canadian storage projects will cause a shift in the monthly flow regime. Lower flows will exist during May, June, July, and August. August is our maximum water-temperature month in the Columbia River. Thus, with Canadian storage, flows will be reduced during the warmest months of the year, as shown on Figure 6.

Figure 7 demonstrates the change in daily flows from hydroelectric peaking. Peaking will play its part in the reduction of flow during the evening hours. Water will be reduced to 50,000 cubic feet per second (c.f.s.) for seven hours at the Dalles Dam on the Columbia River, even though the "average daily discharge" is 120,000 c.f.s. If you have a ther-

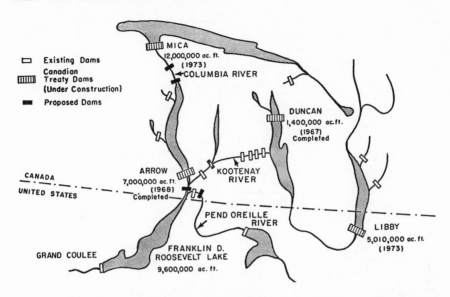

mal-electric plant on the river that is producing firm or constant power and rejecting a constant amount of heat, the reduction of flow will allow for excessive heating of the river.

Figure 8 shows nitrogen gas-saturation in the Columbia River. Nitrogen gas is introduced into the river at the spillway of dams. It is held in solution by the pressure of the deep-end impoundment forebays. The quantity of nitrogen gas found in the Columbia River has been measured for the last three years and has been shown to be above dangerous N_2 saturation levels that are lethal to juvenile salmonids. As temperatures are increased, nitrogen gas-saturation levels are increased from a sublethal to a lethal level. If the temperature in a stretch of the river where sublethal N_2 conditions exist is raised rapidly, lethal conditions are created.

Temperature conditions at the Hanford Reservation during September 1966 are shown in Figure 9. In general, there are two types of temperature increases that can be lethal to fish in this area: (1) sharp increases in the discharge plumes, and (2) an increase in the entire receiving water from one point to another downstream.

Hanford's "N" reactor and the Washington Public Power Supply System's electrical generation plant are shown on Figure 10. The WPPSS generators are run from steam produced at the "N" reactor. The complex has two

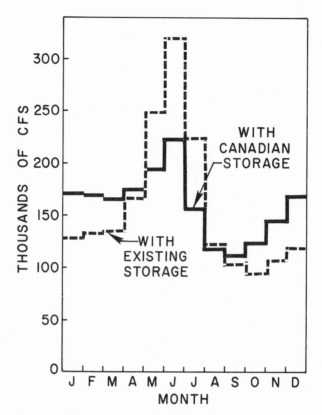

Fig. 6——.Average monthly river flow of the Columbia
River at the Dalles Dam before and after completion
of Canadian storage projects.
(Nelson 1967)

direct and independent cooling-water systems. When the WPPSS plant
produces 860 megawatts of electrical power, the plant discharges 1250
c.f.s. of water into the Columbia River at 38° F. over ambient temperature.
Different operation conditions will produce discharge temperatures from
2.2° to 47.3° F. above ambient river temperatures. Table 2 shows the
predicted amount of heat rejected from the "N"-WPPSS thermal electric
plant at Hanford.

Figure 11 shows a plume of pollution discharge in the upper Columbia
River and illustrates how a thermal plume would dissipate downstream.
Water-temperature increases at Hanford in 1966 are shown on Figure 12.
This is a different presentation of the data introduced by Dr. Nakatani,
i.e., hourly temperatures. Temperature increases began to occur following

TABLE 2. Predicted amount of heat rejected from the "N"-W.P.P.S.S. thermal-electric plant at Hanford, Washington, under electrical and non-electrical production operation.

Computations of heat wasted from "N" reactor: Possible combinations

A. Plant power maintained at 4000 MWth level with electrical production varied.

MWe	BTU/hr utilized for power production $\times 10^9$	BTU/hr wasted into the water $\times 10^9$	Temperature WPPSS only (1250 cfs)	increase (° F.) WPPSS + AEC (2500 cfs)
100	.341	13.311	47.3	23.6
200	.682	12.970	46.1	23.0
300	1.023	12.629	44.9	22.4
400	1.365	12.287	43.7	21.8
500	1.706	11.946	42.5	21.2
600	2.047	11.605	41.2	20.6
700	2.389	11.263	40.0	20.0
800	2.730	10.922	38.8	19.4
860	2.935	10.717	38.1	19.0

B. Plant power not maintained at peak level, but varied to produce different levels of electrical production.[3]

MWe	BTU/hr utilized for power production $\times 10^9$	BTU/hr wasted into the water $\times 10^9$	Temperature WPPSS only (1250 cfs)	increase (° F.) WPPSS + AEC (2500 cfs)
100	.341	1.245	4.4	2.2
200	.682	2.486	8.8	4.4
300	1.023	3.735	13.3	6.6
400	1.365	4.983	17.7	8.8
500	1.706	6.228	22.2	11.0
600	2.047	7.473	26.6	13.2
700	2.389	8.722	31.0	15.5
800	2.730	9.967	35.5	17.7
860	2.935	10.716	38.1	19.0

[3] Assume 21.5 plant efficiency

the strike, on August 25. On Labor Day weekend, increases in the magnitude of 6° to 7° F. were measured from Priest Rapids Dam to Richland, Washington. These temperature differences are between two points that are roughly 35 miles apart; the distance from the outfall of the plants to Richland is about 20 miles. We must assume there is some dissipation of heat from the plant outfall to Richland, however.

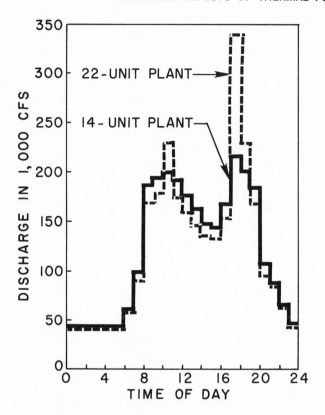

Fig. 7—.Predicted typical daily discharge pattern at the Dalles Dam on the Columbia River upon comple-tion of Canadian storage dams.
(Brown, et al., 1964)

Figure 13 demonstrates U.S.G.S. temperature differences in the Hanford area 9° F. during April. Qualifications were: (1) hot-air temperatures on a weekend, (2) Priest Rapids Dam presented daily average temperatures and Richland records represented instantaneous maximum, and (3) flows were reduced from 80-to-100,000 c.f.s. (normal at this time of the year) to 45,000 c.f.s. This value is considered to be fairly low flow (minimum-flow regulations are 35,000 c.f.s.). This condition was explained to biologists by a statement that the constant rejection of heat by the reactors at Hanford did not cause the problem; instead, it was said to be the unnatural occurrence of low flows. One wonders if it is like this: a man holds a gun aimed at a fish, but he fails to pull the trigger; however, another man walks up and pulls the trigger, and the fish dies. Who is to blame? What is

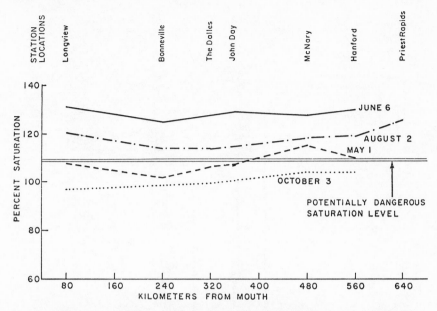

Fig. 8——.Saturation values of dissolved nitrogen in samples of Columbia River
water from Longview to Priest Rapids Dam, May 1-October 3, 1967.
(Ebel 1968)

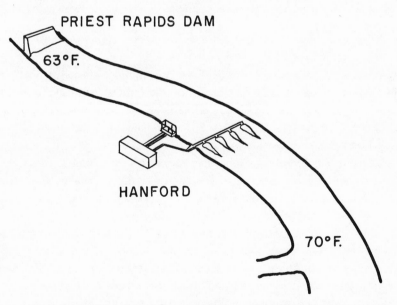

Fig. 9——.Water temperature conditions at Hanford area compound:
nitrogen gas problem.

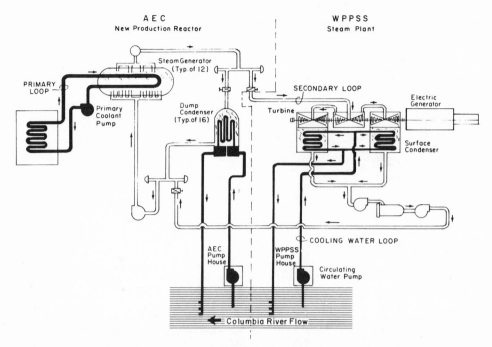

Fig. 10——.Diagrammatic plan view of the "N" reactor and the Washington Public Power Supply System thermal electric plant at Hanford, Washington.

needed here is the realization that a problem exists, and we must develop solutions to the problems.

Fish can be influenced by both direct and indirect effects of heat. It is the indirect effects of heat that are often the hardest to measure and the hardest to define. Some indirect effects can be listed as nitrogen gas-saturation, competitors, predators, diseases, and parasites. Each has an impact on the total biological population.

Preferred temperature on some fish found in the Columbia River Basin are demonstrated in Figure 14. Temperatures of the Snake River today favor the predators and competitors, and work against the valuable ana-dromous fish found in our river systems.

Figure 15 shows nitrogen gas in graduated cylinders and demonstrates the problem: two graduated cylinders are supersaturated with nitrogen gas. One is held at 55° F.; the other has been raised 3° F. to 58° F. Notice how the gas in the cylinder marked 58° F. comes out of solution.

The nitrogen-gas problem is illustrated by placing fish in aquaria with supersaturated gas conditions as shown on Figure 16. The temperature in

Fig. 11—.Sulfite plume at Castlegar, B.C., shown as an example of how a thermal plume might disperse down-river.

one tank is held at 55° F. and the other is raised to 58° F. This condition occurs in the Columbia River at Hanford. You expose the fish in an area of supersaturation of gas to a 3° F. increase, and the gas in the body of the fish comes out of solution. Bubbles in the bloodstream clog the entrance to the heart (Conus Arteriosis) and the fish dies. The fish effervesce like soda-pop.

Figure 17 shows the water-temperature laboratory (floating barge) on the Columbia River. The Bureau of Commercial Fisheries has recently established this facility on the Columbia River near the two proposed sites for thermal-electric plants, i.e., the trojan site at Prescott and the Kalama plant. The barge is approximately 110 feet long and 30 feet wide and has equipment for heating and cooling Columbia River water. We capture fish as they pass this area and test their thermal tolerance in river water.

We have taken some color films of the effect of thermal shock on some species of fish found in the Columbia River.[1] Ambient river temperatures

1. Film was shown, with following commentary.—P.A.K.

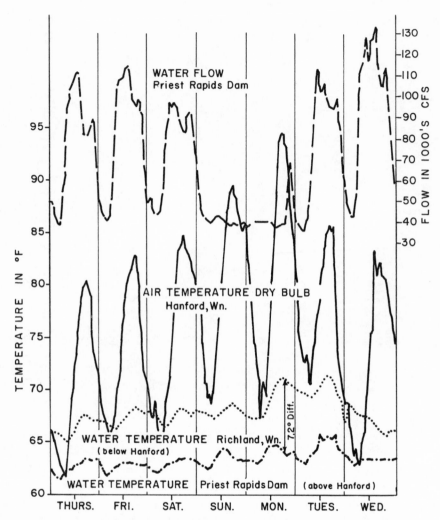

Fig. 12—.Water temperature increases at Hanford, September 1–7, 1966.

are approximately 60° F., and we have increased the water temperature 30° F. to show the effect that the thermal plumes at Hanford could have on these species. A small starry flounder will be shown in the apparatus first.

There is an aquarium with both circulating hot water and a constant air supply shown. A thermograph showing the temperature of the water can be seen under a timer. The starry flounder shown succumbed in 22 seconds, tetanized.

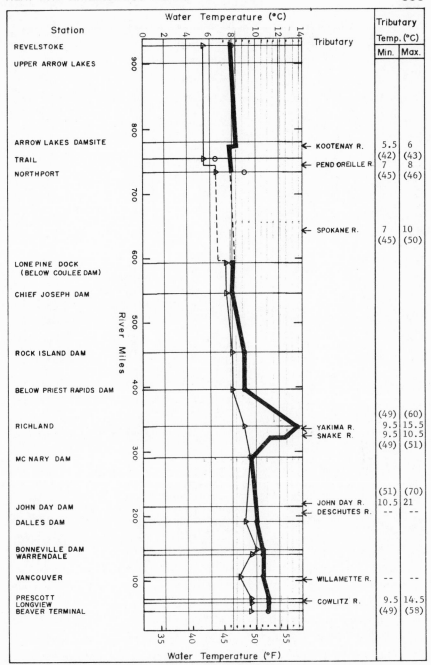

Fig. 13—.Water temperature profile of Columbia River, April 23–29, 1968.
(U.S.G.S., Portland)

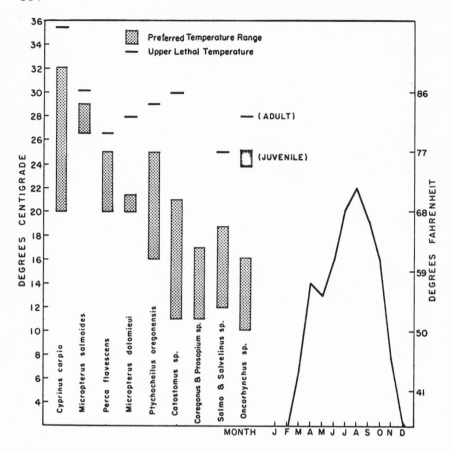

Fig. 14—.Preferred and lethal temperatures of some Columbia Basin fish compared with lower Snake River water temperatures. Temperatures were recorded
at the mouth of the Snake, and monthly average temperatures for 1954 to 1957
were combined.
(From Novotny, unpublished)

The sturgeon is next. This fish is approximately 8 inches long. It is
an anadromous fish, and it is an economically valuable species to commercial fishermen. Loss of equilibrium takes 30 seconds; death comes
in 60 seconds.

The next group of fish are stickleback. These fish succumb to the heat
in approximately 35 seconds.

The next fish is a yearling chinook, a bit larger fish (about 5 inches),
with more mass and a better swimming ability than most downstream
migrants. It is quite probable that this fish could escape the 11-foot

Fig. 15—.Example of nitrogen gas being released from water at a 3-degree-Fahrenheit temperature increase. Water is supersaturated with N_2.

discharge plumes of thermal plants. This fish, however, succumbs in 55 seconds.

The next group of fish are the 0-year chinook salmon, the kind of downstream migrants that are found in the Hanford area. Notice how these fish are in trouble in 5 seconds and are tetanized in 11 seconds. This is only a 30-degree increase in water temperature, a condition that can be shown to exist in the Hanford Reservation.

The next and final portion of the film shows a group of chum salmon. These fish all succumb in 10 seconds. Their gill rakers are all distended and the fish are very stiff, tetanized.

I would like to emphasize the fact that the fish species of major economic value in the Columbia River are anadromous. Adults must move upstream to spawn, and juveniles must migrate to the ocean.

Each of these species has a life cycle and life history that varies; each has a different set of thermal requirements, and this is important, although most of these thermal requirements are yet to be specifically

Fig. 16—.Gas bubbles (N_2) forming on fish and aquarium. Water has been saturated with nitrogen gas and increased 3 degrees F.

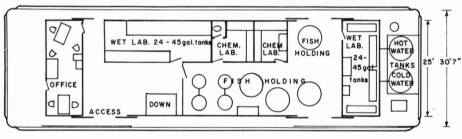

PLAN VIEW

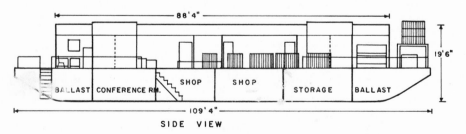

SIDE VIEW

Fig. 17—.Floor plan of water temperature barge presently moored at Prescott, Oregon, the site of the proposed P.G.E. Trojan thermal-electric plant.

described for species in the Columbia River. Each is basically a cold-water species of fish. These facts necessitate the consideration of a different set of water-temperature standards for this area (the Pacific Northwest) than are found in other regions of the United States. We have a different problem here.

The Bureau of Commercial Fisheries has studies under way at the Prescott Laboratory on the Columbia River and at the Mountlake Laboratory in Seattle, Washington, that will aid in the development of criteria for more meaningful water-temperature standards concerning economically important species of anadromous fish.

REFERENCES

Battelle-Northwest. 1966. "Final Report on Study Program to Determine the Potential Impact of Thermal Power Plants on Pacific Northwest Streams to Fish Passage Research Projects." U.S. Fish and Wildlife Service, Bureau of Commercial Fisheries, Seattle, Washington.

Brown, Curtis B., David J. Lewis, Chester E. Mohler, and Warren H. Marple, 1964. "Impact of Recent Power Developments on Multiple Use of the Columbia River." A panel presentation to a public meeting of the Pacific Northwest Field Committee, Dept. Int., Missoula, Montana, October 13, 1964.

Davis, William S., and George R. Snyder. 1967. "Aspects of Thermal Pollution That Endanger Salmonid Fish in the Columbia River." Unpublished manuscript, Bur. Com. Fish., Fish Passage Research Program, Seattle, Washington.

Ebel, Wesley J. 1968. "Nitrogen Supersaturation in the Columbia River: Preliminary Report." Unpublished manuscript. U.S. Fish Wildl. Serv., Bur. Com. Fish., Biol. Lab., Seattle, Washington.

Hurd, Owen W., and Richard T. Richards. 1965. "Hydraulic Features of the Hanford Nuclear Steam Electric Generating Plant." Fifth Biennial Hydraulics Conference, Washington State University, College of Engineering Research Division, Pullman, Washington.

Nelson, Mack L. 1967. "Impact of Canadian Treaty Storage on the United States." *Proc. Sixth Biennial Hydraulics Conference, Univ. Idaho, Moscow, Idaho,* pp. 155–170.

Novotny, Anthony J. 1964. "Importance of Water Temperature in the Main Stems of the Columbia and Snake Rivers in Relation to the Survival of Salmon." In *Fish-Passage Research Program Review of Progress, 1964 vol. 5*: Adaptability of *Salmon to New Environments Created by Dams.* U.S. Fish and Wildlife Serv., Bureau of Commercial Fisheries, Seattle, Washington.

Novotny, Anthony J., and Shirley Miller Clark. 1967. "Preliminary Predictions of Water Temperatures in John Day Reservoir." Unpublished manuscript. U.S. Fish Wildl. Serv., Bur. Com. Fish., Biol. Lab., Seattle, Washington.

Snyder, George R. 1968. "Thermal Plants, Thermal Pollution, and Fish—The Problem in the Columbia River and the Pacific Northwest." Processed manuscript, U.S. Fish Wildl. Serv., Bur. Com. Fish., Biol. Lab., Seattle, Wash.

Snyder, George R., and William D. Parente. 1968. "Selected Temperature-Depth Profiles of Arrow Lakes, British Columbia (1965–66) and Surface Temperatures and Flows of the Columbia, Kootenay, and Pend Orielle Rivers (1964–66)." Manuscript. U.S. Fish Wildl. Serv., Bur. Com. Fish., Biol. Lab., Seattle, Wash.

DISCUSSION FROM THE FLOOR

Roy E. Nakatani: Mr. Snyder, I enjoyed your film on the fish-fry. In fact, it was very dramatic. Because of limited time, I didn't discuss the problem of the relationship between the behavior of the fish and the hydraulic characteristic of Hanford's discharge plumes. However, I did mention that Dr. Chuck Coutant and his colleagues have experimented with drifting young salmon in live-boxes through reactor plumes in an effort to obtain deliberate kills. We recognize that a fish in a live-box is confined and is not a free-swimming fish; nevertheless, no significant numbers of salmon were killed in the drift study.

The reason Coutant did not see kills is obviously because the fish in the live-box did not experience extreme temperatures like Mr. Snyder's experiment. That is, sudden exposure to delta T of about 30° to 40° F. from the base temperature.

I believe Dr. Coutant should really make some comments, since he carried out the field work. In one case, we have Mr. Snyder's very dramatic laboratory experiment; and in the other case, we have actual field observations. The hydraulic conditions near the discharge are such that Dr. Coutant had a difficult time keeping the thermister probe right in the upwellings of warm water being discharged. It was necessary to weight the probe to hold the sensor in warm waters. The temperature experience was not like Synder's experiment.

Charles C. Coutant: Maybe I should go back just one step to temperature changes close to the outfall before I comment on holding temperature probes at the end of the pipeline. It turns out that the temperature in the plume (for sake of discussion, we can use Mr. Snyder's 30-degree initial temperature) has generally lost about 80 percent of the delta T within less than five seconds of the release of water from the end of the

338

pipe out in the river. We measure this at the river surface. This loss occurs because of turbulent mixing with cool river-water. If a fish is travel-
ling downstream at the rate of river flow, it generally would not
experience the constant 30 degrees of temperature in the plume area for
even the five seconds that it took for Mr. Snyder to show fish rolling over.

I don't deny the importance of the type of experiment that Mr. Snyder
is doing, however; we're doing them, ourselves. A reason for determining thermal-resistance times at constant temperatures, we believe, is to
help us predict the likelihood of fish death in a plume where the water
temperature is dropping from this 30 degrees to nearly base river-
temperature. It is proving very useful for such predictions to do the type
of experiment that George and I are doing, where we find a particular
time of death for a particular temperature in the laboratory. This lets
you take the plume-temperature curve and determine from lab data a
certain rate of dying for each short-time interval in the plume curve at
an average temperature for that interval. You can then plot the rates of
dying and time of probable death of a fish in this decreasing temperature. While our estimates are still somewhat crude, further development
of a thermal-death model will, it is hoped, provide us with accurate
predictability. So let me say again that the type of experiment that
George is doing is not incorrect, by any means: it just requires some
interpretation to be applicable to the field conditions which we have
observed.

This plume-temperature curve assumes that a fish could, in fact, start
at the point of discharge and move through the center line of the plume
from the 30 degrees to the point where you are essentially at river temperature, plus whatever increment you have after mixing. It is this assumption that we have been making in trying to look at fish fatalities in
Hanford plumes.

Actually, our experience has been, as Roy has mentioned, that it's
very difficult to force a very tiny thermister probe into an area of this
30-degree initial-outlet temperature. We get into a question of hydraulics
which is far removed from my training. We need a good hydraulics
engineer to comment on turbulent outfall mixing—possibly, after we've
spent some money and time on some studies.

There's a great deal of turbulent mixing near the outfall, as I mentioned. The water is shooting out of these open-ended pipes, if this is
what we assume they are, at a very high velocity. If you wish to hold
a probe in this discharge temperature, as I have tried to do, you find

that you can't do it. In fact, it's very difficult to get increments more than half the delta T of the outfall, even with a lot of effort.

Because of this difficulty, there is some real question in my mind whether a fish could, in fact, get into the center line of the plume to experience the types of temperatures that we might ascribe to it. This is an area that needs a great deal of observation. The practical question, if we assume that fish are out there running a collision course with the plume, is whether they will go through the kinds of high-temperature experiences that we can sketch out on a paper for a nice, hypothetical plume. I invited Mr. Snyder to comment on this, because I know he and I have both worked on the same problem.

George R. Snyder: I think it is important that we first define the problem. Presently, I have to calculate temperature increases. Normally, I think an engineer should tell me what temperature increases are planned for a discharge area. From this point, biologists can tell engineers what effect those specific temperatures will have on fish populations.

The characteristics of the "N"-reactor plume most certainly are needed. We need to physically define not only the flow, but also the temperature characteristics of that plume. I think it's necessary to define the plume physically, because we need these data to apply to other areas.

I would also like to mention that Bob McConnell and Ted Blahm are responsible for the studies that are going on at the Prescott Water Treatment Barge.

Rex A. Elder: I am a hydraulic engineer, although I'm not sure that I'm going to admit to being the good one that he wants. I will make a few remarks from the engineering standpoint.

First of all, addressing to Dr. Coutant's remark and to George Snyder's, I take just a little opposite stand to George. He wants *us* to tell *him;* we want *him* to tell *us!* I'll explain *why* we want him to tell us.

The field of hydraulics is not quite like the field of structures, where you can put in fairly well-known values, a big enough safety factor, and be pretty well assured the bridge is going to stand up. In a lot of our designs, we can't be 100 percent sure of just what's going to happen. However, they can be researched. We do this in our laboratories, right along; and we can learn a lot of things, if we know what we're looking for. This is where we need the data from you botanists and biologists and what-not involved in the flora and the fauna—because, until we know, or at least have some idea of what you want, we don't really know what to look for.

Now, this matter of diffusers is not a very difficult matter; however, it takes some time and considerable money. We more or less can do quite a wide range of things. But we certainly can't do it if we don't know what we're trying to achieve. So some place here we have to stop sitting and waiting for the other guy; we have to get together and talk about what we know. This is where this meeting is very important and why, as an engineer, I'm learning things, because now I'm commencing to get some idea about what you biologists think we need, and therefore I have a better idea of what to go looking for and how we're going to design these things.

The other remark I'd like to make concerns the Columbia River. The Columbia River is a majestic, very fine river system, and it is certainly not typical of all of the river systems in the United States. I think there is a considerable tendency to take many of the problems that are occurring and have occurred on the Columbia River, and because of our ignorance, transfer these over to many of the other rivers in the United States. Here we have to be very careful, because most of our other rivers do not have anadromous fishes or at least not to the extent they are here; and most of our other rivers are entirely different, from a temperature standpoint. The temperatures occur at different times of the year, and they will approach equilibrium temperature, where the Columbia River never does. There are so many factors that are different in the majority of the rivers in the United States as compared to the Columbia that we have to be very careful in trying to translate the problems of the Columbia to other areas.

Robert Phillips: I felt that Dr. Nakatani implied that the fall chinook were spawning in the area of the Hanford outfall and were thriving there. I have two questions. First of all, what was the intra-gravel water temperature from the time of spawning through incubation? And second, what was the survival, at least up until the time of hatching, of the eggs deposited there?

Nakatani: We don't have data on an actual, natural, salmon nest. We have considered getting scuba divers down to make direct observations, but the more experienced Washington State Fisheries divers told me that if you dive in water with a velocity of about six to seven feet per second, it's extremely difficult if not hazardous. The face mask just rips off, unless you ride a sled to shear off the fast water. The Columbia River chinook salmon, which may average 20 pounds, spawns in relatively deep waters, as deep as 20 feet.

Don Watson of our staff provided me with some temperature data from a simulated salmon nest located just downstream from the outfall of the 100-F reactor. These data are not from a natural nest, but they may be a partial answer to your questions. Watson drove a special pipe with sensors into the gravel to obtain temperature at different gravel depths. I deleted this information from my presentation but I can give you some data now.

The data collected during November 1954 to May 1955 at gravel depths of 0 inches, 6 inches, 15 inches, 21 inches, and 30 inches seldom showed more than 1° C. or 1.8° F. difference between the river gravel and the water just above the gravel. There was generally a lag in the temperature change between the water below the gravel and the water just above the gravel, and the temperatures ranged from 1° to 4° C. higher in the simulated nest than in the river water immediately upstream from the effluent outfall.

If you recall the slide of the map of Hanford, it showed our experimental hatchery located at 100-F is downstream from all the reactors. Clearly, the hatchery receives water with elevated temperatures because of reactor effluent. Yet, after some 20 years of reactor operations, the hatchery experienced no fish kills attributable to reactor effluent. One year, when the 100-F reactor was down, some chlorine was accidentally spilled right into our hatchery intake and killed many of our fish. There was quite a bit of excitement, until we were able to determine the cause of the fish kill.

In brief, despite the potential hazards of reactor effluents, we have reared salmon and trout satisfactorily for many years. We have the only hatchery, to my knowledge, that uses untreated Columbia River water. Our hatchery rears fish for our research and also functions as a built-in continuous monitoring station. I believe it is important to remember this point.

Robert Smith: Dr. Nakatani, I was very much interested in your slides which showed the spawning grounds, right in the discharge of the heated effluent. I think the hydrodynamics problem is one of the most critical problems in any of these situations.

Assuming that the temperatures were as high as has been suggested, could someone say what the water depths were in the vicinity of the discharge pipes, how deep the discharge pipe was from the surface, and whether or not is was pointed upward? Are there any figures on flow-rates, both the current velocity downstream past the discharge, and the discharge flow-rate from the pipe? Because the initial pollution problem is the starting point, if the initial pollution is high, then so much the better for organisms downstream. If the initial pollution is low, other things are important. For example, the higher the temperature difference between the surface-water temperature and the temperature of the effluent, probably the smaller the degree of mixing in the vertical. In this case, the horizontal mixing becomes that much more important. When the horizontal mixing is emphasized over the vertical mixing, then the loss of the heat to the atmosphere is enhanced. Does anyone have numbers to quote?

Nakatani: I'm not a hydraulics engineer, and I'm sorry, I don't have the numbers you ask. However, I can comment on some observations on the response of salmon to temperature and velocity gradients. In salmon, we are dealing with an extremely sensitive, complex biological system. I recall working with hydraulics engineers at the University of Washington on a rectangular flume designed to provide a constant 3-feet-per-second water. A group of small salmon was introduced into the test flume and the fish were expected to be swept down, because these little fish can hardly swim against 1-foot-per-second water. However, the fish were not swept down, because each fish found small standing waves and swam in these small areas with low velocities. The fish may be a better engineer in terms of picking out small areas of low velocities.

I believe the salmon are also very responsive to temperature gradients. I find it difficult to believe an upstream adult salmon will continue to swim into a reactor plume and find itself in extreme lethal temperatures. Because the fish are responsive to temperature gradients, the inference is that the fish will not experience extreme temperatures if they can avoid it.

With reference to questions of Mr. Smith about the hydraulics characteristics and the depth of discharge, I need to ask someone who knows more about it. Dr. Foster might know about the depth of the discharges.

Richard F. Foster: I think that it would be fair to say that the order of magnitude of the depth of water we're talking about is something like 40 feet. The pipes can be considered as on the bottom, discharging in a direction somewhat toward the surface. They are discharging into water velocities of something on the order of 5 feet per second.

The characteristics of these plumes have been studied extensively by both the Douglas United Nuclear people and by Battelle Northwest. We do know that the plume is such that there is immediate vertical mixing. The effluent is distributed between the bottom and the top of the river quite uniformly in an extremely short distance. However, as the imagery of the plumes indicated, the horizontal mixing is comparatively slow and does not extend completely across the river until several miles downstream.

James Adams: First of all, I would like to discuss some possible misconceptions. George Snyder showed a picture of the Troy site of Portland General Electric and then he talked about the 30-degree temperature differentials and he talked about the 38-degree temperature differentials up at the Hanford facilities. The Hanford reactors are typical. They are plutonium production reactors and I don't know of any commercial nuclear-thermal power stations in the United States, either being built, in operation, or planned to be built, that are talking about this kind of temperature differential. Most of them are talking about 16° to 24° F. temperature differentials.

Second, on the Sacramento-San Joaquin Rivers, we have 2500 megawatts of thermal generation which were first installed back in 1952–53. We also have the largest king salmon fingerlings, year class zeros, through the condenser tubes of the operating plant with a 16-degree differential, with no mortality. I have a few slides here to demonstrate this.

Figure 1 shows the Delta section of the Sacramento-San Joaquin Rivers. The two arrows on the bottom are these two plants. The Pittsburg plant is 1340 megawatts, and the Antioch plant is 1280 megawatts.

The Sacramento River is coming in from the upper right-hand corner, the San Joaquin over on the other side. The Sacramento and the San Joaquin have both been extensively dammed. They used to have over 3,000 miles of spawning area; now they have less than 180 miles; however, the system still has the largest king salmon run.

Figure 2 shows a picture of a typical temperature distribution at the Pittsburg plant on May 20, 1964. It operates at a load of 990 megawatts, the cooling-water flow being 1600 second-feet. This is at the time of the downstream migration of the king salmon and you can see the temperature pattern. Ambient river temperatures are about 65° F. This is a surface-temperature measurement, determined with I. R. temperature devices, supplemented by boat work with thermister probes and bathythermographs.

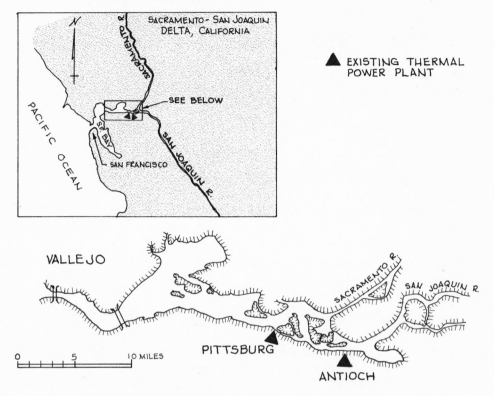

Fig. 1—.Delta section of the Sacramento-San Joaquin Rivers, California, showing location of Pittsburg and Antioch Power Plants.

Figure 3 shows the Contra Costa plant on the San Joaquin River, a much lower megawatt generation at this time (471), and a 16-degree temperature differential. The cooling-water flow is 845 second-feet, and note that this is a much shallower area.

If you look at the temperature-depth stations, the second one from the left (Number 2), you see a thin line which shows the heat distribution with depth. Most of the warm water is carried in the surface film.

Figure 4 shows an ocean-side conventional steam plant. This is about 880 megawatts, at this particular time, although Morro Bay is a 1000-megawatt station. Observe the surface-temperature canopy at this particular plant. The cooling-water flow is 1050 cfs, the temperature differential 20° F., and in this picture, what you can really see is about an 8-degree differential.

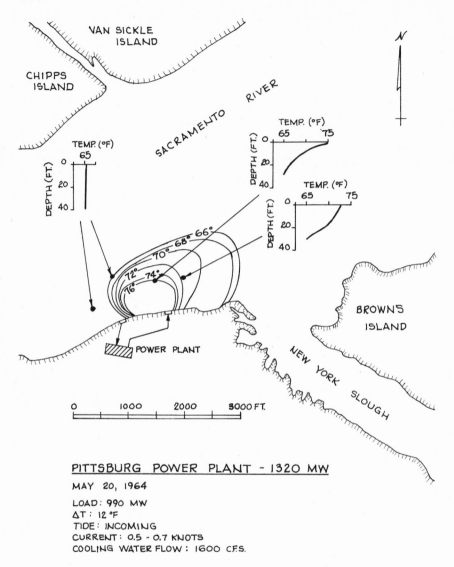

PITTSBURG POWER PLANT - 1320 MW

MAY 20, 1964

LOAD: 990 MW
ΔT : 12 °F
TIDE: INCOMING
CURRENT: 0.5 - 0.7 KNOTS
COOLING WATER FLOW: 1600 C.F.S.

Fig. 2—.Temperature distribution at Pittsburg Power Plant, California, on May 20, 1964.

Figure 5 shows some surface-temperature measurements at the Morro Bay power plant relating the isotherm temperature rise to the acres enclosed within the isotherm. You can see the different measurements; and by drawing a line over on the left-hand side on the graph, it can be seen that a 2-degree increase affects about 100 surface acres, or maybe

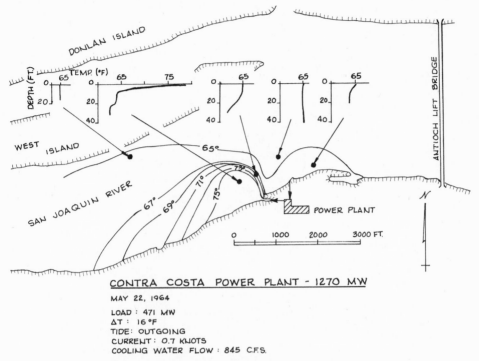

CONTRA COSTA POWER PLANT - 1270 MW

MAY 22, 1964

LOAD : 471 MW
ΔT : 16 °F
TIDE: OUTGOING
CURRENT : 0.7 KNOTS
COOLING WATER FLOW : 845 C.F.S.

Fig. 3—.Temperature distribution at Contra Costa (Antioch) Power Plant, California, on May 22, 1964.

slightly more. Looking at a 12-degree temperature rise, the worst we can see is about one acre.

Figure 6 shows the same temperature scale on the bottom, but distance on the other axis. From this figure, we can get an idea of what the upper limit is. A variety of conditions are shown to give an idea of what these things look like. The point I want to bring out is that you can build 1000-megawatt generating stations on the coast or in estuarine areas and you're not going to warm up the whole coastline. You could build 2000- , 3000- , or even 8000-megawatt units, as we are now planning in California.

We have a neat thing in California, inasmuch as Point Conception is a normal dividing line between so-called cold-water-warm-water forms. Species which are normally found in the cold-water regime north of Point Conception are called cold-water forms. If they're usually found south of Point Conception, or the warmer-water areas, they are called warm-water forms.

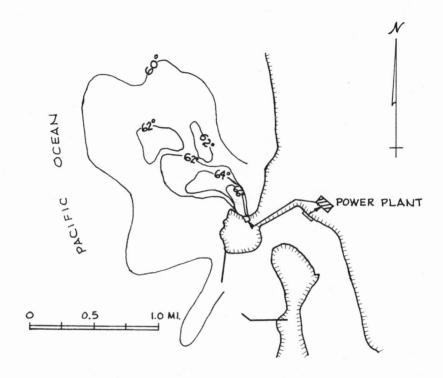

MORRO BAY POWER PLANT - 1030 MW

SEPTEMBER 12, 1963

LOAD: 833,000 KW
ΔT : 20°F
COOLING WATER FLOW: 1,050 C.F.S.

Fig. 4—.Temperature distribution at Morro Bay Power Plant, California, on September 12, 1963.

At Diablo Canyon, where we're building a 1000-megawatt station, we're already talking about our second unit. We find 66 species in a comparable area, 47 percent of them cold-water forms. When we consider all of Diablo, it runs about 65 percent cold-water forms (Diablo presently has no power plant).

In general, what I'm trying to say is that you have a very definite biological response to these discharges. I think you could predict what this biological response is by the use of temperature-prediction equations, I just want to dispel this idea that a 1000-, or 2000-, or 3000-

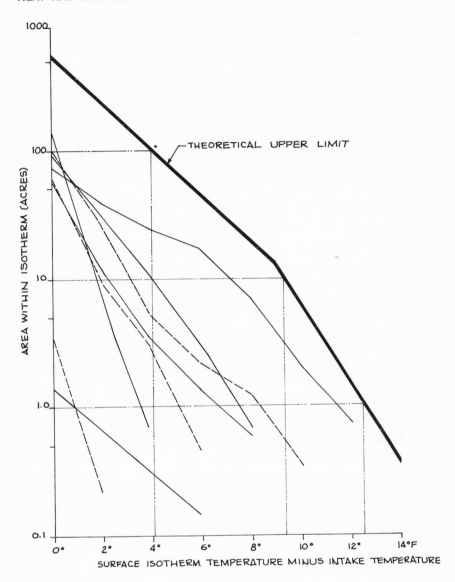

MORRO BAY POWER PLANT
RESULT OF NINE RUNS

Fig. 5——.Relationship between surface-temperature elevation and surface area affected for 9 different surveys at Morro Bay Power Plant, California.

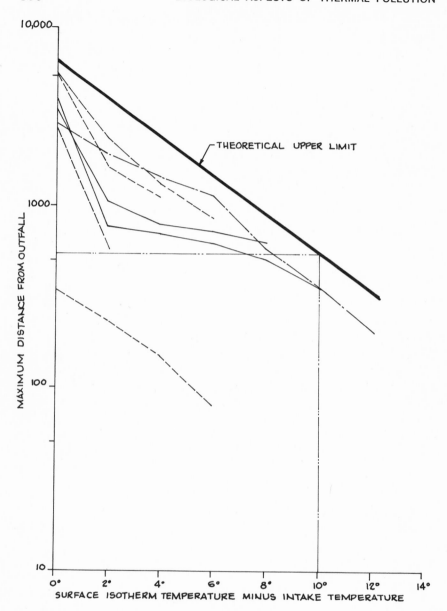

MORRO BAY POWER PLANT
RESULTS OF NINE RUNS

Fig. 6——.Relationship between surface-temperature elevation and maximum distance at which temperature was measured for 9 different surveys, Morro Bay Power Plant, California.

TABLE 1. Invertebrates Identified at Morro Bay Power Plant and at Diablo (No Power Plant)

Location	No. of Species	% Coldwater Forms
1) Morro Bay P.P. Discharge Canal	21 sp.	5% C.W.
2) Morro: 300 ft. out	34 sp.	33% C.W.
3) Morro: 500 ft. out	54 sp.	61% C.W.
4) Diablo: No Power Plant Similar area	66 sp.	47% C.W.
5) Diablo: All Areas	—	65% C.W.

megawatt plant is going to warm up considerable areas of the coastline. It isn't going to happen on the Pacific Coast.

Gerald R. Bouck: I was just talking to someone who seemed to be a little confused on a couple of points concerning the Columbia River. I think perhaps I can clear up some of them.

The Hanford area is the last remaining relatively free-flowing section of the Columbia River, and it is about the only section of the river where salmon spawn. There is a dam proposed, to inundate this reach; and this will, possibly, be the end of salmon-spawning in the main-stem Columbia River.

With that in mind, there are proposed to be many (38 ?) thermal-electric stations on the Columbia River; and none of these would be discharging into a free-flowing river with the hydraulics characteristics that have been discussed here, but rather they would be discharging into impoundments. Now, whether this would worsen the situation or not, I don't know, but I thought it ought to be pointed out.

Second, I'd like to ask Roy about the dissolved-nitrogen problem on the Hanford region of the Columbia. Mr. Snyder has implied that, if the plumes of hot water don't kill these small salmon as they go downstream towards the ocean, when they reach the thermally-mixed area of three or so degrees warmer, the nitrogen supersaturation problem will kill them. I understand that the intake for your hatchery is located down below the reactors. Do you have a problem with nitrogen supersaturation in your Columbia River hatchery water?

Nakatani: We certainly have nitrogen-supersaturation problems in our hatchery, unless we take precautionary or corrective measures, especially when we heat the water for temperature-controlled experiments. Obviously, when we heat our hatchery water, we get supersaturation;

but the heated water is agitated over a splashing device to bring the dissolved gases to air saturation.

In our hatchery, on occasions, we have experienced the so-called gas-bubble disease in young fish. There seem to be at least two situations when we talk about "gas-bubble" disease: on one hand, we observe the disease when fish are exposed chronically to supersaturation like in our hatchery fish; and on the other hand, I think what Mr. Snyder is concerned about is more of an acute situation, where fish in supersaturated waters are suddenly brought to the surface. The Bureau of Commercial Fisheries is experiencing this latter type of problem in sampling young fish in gate-wells of Priest Rapids Dam above Hanford.

If the supersaturation of dissolved gases causes a significant problem or damage to small fish, then this problem is further aggravated by a sudden elevation of temperature. We haven't investigated the interaction of elevated temperatures and dissolved gases, but we are planning some work with adult salmon this year.

Several years ago, we proposed to investigate the etiology of the "gas-bubble" disease developed under a chronic situation for basic scientific interest. We had observed young chinook salmon being reared in supersaturated waters at Rocky Reach Dam and carried out some histopathology on the diseased fish. We have this information, but I believe Mr. Snyder is concerned about a different situation. Further, I believe the projected plans on water management call for all the flow going through the turbines with little or no spills. Hence, the problem of supersaturation of dissolved gases is with us only when much water is spilled because the supersaturation is related to the manner in which the water is spilled. Is this correct, Mr. Snyder?

Snyder: Yes, this is right. Let me say a word about gas-bubble disease. The effect of nitrogen gas on salmon appears to be aggravated by temperature increases. This, in effect, is what happens when fish come up towards the warmer surface. This is where the Bureau biologists first witnessed gas-bubble disease in the gate-wells of hydroelectric dams. We dipped fish out of the gate-wells, put them in the holding containers for marking, and they effervesced.

I am suggesting that, if the fish do not enter the heated portion of the plumes at Hanford, they are forced to the surface of the water by hydraulic action; and if the sharp 30° F.-increases do not kill them, the lesser temperature increases plus the N_2 effects will kill them.

Bouck: I think it's important to recognize the fact that any increase in temperature will also increase the nitrogen partial pressure and that this will be a characteristic of any river. Hence, the heating of river water may cause nitrogen-embolism problems that could be acutely lethal or chronic and cause hyperplasia from hypoxia, and this may very well be one of the main, characteristic dangers in any river that is heated to sublethal temperatures.

There's one other thing that I think someone should comment on: namely, that it has been implied that the increase in the salmon-spawning population in the Hanford reach in recent years has been due to some benefit or at least would appear to be due to some benefit from Hanford. In contrast, it is also alleged that decreased commercial fishing in the lower Columbia actually allows this increased spawning population.

Chapter **11**   John S. Alabaster

EFFECTS OF HEATED DISCHARGES
ON FRESHWATER FISH IN BRITAIN

IT IS obvious that conditions in Britain must be very different from those in America, or even parts of Europe, with regard to the normal temperature of rivers, the temperature and discharge of heated effluents and the species composition of aquatic life. Yet some results that emerge from the work carried out are of general interest and these can be stressed rather than the quantitative detail, much of which is already published elsewhere (for example, Alabaster and Downing, 1966); at the same time, some gaps in our knowledge can be mentioned.

Water Temperature

Two extreme examples of normal seasonal river temperatures, taken from data in the *Surface Water Year Book of Great Britain,* which is published annually by the Ministry of Housing and Local Government, are shown in Fig. 1, the lower curve being for the River Tummel in Scotland during 1958, a wet year, while the higher one is for the River Medway in the south of England, a region of low rainfall, taken in 1959 when the summer was exceptionally warm and dry. Values for most other rivers would come within the range for these two examples, with a minimum less than 5° C. and a maximum seldom much above 20° C. The weekly range of temperature is generally less than 8 deg. C.[1] and the daily fluctuation

Crown copyright. Reproduced by permission of the Controller of H.M.S.O.

The author wishes also to acknowledge, gratefully, the considerable help given by the Central Electricity Generating Board, River Authorities, and the Water Pollution Research Laboratory in the field and laboratory studies.—J.S.A.

1. The abbreviation "deg. C." is used for the differences and ranges of temperature; "° C." is used for actual temperatures.

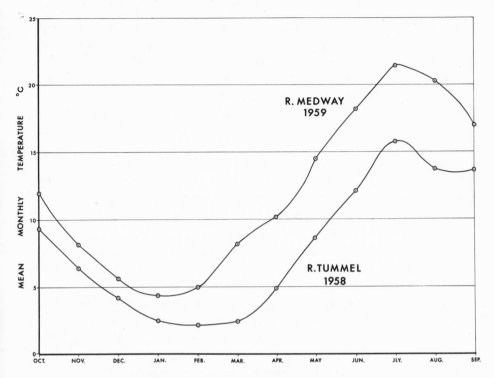

Fig. 1——.Examples of normal river temperatures in Britain.

less than 3 deg. C. On such background values is superimposed any increment derived from heated discharges.

By far the greatest quantity of heat discharged is in cooling-water from
the electricity-generating industry; and as the water passes through condensers, it is normally warmed 6 deg. to 9 deg. C., and in winter perhaps
10 deg. C. Future trends are for differentials of nearly 13 deg. C. in large
plants using estuarine or seawater for cooling purposes, while in some
other parts of Europe the increase is normally 7.5 deg. to 11 deg. C. and,
in winter, 12 deg. to 16 deg. C. (E.I.F.A.C., 1968).

About half the plant uses estuarine or seawater with direct cooling systems and some is located at the mouths of salmon and sea-trout rivers. The
remainder relies on river water for cooling purposes, just over half using
the direct system exclusively, nearly a third recirculating the cooling-water
through towers in which it is cooled before reuse (the "indirect" or "closed"
system) and the rest employing both systems in combination, using the
direct system when river supplies are adequate and cooling towers in ad-

dition whenever necessary. This is the "mixed" system in which the water circuits may be either separate or interconnected. In the latter case, the mixed water may be warmer before and after passing through the condensers than that used in direct systems because the effluent from the towers is generally warmer than normal river water. The total amount of heat to be dissipated is much the same per kwhr whether discharged to a body of water, as with direct cooling, or mostly to the atmosphere, as with indirect cooling.

The demand for electric power is highest in winter and during daytime. Peak requirements in summer can therefore coincide with maximum river temperatures and times when the effectiveness of cooling towers is reduced because of high air temperatures. Only a minority of the stations using fresh water generate continuously to supply part of the base load of electricity. The majority generate mainly at times of peak demand and are subject to wide variations in load, often being shut down during part of the night and at weekends; their effluents are therefore discharged largely during the day, and may vary in temperature with variation in load, since the rate of flow of cooling-water may not match the loading. Temperatures can also become abnormally high when effluent accidentally recirculates in a river back to the cooling-water intake; this may occur even when the point of intake is upstream of the outfall and the total river flow exceeds that of the cooling-water. On the other hand, such recirculation does not necessarily occur, even when the outlet is upstream of the inlet. Recirculation in towers dissipates heat to the atmosphere and, in mixed systems, reduces the volume of effluent discharged but increases its temperature; fluctuations in effluent temperature are therefore related to the proportion of cooling-water being recirculated.

The extent to which heated effluents increase the temperature of the receiving water depends not only on the relative temperature and volume of heated and unheated water but also upon the degree of mixing. In the rare circumstances when most, if not all, of the river flow is taken for cooling purposes, temperatures downstream of the outfall may be uniformly high throughout the water column and close to the values expected from the temperature and flow of effluent and river, though steep temperature gradients may occur in the immediate vicinity of the outfall. Such a situation is often found in the River Trent and examples of its temperature at Shardlow, three miles below Castle Donington Power Station, and at Trent Bridge, much further downstream, are shown in Figure 2 together with temperatures of a tributary upstream. The pattern of tempera-

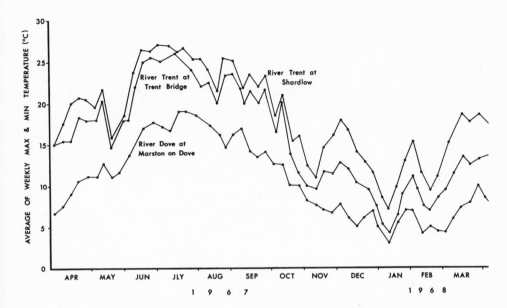

Fig. 2——.Water temperature in the Trent River Authority area.
Data supplied by the Trent River Authority

tures over the year below the Power Station generally follows that upstream but fluctuations are greater and the increment in temperature in winter—for example, at the end of November—is sometimes higher than in summer; the weekly ranges of temperature (not illustrated) are also much greater at the two lower points in the river than in the tributary.

Where, as sometimes happens, cooling-water flows through a long channel before discharging into a river, temperatures may be high because there is no dilution with cooler river water and, furthermore, they may remain relatively high even when the discharge stops, since ponded-up warm water is not flushed away. Such a situation is found at Peterborough on the River Nene, where the channel is over a third of a mile long and where the increment in temperature over ambient river values is generally higher than for the Trent and often more than 12 deg. C.

Fairly uniform temperatures may also occur in parts of large estuaries where mixing is good; in the Tyne, the increment has been shown to be about 3 deg. C. under conditions of low fresh-water flow (Swain and Newman, 1962) and in the Thames, the estimated increase is about 4 deg. C. (Gameson, Hall, and Preddy, 1957).

In most situations, however, in both rivers and estuaries, the hottest

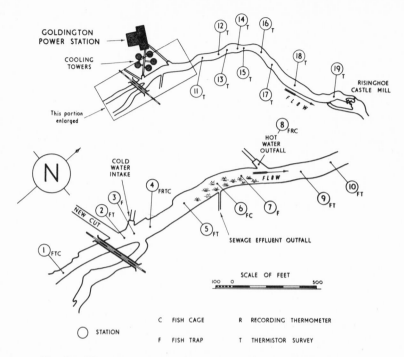

Fig. 3—.Diagram and map of the River Ouse at Goldington.

water persists downstream as a layer on the surface, leaving the stream-
bed either unaffected or warmed only a few degrees above normal, as in
the River Ouse, Bedfordshire, one of the sites of which is shown diagram-
matically in Figure 3. The distribution of temperature at the surface and one
foot from the bottom is shown in Figure 4 and is typical of places where
only a fraction of the river flow is used by the generating station and also
where the effluents are discharged intermittently.

Though water is often stratified in this way near outfalls, the layers tend
to mix together further from the effluent source. Moreover, weirs are often
found within a few miles downstream, and any stratification of the water
above the weir may be destroyed below. Other factors, such as tributary
and infiltration water, solar radiation, ambient air and ground temperatures, and distance downstream from the power station, influence river
temperature. In estuaries, such as that of the River Usk studied by Swain
and Newman (1957), vertical and horizontal stratification may be evident
only close to the effluent outfall at slack water. Figure 5 illustrates the
tendency for water in excess of 20° C. to remain in the immediate vicinity

(a)   SURFACE   TEMPERATURES

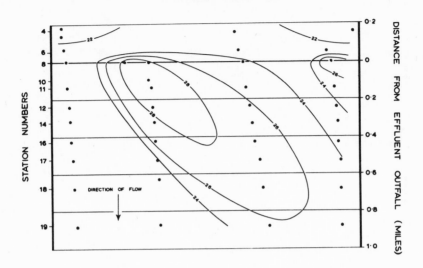

(b)   TEMPERATURES   ONE   FOOT   FROM   RIVER   BOTTOM

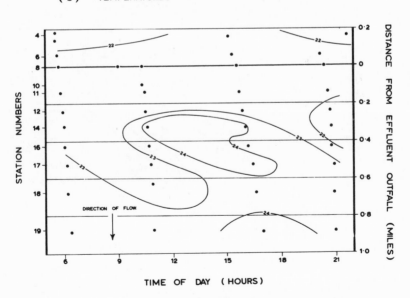

—24—   24° C. ISOTHERM   •   POINTS   WHERE   TEMPERATURES   WERE   MEASURED

Fig. 4—.Temperature distribution in the River Ouse at Goldington,
20 August 1959.

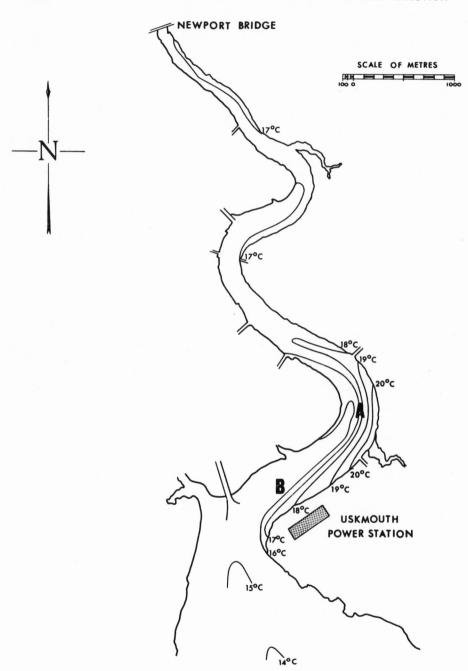

Fig. 5—.Temperature distribution in the estuary of the River Usk.

of the outfall at slack low water, while temperatures as low as 16° C. occur on the opposite side of the estuary. These conditions are temporary and the influx of seawater quickly disperses the hot water until, at high water, stratification is almost nonexistent and temperature differences in the whole of the lower estuary are no greater than 1 deg. C. The transient effect of the heated discharge is illustrated in Figure 6, where the difference between average temperatures at positions A and B (shown on Fig. 5) is shown for several tidal cycles. The slight change in pattern after 1961 may be accidental but coincides with the introduction of a mid-stream outfall.

From the foregoing, it must be clear that conditions in rivers and in estuaries receiving heated effluent are extremely diverse and variable.

Lethal Temperatures for Fish

Laboratory studies help in assessing the likely effect of heated effluents on fish. Various authors have shown that, when temperatures are gradually raised, the levels ultimately lethal are close to 30° C. for Atlantic salmon, brown trout, and sea trout; about 33.5° C. for roach; and probably at least 35° C. for tench. Cocking (1957) found that roach cannot be kept in good condition in the laboratory unless temperatures are lower than 30° C. With the onset of summer, the temperature of rivers where heated effluent is continuously discharged may gradually reach but seldom exceed 30° C. If this were to be maintained for long, one would expect trout to be killed and some species of coarse fish to be adversely affected.

In the field, however, temperatures in the vicinity of effluent outfalls may fluctuate sharply. The maxima, though lower than the ultimate lethal levels determined when temperatures rise gradually, may kill fish more or less quickly, depending upon the temperature to which the fish have been acclimated beforehand. As an example, the resistance of salmon and sea-trout smolts is illustrated in Figure 7; temperature changes corresponding to 50 percent kill of fish at 100 and 1,000 minutes are plotted against acclimation temperature and shown as two diagonal hatched strips representing the survival of fish transferred from cool fresh water to heated fresh water and heated seawater, respectively. The rise in temperature expected across condensers is shown as a horizontal hatched band, potentially lethal conditions being indicated where this cuts and overlaps the diagonal areas. This shows that smolts acclimated to temperatures prevailing during their migration withstand the normal increase in temperatures expected across condensers except during the latter part of the period of migration. The risk would be greatest where fish were transferred from

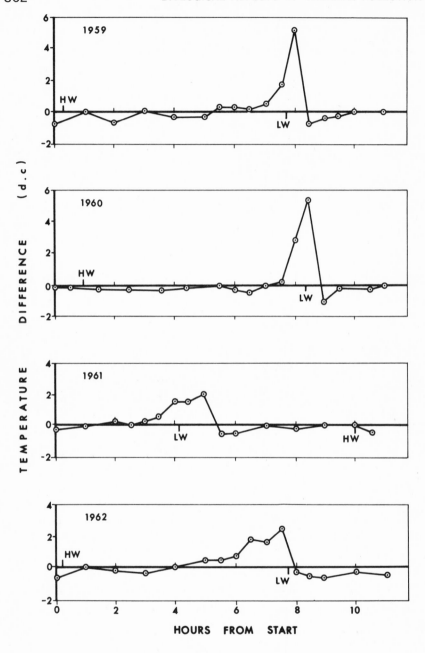

HW : HIGH WATER   LW : LOW WATER

Fig. 6——.Difference between mean temperature at positions A and B in Fig.
5 (B-A) in the River Usk.

MEDIAN LETHAL TEMPERATURES

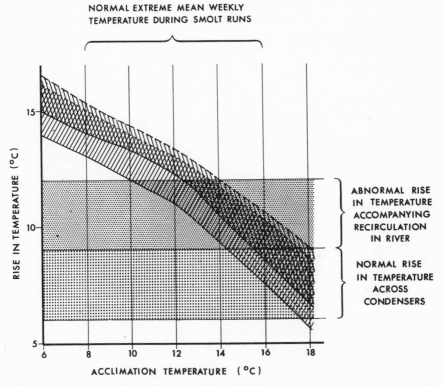

Fig. 7——.Relation between lethal and acclimation temperatures for salmon and seatrout smolts.

cool fresh water to warm, full-strength seawater, as might happen where power stations are located near estuarial barrages; the fish would show maximum resistance to increased temperatures if transferred to 30-percent seawater; and where exposed to seawater of full strength, some hours before being subject to a rise of temperature, they would be more resistant than fish experiencing a simultaneous increase in temperature and salinity. More needs to be known about acclimation to both seawater and temperature and the interaction of other factors, such as day length.

Other species are more resistant than salmonids; thus, for example, acclimation temperatures at which a change of 8 deg. C. is lethal in 1,000 minutes or less is 18° C. for trout, 22° C. for perch, and 26° C. for tench. Assuming fish to be acclimated to normal summer river temperatures (generally no more than 20° C.), sudden exposure to effluent heated 8 deg. C. higher than this would be lethal to trout but would seldom kill coarse fish.

It is therefore to be expected that, in summer, most heated effluents would be lethal to trout acclimated to normal river temperatures, and that some might also kill the sensitive coarse fish. Field tests with caged fish in summer have indeed shown that all trout and a substantial proportion of the roach, perch, and gudgeon, but none of the tench and carp, died within 24 hours of transfer from ambient river temperatures to heated effluent. Such tests are not strictly comparable with those conducted in the laboratory, because temperatures are rarely constant and the rate of mortality is sometimes irregular. If anything, survival in the field appears to be underestimated, even after allowing for the presence of other toxic factors in the water. Probably maximum temperatures are limiting, but it is also possible that handling procedures in the field are stressing the fish more than they do in the laboratory; more needs to be known about these aspects.

In winter, the possibility arises of fish being acclimated to temperatures higher than ambient, as in the Trent and the effluent channel at Peterborough, and then being exposed to unheated river water and a sudden drop in temperature when stations come off load, though at present this seems unlikely to occur often. Fish kills have not been reported, but they may be less obvious than those caused by a sudden increase in temperature if they are long delayed and individual times of survival are spread over a long period. This aspect of thermal pollution has not been specially studied, and perhaps more attention should be given to it.

Observations on fish caged in heated effluents and in receiving streams can detect conditions which are potentially, though not necessarily, lethal to free-living fish. Furthermore, where the water is thermally stratified, temperatures near the stream-bed may not be lethal, at all; and in addition, fish exposed to heated water near outfalls may have become partially acclimated to temperatures above normal, becoming more resistant to lethal high values; at Goldington, for example, (Fig. 3) perch caught downstream of the generating station survived in cages in the effluent for 140 minutes on one occasion, whereas those transferred from upstream survived for only 48 minutes.

There are occasional reports of fish kills in summer, but I have observed a mortality among free-living fish only once under normal operating conditions when the temperature in an effluent outfall rose exceptionally from 30.5° C. to 36.5° C. in three hours. Relatively few fish were killed, and many of the larger individuals appeared to escape successfully. The only other instance of a mortality occurred during an experiment at Goldington when, having no flow through the cooling towers, recirculation of cooling-water was induced in the river, causing effluent temperatures to rise above normal. The effluent began discharging at 0700 hours, reached 30° C. in 2 hours, and by noon, river temperatures were lethal between Stations 10 and 6 (see Fig. 3) and at the surface at Station 4, as demonstrated by cage tests and by the presence of dead fish in traps which had been emptied at 0700 hours. Dead fish appeared at 1230 hours on the surface of the water between the outfall and Station 6 and, altogether, several hundred small gudgeon and a few other large fish were gathered. Since only a few large fish were found, even where those in traps were killed by the afternoon, it seems that others present near the outlet must have moved away to avoid the lethal conditions which developed there.

Fish Behavior in Temperature Gradients

Field observations suggest that fish may escape lethal conditions near heated effluent outfalls and this conclusion is supported by the results of experiments with fish in temperature gradients, an example of which is shown in Figure 8. Two batches of bream, one acclimated beforehand to cold water and the other to warm water, were placed in the cold and warm ends of a 500-foot-long channel in which a temperature gradient had been established. During the course of a day, the distribution of temperature was altered, and the two groups of fish, which remained in separate shoals, kept to the cold and warm water respectively. Similar results have been obtained with grayling, trout, bleak, perch, tench, and roach. Only very few large fish (1 out of 13 tench and 3 out of 38 bleak) but a large number of small gudgeon (359 out of 607) were killed in these experiments. The reason for the vulnerability of gudgeon under both field and laboratory conditions is not known.

At the end of these experiments, the fish have tended to occur at tem-

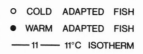

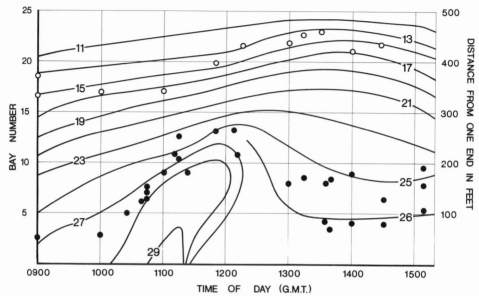

Fig. 8—.Positions of bream shoals and distribution of temperatures in a winding channel 500 feet long and 4 feet wide.

peratures higher than those selected initially, suggesting that some acclimation to temperature had taken place. With bream exposed to increases of up to 8.3 deg. C. within several hours, final selected temperatures increased on average 1 deg. C. for every 4 deg. C. of change experienced. One would therefore expect their resistance to lethal temperatures to increase (as found with perch in the field) and perhaps their avoidance of a given change of temperature to be reduced if they were continuously exposed to temperature fluctuations in the neighborhood of heated-effluent outfalls.

The effect of more prolonged acclimation has been studied by keeping roach for several weeks in the channel in extensive gradients of temperature; the temperature selected increased steadily, reaching between 20° and 25° C. after two weeks. This suggests that, when normal river temperatures do not exceed 20° C., heated effluents would attract this species, especially as its final preferendum could be higher than the values found in this relatively short experiment; carp would be similarly attracted, but trout are likely to be repelled except perhaps during the wintertime.

It appears from this that, in general, a sudden increase in river temperature of 6 deg. to 9 deg. C., caused by the intermittent discharge of a heated effluent, would generally suffice to drive the majority of fish away from lethal conditions. No information is available on the temperatures avoided by migratory stages of Atlantic salmon and sea trout, but it seems likely that smolts and adults, in spite of their migratory urge, would also react to and avoid sudden temperature changes. At the same time, there would be a tendency for fish to become acclimated to temperatures above normal and for them to select preferred temperatures in the neighborhood of effluent outfalls, provided temperatures were not too high or too variable.

Natural Fish Populations

Whatever inferences may be drawn from laboratory and field experiments, natural fish populations must also be studied to assess the effect of heated effluents on fish and fisheries. In general, too little is known about them. Direct counts can sometimes be made in rivers containing migratory species; but often, indirect sampling methods must be used, some of which may depend on the activity of the fish, which in turn may be affected by temperature. Sometimes the efficiency of sampling is so low that catch-effort or even mark-recapture methods of estimating populations cannot be used.

The distribution of coarse fish has been studied by several methods in some rivers receiving heated effluents. As expected, no significant difference was found between the populations above and below the outfalls at those stations, including Goldington, where the river water was markedly stratified and temperatures near the stream bed were the same as or close to ambient values. At other sites—for example, those on the River Trent, where the water was not stratified—significantly more fish were caught in traps in heated water in 1958 and in unheated water in 1959.

In Figure 9, the catch in traps at all stations at normal temperature and that at all stations above normal at both sites, expressed as a percentage of the mean catch at each site in each year, has been plotted against the mean water-temperature. Most roach and gudgeon are caught at temperatures between 21° C. and 27° C. and fewest at temperatures outside these limits. These figures probably reflect the abundance of the populations since they compare favorably with the size and composition of

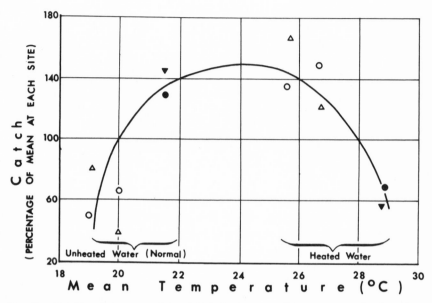

Fig. 9——.Relation between catch of fish in traps and temperature in the River Trent.

catches in seine-net samples and with estimates of numbers based on the recapture of marked fish. Such a distribution of fish in the Trent would be expected from the results in the gradient channel, where roach selected a temperature between 19.5° C. and 25° C. when given a choice in a temperature gradient; it is also consistent with Cocking's (1957) observation that temperatures must be less than 30° C. for roach to live healthily.

The Trent River Authority have continued to study the fisheries of the Trent; and although there is no further information on population sizes, recent observations show that the growth of roach is generally above average for Britain, at least until the fifth year of life.[2]

Conclusions about other species are not so clear-cut; but bream, tench, and carp also tend to occur in the warmest water available, bream being particularly abundant in the effluent channel at Peterborough. No un-polluted trout streams receive heated effluent, but there is little doubt that the fisheries would be destroyed if they did.

2. D. G. Holland: personal communication.

Estuaries containing migratory salmon and sea trout are so little affected by heated effluents that no effects on the passage of fish would be expected; this is confirmed in the River Usk, in which catches of salmon have remained constant over a number of years during a period when the power station has come into operation.

Indirect Effects

In rivers to which sewage and industrial wastes as well as heated effluents are discharged, the indirect effects of increased temperature are likely to be important; for example, the lethal effect of low concentrations of dissolved oxygen increases with temperature and the concentration of dissolved oxygen in solution is in turn determined by its solubility and an equilibrium between rates of oxidation, respiration, photosynthesis, and aeration, all of which are temperature-dependent. Furthermore, the time of survival of fish in toxic solutions is reduced at high temperature, yet threshold concentrations for survival may be increased at high temperature, that is, toxicity is reduced; thus, in polluted streams there may be advantages in maintaining a high temperature in winter. Finally, where a large proportion of a stream or watershed is abstracted to make up for evaporative losses in cooling towers and to prevent undue concentration of minerals in the cooling-water, there may be a significant loss of dilution water for polluted sources situated downstream.

Summary

The increase in temperature of the water used by electricity-generating stations for cooling purposes is often potentially lethal during the summer to trout and to the more sensitive species of coarse fish, such as roach, perch, and gudgeon, though not to tench and carp. Coarse fish are present in most rivers receiving these heated discharges but are rarely killed even close to the effluent outfall, probably because they avoid lethal conditions which are often confined to the surface layers, and also because those fish exposed to heated water in the neighborhood of outfalls have become partially acclimated to temperatures above normal and thus become more resistant. When fish are killed, effluent temperatures are higher than normal, either because of exceptional operating conditions or intentionally for experimental purposes, but many fish apparently escape successfully. Most danger occurs near the outfalls, particularly when effluent recirculates both in the river and through cooling towers. Effluent temperatures are not always potentially lethal, even in summer, and may attract roach, gudgeon, carp, tench, and bream. Indirect effects of heated effluents on fish in polluted rivers are likely to be important.

REFERENCES

Alabaster, J. S., and A. L. Downing. 1966. "A Field and Laboratory Investigation of the Effect of Heated Effluents on Fish." Fish. Investig. Series I 6(4).

Cocking, A. W. 1957. "Relation Between the Ultimate Upper Lethal Temperature and the Temperature Range for Good Health in the Roach (Rutilis rutilus)." *Nature,* London 180:661–662.

E.I.F.A.C. 1968. "Report on Temperature and Fisheries Based Mainly on Slavonic Literature." European Inland Fisheries Advisory Commission, Technical Report No 6. F.A.O., Rome (32pp.).

Gameson, A. L. H., H. Hall, and W. S. Preddy. 1957. "Effects of Heated Discharges on the Temperature of the Thames Estuary." *Engineer,* London 204:816–819; 850–852 and 893–896.

Swain, A., and O. F. Newman. 1957. "Hydrographical Survey of the River Usk." Fish. Investig., Series I 6(1):34pp.

Swain, A., and O. F. Newman. 1962. "Hydrographical Survey of the River Tyne." Fish. Investig., Series I 6(3):45pp.

DISCUSSION/ Edward C. Raney

JOHN ALABASTER's studies on the effects of heated effluents have been published (see Edward C. Raney and Bruce W. Menzel, 1967 [or revised edition 1968], A bibliography: "Heated Effluents and Effects on Aquatic Life with Emphasis on Fishes" for references). His summary today may elicit questions from the floor later. He informed me that no further work of this type is presently being done in Britain.

In the United States and Canada, we have more species of fishes and many research biologists who have produced numerous publications of investigations on the relation of temperature and fishes and other aquatic organisms. Such studies limited to salmonid biology probably number at least 400 references. More than 1200 references were included in the Raney and Menzel bibliography mentioned above. A new edition planned for October 1968 will have 400 additions. Other available bibliographies include J. A. Mihursky and V. S. Kennedy, 1967, "Bibliography on Effects of Temperature in the Aquatic Environment" (1,200 entries), Univ. Md. Nat. Res. Inst., Contrib. no. 326; and H. Bruce Gerber (and ASCE Committee on Thermal Pollution), 1967, "Bibliography on Thermal Pollution" (863 entries), Jour. Sanit. Eng. Div., Proc. ASCE.

The principles regarding temperature requirements of fishes have been elucidated through the efforts of many biologists, particularly, E. C. Black, J. R. Brett, H. O. Bull, J. Cairns, Jr. P. Doudoroff, F. E. J. Fry, W. S. Hoar, O. Kinne, F. B. Sumner, and F. J. Trembley; and, more recently, by C. C. Coutant, Clark Hubbs, J. A. Mihursky, K. Strawn, and D. E. Wohlschlag.

These principles are:

1. Fishes are cold-blooded or heterothermic. The body tissues quickly change to the temperature of the surrounding water, because the gills must

be in contact with water to permit the passage of oxygen and carbon dioxide.

2. Fishes can perceive minute differences in temperature—much less than 1° F.

3. Fishes have a lower and an upper lethal-temperature limit or tolerance. This is specific for each species and the requirements may differ for different stages in life history, such as spawning, development of eggs, and development of young. These limits permit adjustment to differences in seasonal temperatures, as well as minor fluctuations in temperature. The lethal temperature could be modified by other factors, such as oxygen, carbon dioxide, salinity, and ions of heavy metals. In some widely distributed fishes, infra-specific population differences in lethal limits are expected.

4. Fishes are able to acclimate to a given temperature (within the lethal limits) and do so regularly with temperature changes accompanying the seasons. A fish, when permitted to acclimate, may finally choose a preferred temperature.

5. Once acclimated to a given temperature, fishes acclimate more readily to an increase than to a decrease in temperature.

6. The activity of fishes and other cold-blooded animals usually increases with a rise in temperature. However, these relationships can be complicated. Some trouts, for example, show a maximum activity at a lower temperature. For example, the lake trout has an upper lethal-temperature of about 74° F., while its maximum maintained swimming speed is displayed at 61° F. (Gibson and Fry, 1954).

7. It has been shown that fishes can and will follow a gradient. They normally follow this gradient to or toward their preferred temperature.

An understanding of these principles will assist in understanding fish behavior and may be helpful in planning structures to guide and protect fishes.

An important service may be rendered by an experienced aquatic biologist in the early stages of the search for a site for a nuclear power plant and perhaps for almost any power plant. Many of the companies in the western part of the United States have young biologists on the staff, but most of the eastern companies do not. Many companies prefer to retain a consulting biologist. Often, after a few days' study, the biologist may prevent unnecessary large expenditures by making suggestions about where cooling water should be taken and where the effluent should be discharged.

Certainly in these times, no power plant with a substantial heated effluent should be built without a biological study and a thermal study on a model. The former study should cover at least three years before and continue for four or more years after operations begin. Probably the aquatic habitat near large nuclear plants will require continuous study after operations have started. In some heavily used navigable rivers, it might be possible to have a river laboratory set up on a barge like that presently employed on the Columbia River.

John Alabaster alluded to the proposition that many warm-water fish populations might do better if fewer individuals were present. I'm studying a reservoir now that has a large population of channel catfish, white crappie, and other warm-water fishes and which would probably serve much better as a fishery if only three-fourths as many individuals occurred. However, dead fishes attract attention, whether caused by natural mortality or otherwise. Advanced study in the area often assists with a rational explanation of lack of fishes or of fish kills. As far as we know, there is little of a problem, at the moment, of direct killing of valuable fishes at outfalls discharging heated effluents. What the sublethal or indirect effects are on the aquatic populations is still largely unknown.

At large nuclear-power stations where much cooling water, or a combination of cooling water and cooling towers is needed, the biologist can be helpful to engineers. Many cooling-water intakes use the same vertical traveling screen which was designed many years ago. The screen is often put in the wrong place. Some suggestions follow.

Stationary screens should be avoided. They are impossible to keep clean and fishes beat themselves to death upon them. If stationary or the usual vertical traveling screens are used, they should be placed at the entrance of the canal with escape pathways provided both up- and downstream, and the in-take velocity before the screen should be reduced to less than one foot per second. Combinations of the above with electricity, air bubbles, lights, sound, and water jets have been tried but with little success and are not recommended for large installations. The screen should be placed so that recirculation of warm water is less than 1° F. If possible, the screens should be placed in the canal at an acute angle, with a bypass arrangement. When heating screens for de-icing in the winter, use care not to create conditions which will attract fishes.

A very promising new type of screen is now in an advanced stage of study and development by Daniel W. Bates of the U.S. Bureau of Commercial Fisheries, Portland, Oregon.[1] It is a horizontal traveling-matching-current (velocity) screen which, when set in a canal at an angle of 30 degrees and provided with a by-pass, is self-cleaning and virtually 100 percent efficient in screening against fishes and other larger aquatic organisms.

1. Daniel W. Bates, Program Leader, Freshwater and Estuarine Research, U.S. Bureau of Fisheries, Portland, Ore., gave a paper on the horizontal traveling screen at the Nashville Symposium on August 15, 1968. See pp. 225–242: Parker, Frank L., and Peter A. Krenkel, editors. 1969. *Engineering Aspects of Thermal Pollution: Proceedings of the National Symposium on Thermal Pollution, Sponsored by the Federal Water Pollution Control Administration and Vanderbilt University, Nashville, Tennessee, August 14–16, 1968.* Nashville: Vanderbilt University Press.—P.A.K.

DISCUSSION FROM THE FLOOR

Kirk Strawn: What light conditions did you use in your temperature-choice experiments, and what was the photo period?

John S. Alabaster: We worked during the wintertime, using natural daylight, and the fish had been exposed, prior to the experiments, to the natural-light regime.

George E. Burdick: Is the mixing zone a concept in England on heated wastes? And if so, what are its parameters?

Alabaster: The conditions of consent to the discharge of heated effluents are based upon effluent temperatures and upon the river temperature downstream of the discharge point.

Fred March: Being an attorney, I'm just sort of sitting in here, but I work with water-pollution problems and related matters in the Interior Department. I'd like to ask the gentleman from England: Does England license its nuclear plants? Do you have an authority, that, say, permits applicants to build nuclear plants, or how are these nuclear plants built? Who passes judgment on whether or not a nuclear plant should be built in a certain siting location, and what steps are taken to prevent what you might call damage to fish and wildlife and other types of water resources or the environment?

Alabaster: With regard to water pollution, the authority proposing to build a power station must make a formal application to the River Authority to discharge an effluent of specified characteristics. The River Authority is empowered to lay down conditions of consent to the discharge, which, if disputed, may be subject to appeal to the Minister of Housing and Local Government.

In practice, we have not found this system breaking down, since most questions can be resolved by informal discussion at an early stage. It has worked well in the past, and I believe it will work in the future, too. Does that answer your question?

March: I certainly appreciate your information. In the United States, the Atomic Energy Commission doesn't actually license the many aspects of this that relate to thermal pollution. They will give you a license to build the plant, but they don't take into consideration, under persent law, many of the facts that may result in damage of the waters and related natural resources of navigable rivers.

This brings up the question, do you think it's desirable that some sort of national legislation should be enacted in the United States that would have one central authority to pass judgment on all aspects of all applications to build nuclear plants?

Alabaster: I have no experience of American conditions, and I certainly wouldn't venture an opinion on them. In Britain, our statute-law makes general provisions whereby the River Authority is given power to lay down conditions of consent to the discharge of effluents.

March: I appreciate your remarks very much; and in passing, I would like to state that there is legislation in this country, now pending in Congress, that attempts to face, among other things, the problems of thermal pollution from nuclear power plants: S. 3851 and H.R. 18667, both of the 90th Congress, would, among other things, add a new Section 111 to the Atomic Energy Act of 1954, as amended, which would give jurisdiction to the Atomic Energy Commission to consider and determine the thermal-pollution problems associated with nuclear-powered facilities constructed and operating under AEC license. S. 3330 and H.R. 16934, identical bills of the 90th Congress, known as the "Electric Power-plant Siting Act of 1968," would direct the Federal Power Commission to survey possible thermal and nuclear electric-generating-plant sites in the United States and to publish a plan of their location and would require the AEC to observe any national power-plant siting plan adopted by the federal government to implement this legislation.[1] And I might say that those who are biologists also must remember they have to work within the political system and that what they know about fish and wildlife and other animals and how thermal pollution affects these natural resources, they should make known to their congressmen and senators: to know about these particular problems that biologists know are of concern to life in river systems is the only way you can

1. See remarks of Senator Edward Kennedy, explaining the need for enactment of S. 3330 in the *Congressional Record,* April 17, 1968, pp. S4150–S4152.

protect the resources—if you are able to get the necessary tools put into law that would protect them.

Alabaster: I personally am against over-all standards which are aiming to cover large areas. We had a provision in the Rivers Prevention-of-Pollution Act, 1951, for the river authorities to make by-laws laying down standards for effluents. This section has been repealed, but we retain the section which gives them power to issue conditions of consent on individual discharges, bearing in mind local conditions, particularly volume and quality of dilution water. I'm very much in favor of this principle.

I'm also rather against the idea of limiting the temperature of an effluent or of a river to a single maximum value; one might consider the possibility of having different conditions at different times of the year. Certainly, young stages of some fish are much more sensitive than adults.

We have an example in Europe of a species, *Lota lota,* which is extremely sensitive in its young stages to temperature change greater than two or three C. degrees above zero, and for which a very stringent standard would be required in winter. So we might consider different temperature standards at different times of the year.

John G. Wilson: I note your use continuously of the term "River Authority"; and of course, being a layman, I would like clarification on this. Is this "River Authority" you speak of a national agency or is it, rather, a commission of some type concerned only with one specific river basin, or with many multi-river basins?

Alabaster: The River Authorities were set up as River Boards, following the passing of the River Boards Act, 1948, in which details of their composition, function, and powers can be found. They became River Authorities following the passing of the Water Resources Act, 1963, and are responsible for fisheries, pollution prevention, land drainage, and water conservation, in whole watershed areas.

They comprise mostly representatives of local government whose councils supply funds from a precept on local taxes, together with several members appointed by the Minister of Agriculture and Fisheries to represent fishing and land-drainage interests, and by the Minister of Housing and Local Government. The Authorities appoint their own technical officers to carry out their statutory functions.

Wilson: I noticed you used the word *authorities* in the plural. Does that mean that *one* River Authority has jurisdiction over *one* river basin?

Alabaster: In general, yes, but sometimes one authority would have several watershed areas under its contol. We have just over thirty authorities covering the whole of England and Wales.

Wilson: Of course, our system in the United States is quite different from that.

Alabaster: The system in Scotland is quite different, too.

Leon Verhoeven: This is rather a general comment I wish to make. Yesterday, Max Katz said, in effect, "We biologists are missing the boat if we do not take a definite stand on thermal pollution. It is all right for us to admit that we do not know everything we would like to know about the effects of above-normal water temperatures on salmon and steelhead."

It seems completely illogical, to me, for us to suggest that we should not oppose the addition of heat to anadromous-fish streams such as the Columbia River. We know that unfavorable temperatures cause fish to be moribund. We also know that moribund fish are more susceptible to disease and predators. We know that these increased temperatures may favor populations of predators, such as squawfish.

We know that these increased temperatures cause avoidance and sometimes straying. For instance, we have an example of a steelhead entering the Columbia River last summer, ascending McNary Dam, being tagged, dropping back downstream over McNary Dam, and then entering the John Day River, sixty or seventy miles downstream. Supposedly, at the time this steelhead came up the river, the John Day Dam was too low and too warm for the steelhead to enter, so it lay in the cooler, main-stem Columbia.

We have an example of chinook salmon by-passing the fishways at the University of Washington hatchery because the temperature was one or two degrees above normal. These fish strayed and spawned, unsuccessfully, on the shores of Lake Washington.

We have an example of sockeye going up the Columbia River before Wells and some of the other dams lower downstream from Wells were built, where these fish would come to the Okanogan River, which might be too warm at the time for them to enter; they would lie below Chief Joseph Dam, in the cooler water, until the Okanogan cooler off, and then they would ascend the Okanogan.

We also know that the chinooks at Hanford choose the bank opposite the discharge of heated effluent from that project.

The cost to the consumer to protect the environment will be small, but it will cost more, later; and the damage caused by projects that are built in the interim cannot be entirely compensated for. There are going to be more projects, because the ones that are being proposed now are just the beginning.

Fred Limpert: I think Dr. Bouck this morning left the impression that there would be thermal power-plants aggregating 26,000 megawatts' capacity located on the Columbia River. I don't think that there's anything in the present planning horizon that would put anywhere near this number on the main-stem Columbia. It is true that there may be that many sites; but even if there are, certainly many of the plants, even if they were put on the river, would not be once-through cooling.

Dr. Raney pointed out this morning the necessity for planning coordination to obtain proper siting. Bonneville Power Administration markets the power from all the federal projects in the Pacific Northwest through its extensive transmission system. We are also the primary agent for wheeling electric power from both public and private groups and responsible for major power contracts with the Pacific Southwest utilities.

Most of the economic hydro-sites in the Pacific Northwest have been used up; and in order to meet the estimated 15,000,000 to 16,000,000 kilowatts of thermal generation that will be needed in the Northwest in the next twenty years, we requested Battelle-Northwest to make a siting study of this area. They investigated the unique aspects of nuclear power-plant siting in the Pacific Northwest in six major areas, mainly Puget Sound, the lower Columbia, Western Oregon, the Middle Columbia, Northwest Montana, and Southeastern Idaho.

Example sites were selected within each one of these areas to illustrate some of the various problems and siting factors. They also examined the important economic factors incorporating nuclear power-plants into the predominantly hydro system that we have at the present time.

A 700-page report resulted, and included in this report is an examination of some of the environmental effects of the siting of these plants in various areas of the Pacific Northwest and some of the proposed means for taking care of these problems, including cooling towers. Because this is a 700-page report and was quite costly to produce, Bonneville is getting $10 a copy for it, but there are copies available, if people in the area would like to get them.

Joel W. Hedgpeth: In the course of the last couple of days, my name has been taken in vain a couple of times, and also I was credited with some of Dr. Strickland's fine remarks. But I have gotten the impression, whether rightly or wrongly, that insofar as disposal of heat on the Pacific Coast is concerned, the ocean is regarded as the most economical way to waste an awful lot of heat. I won't argue with Mr. Adams. I suspect that there is not going to be very much temperature effect, more than locally, in the vicinity of large reactors on the open, rocky shore. Of course the matter of estuaries or situations with funny little current problems is another matter, and each case must be studied individually.

We may be reaching a stage in the future when we can't afford to throw all this heat away. It might pay the power companies, expecially the Pacific Gas & Electric Company, which dearly loves to have us think of it as an eleemosynary and angelic institution, to consider some pioneering in this respect. They have a reactor in Humboldt Bay, and another one at Morro Bay, where they might want to reverse the currents, sometime, and combine power generation with experimental fish hatcheries.

Of course, one of the problems at this time with all our aqua-culture is that we are specializing in luxury foods, like oysters. We haven't really gotten to the point of wholesale production of essentially cheap food for the peasants. Whether we ever will, I don't know.

I did notice with interest that Dr. Alabaster managed to use some left-over power-plant facilities for a laboratory.

I really don't consider myself as an apostle of negative inaction, and I did try to say that there are some limits we know we can work within for certain problems. But we are, as biologists, confronted with a problem that engineers don't have. They can try new material out to build a bridge, and if it falls, then they can go back to some old material and build another bridge. But biologists have the problem of time. This is one thing that I think perhaps should also be emphasized in Dr. Allen's talk. Biological processes quite often consume more time than the engineers seem to think we ought to have. When you reflect on it, you must realize that it's only been since the 1940s, a very short time in terms of life processes, that we went into the exponential phase of polluting our environment; and in this short time, we have introduced atomic technology and its by-product of radioactivity, contributed by the physicists; pesticides and detergents, contributed by the chemists; and antibiotics, from the biochemists. I think we should make it plain that the biologists who are supposed to do something about pollution are not responsible for the planetary population explosion. Mr. Katz did seem to blame the biologists for the industrial revolution, at one point in his oration. The nature of biological processes is such that we do not have the safety factor that engineers have, and sometimes we get the warnings rather late. How did DDT get into penguins in the Antarctic? It got there much faster than through any possible food chain we can think of, and we should take this episode as another warning that we may indeed have run out of space for this sort of thing, as George R. Stewart said in his book, *Not as Rich as You Think.*

So there is no end to the problems; however, I do see some hope.

I don't think there are quite as many biological Philistines as Dr. Mount suggested. I also think that most of them are in the fish-and-game departments of the newspapers. I collect clippings, and I'm heartened to notice the general interest in conservation problems and in what we might call "marking the sparrow's fall" increasing in the newspapers. Many of our politicians are now discovering that you can make a career of conservation, and it's happening in the state of Oregon in a very spectacular way, as most of us know.

In fact, it had some of us worried, because one of these individuals is rumored to be concocting his own estuary bill for Oregon, without asking anybody else what should or shouldn't be in it, and that, of course, is potentially dangerous. But that's a local Oregon problem.

It certainly would be nice if we could have a nice, tight little situation, as Britain has, instead of our anarchy of authorities and county boards and everything else that approved things behind each other's backs. But this is the way we have it, so I guess we'll have to live with it.

Chapter 12   J. Frances Allen

RESEARCH NEEDS
FOR THERMAL-POLLUTION CONTROL

THERMAL pollution, although predicted as a critical problem in water pollution, has only recently received appropriate emphasis. A report to the Committee on Public Works in 1963, entitled "A Study of Pollution— Water," points out that, since 1900, electric power production has approximately doubled every 10 years and is expected to double again by 1970. The report further states that, unless controls are effected, this could mean an increase in heat pollution of more than 100 percent by 1973. This statement does not take into account the rise in water temperatures that will accompany the increase in the number of impoundments for hydropower, irrigation, navigation, flood control, and water-supply purposes. Since increased temperature reduces the amount of oxygen which water can hold in solution, the introduction of heat obviously results in additional pollution effects. Fortunately, there is an increasing expression of concern regarding the effects of thermal loading upon ecosystems. Before controls can be effectively instigated, it is essential that pertinent information be available for the development and establishment of controls. These comments will be restricted to the research needs of the biological aspects of thermal-pollution control, including mention of the related physical and chemical environments.

Whether we speak of thermal pollution, thermal effects, thermal enrichment, thermal alterations, or thermal additions, we are considering the same problems, the same research needs, and the same impact on our environment. From the presentations made at this symposium, there is no question that critical needs exist for knowledge and understanding of thermal effects on aquatic biota and their associated environments. Prob-

382

lems are recognized, but to define research needs to resolve these problems is another matter. When individuals were asked to express their views on approaches, there was some agreement but also some contradiction. Each sees the problem from his own experience and broad considerations are difficult to come by. Research needs should be recognized as an effort to find answers to thermal problems related to both beneficial and harmful uses of water. The better we understand the effects of heated effluents on the organisms and their environments, the better our position for controlling thermal effects, at the same time taking advantage, wherever possible, of thermal additions.

Research needs in the area of thermal effects, however, appear to center around two points: the direct and indirect effects of thermal pollution on the organisms, and the effect of thermal alterations affect organisms directly and indirectly is well established; but less understood is the effect of thermal alteration on the environment. Whether concern is with the effects on individual species, populations of a species, community relationships, or the entire ecosystem, one cannot treat temperature as an isolated factor. Temperature must be treated in relation to other environmental features, for it is indisputably established that ecological conditions tend to operate in conjunction with other conditions to produce their effects. Studies of temperature as a single limiting factor are often inconclusive, leading to difficult and questionable interpretation of the resulting data. When considering thermal effects, attention often appears to focus only on fishes; emphasis on the food-chain organisms is neglected. Larvae of shellfish may represent an item in the food chain of other animals, but concern also lies in the food chain for this same molluscan or crustacean larval form. It is apparent from a general survey of the literature that the majority of the research reported has been on temperature effects on fishes, and certainly this has been limited. Much less has been achieved on organisms in the food chain, animal or plant. A review of the subjects on the symposium's program emphasizes the necessary concern for thermal influences on all groups of organisms, regardless of how they are categorized—plankton, nekton, benthos, or marine, estuarine, freshwater forms—for each one represents an integral part of the over-all picture.

Pollution in an ecosystem is now measured only by gross observations, such as odor, fish kills, and excessive plant growth. The measure of pollution is *comparison,* before and after pollution has occurred. Results of thermal pollution cannot be predicted, unless conditions of the environment prior to heat addition are known. To establish necessary base lines, ecological surveys must be made of areas free from thermal pollution but where thermal loadings are a potential. This is not meant to infer that areas already receiving heated effluents should be excluded. Surveys should encompass diverse geographic areas so that temperate, subtropical, and tropical environments are included. The temperature requirements and tolerances of their biota are diverse, susceptibility to excessive heat being quite different for animals and plants in these respective environments. Such investigations should identify the organisms, determine their occurrence and distribution, their interaction (inter- and intra-specific), delineate the characteristics of the physical and chemical environment, and ascertain the reaction or interplay of the biological, physical, and chemical environments with each other. As has been pointed out by Patrick (1968) the functioning of natural aquatic communities must first be understood if we are to know how disturbance or alteration of the environment affects them. A large share of research has dealt with the effects on individual species and not on the biotic communities or ecosystems. Alteration should be directed toward knowing how a natural community is established and how natural aquatic ecosystems are structured and function. As further stated (Patrick, 1968), studies must include the relative values of available species' pool, invasion rate, size of area, diversity of habitats, and variability of the environment in the development of the community. The introduction of warm water into an ecosystem is known to affect the metabolic rate of organisms; but unless the functioning of a natural community is understood, it is difficult to evaluate these effects and relate them to methods for maintaining a natural community. Although Patrick's discussion is concerned particularly with lakes and streams, the same general principles hold for the entire aquatic environment. Regardless of its quick "indicator" contribution, we can hardly rely on SCUBA-diving techniques to determine the *subtle* changes in the environment which may not be immediately detectable. Sampling techniques, data collection, and analyses must be suitably designed. Concomitant laboratory and field studies are essential. No reliable scientist will predict, at least with assurance, the effects of heat addition on a specific body of water, on particular species or communitits, or on any ecosystem, without knowing what condition existed before the effect actually occurred. Effects may not be detrimental. They may be turned to beneficial use; but prediction of effects is essential to the initiation of correct thermal controls.

To attempt to evaluate biological effects by utilizing a single environmental variable, particularly temperature, as already mentioned, is unrealistic. As a classic example, temperature and dissolved oxygen are so interrelated that it is not feasible to study temperature without considering its influence on the amount and availability of dissolved oxygen. The same applies to temperature, salinity, and density. It has been found that the rate of cell division in certain phytoplankton is critically dependent upon temperature and salinity. We must not neglect to consider the large number of variables in any environment which are bound to subject the community members to diverse degrees of stress. Multivariate studies are coming close, in many instances, to duplicating naturally occurring conditions (Mihursky and Kennedy, 1967), a feature we should continually seek. Although such designs are important in focusing on a single parameter and determining important principles, experiments must eventually consider other variables (Mihursky and Kennedy, 1967).

Many more species of aquatic life must be subjected to investigation of thermal effects. Virtually nothing is known of the temperature requirements of the great majority of planktonic, benthic, and pelagic fauna and flora. Perhaps the least is known about the benthos and the most about fishes and phytoplankton, much of which is frankly fragmentary. According to Mihursky and Kennedy (1967), of the approximate 1900 species making up the American Fisheries Society's list of fishes from the United States and Canada, 1960, less than 5 percent have been specifically studied for their response to temperatures. To determine requirements for more than a relatively few species, as far as total numbers of plants and animals go, is realistically a hopeless task. A critical review of the literature verifies that the research on temperature effects is entirely inadequate. Although a myriad respectable projects are reported, none gives a complete treatment for species. Much data are limited to adults, some to larval forms, some to embryonic development; but there is no way to tie these together, for the work has been done on a diversity of species.

Selection of species as experimental animals is thus of critical importance. Such selection may be considered as part of a design for solution of a problem, rather than a research need, depending on one's point of view. It is difficult also to rank the species which require priority of attention for thermal effects. In general, criteria for selection and priority consideration are based on the economic importance, sport or commercial, and whether or not they are significant food organisms which support these populations as a part of the food web. To determine legitimately which is of greater importance—freshwater, marine, or estuarine life—is difficult. Thermal additions to the environment affect all three categories. Neither anadromous nor catadromous species should be neglected. It is more difficult to determine what species are the significant elements in the food chain, for feeding habits are not well known. Whether fishes are really exclusively selective in their feeding is questionable, although they may exhibit a preference. In this aspect, other biologists have much to offer to the group or individual designing research programs on thermal effects of heated waters on food-chain organisms.

As implied previously, knowledge of individual species is, in general, incomplete, as the facts known are usually restricted to only one or two life-history stages of the taxon under consideration. Too often, information is limited to a single stage, which may not be the most critically sensitive stage in the organism's development, particularly in regard to temperature and sources of nutrition. Practically speaking, the one best known may be the least sensitive stage: it may be the one easiest to maintain in the laboratory and/or the most readily identifiable, the reasons for its selection as an experimental organism, in the first place. Thermal effects on, and requirements of each of the life-history stages—for example, eggs, larvae, nymphs—are needed. Even effects on gametes in vivo should not be ignored. In fishes, rises in environmental temperatures bring about increases in metabolic rates and oxygen requirements, sensitivity to toxic materials, reduction in swimming speed, and increased avoidance reactions (Everts, 1963). These should receive attention, and as appropriate, include invertebrates and plankton, as well as fishes.

Lethal effects of heat addition are, of course, important; but of more fundamental significance are the sublethal effects. Heat-death is the obvious result of excessive exposure. The President's Science Advisory Committee's "Environmental Pollution Panel Reports: Restoring the Quality of Our Environment" (1965), states that the development of more effective techniques to measure the tolerance levels of different organisms to pollutants and to identify and assess the changes in abundance and distribution of organisms making up natural and man-controlled biological communities under pollution stress should be investigated. This statement can be applied to thermal additives, as well as to other kinds of undesirable or questionable additions to the environment.

Temperature tolerance, reflecting only survival capacity, may overshadow the long-term effect of excessive temperatures, which have had a definite impact on metabolism, rate of growth, normal development, spawning, and reproductive potential, as well as possible genetic influences. Tolerance is associated with acclimatization to temperature changes, whether they be sudden or gradual, higher or lower, over short or long periods of time. Thermal shock is an obvious outcome of an abrupt change in which the individual has had no opportunity to adjust. Research is needed to determine the capacity of the organisms to adjust to thermal changes and to determine whether acclimatization is entirely physiological. Research programs should be developed to determine the effect of temperature modifications over a span of several generations. This should be accomplished for important components of aquatic life.

Acclimatization involves survival at extreme temperature; reduced rates of metabolism, when exposed for some time to high environmental temperatures, and seasonal acclimatization of metabolism dependent on natural temperature changes (Nicol, 1967). Some organisms require the equivalent of a natural seasonal change. More information is needed on the effects of increased temperature on metabolism, so that the results will be known for all of the physiological processes. It has been observed that the metabolism of isolated tissues sometimes reflects the same temperature influence measurable in the intact animal (Nicol, 1967). Exploration of the effects of increased temperature on the breeding and reproductive potential of aquatic life has not been exhausted. Naylor (1965) points out in his review that available evidence indicates that eurythermal species can breed within a more narrow range of temperatures than they can tolerate as adults. The same writer suggests that temperature increase from heated effluents may prevent breeding and, thus, maintenance of the population would necessitate continual recruitment from the outside. Thus, effluents which narrow annual temperature range in a discharge area might well have biological effects. Gametogensis may be complete, but temperature may never be suitable to trigger spawning. The influence of augmented temperature on sterility has not been adequately explored. Research on thermal impact on breeding, the reproductive cycle, and potential is critical, needs further attention, and should be emphasized.

It is well established that some species grow faster and attain greater growth when the growing season is warmer and longer. In some forms, the rate of growth eventually decreases, while species in colder habitats reach a greater length than those whose growth initially is more rapid. In general, it may be that heated effluents would lead to increased initial growth rates and precocious maturity, resulting in a decrease of adult size and perhaps shortening the life cycle of some species (Naylor, 1967).

Temperature changes regulate the riverward and seaward migration of fishes. Thus, a thorough understanding of thermal stress on migratory populations must be developed; and this we do not have, except in bits and pieces. To determine the effects of increased temperature on the vertical and horizontal migration patterns of invertebrates and plankton is a desirable objective. Larval invertebrates are known to migrate, but what factors influence this behavior? Some fishes congregate at outfalls. Is this a response to temperature—or is it the current to which they are responding; or is it something else?

Temperature variations on predator-prey relationships should be investigated. This type of problem can well be studied as a part of understanding community relationships. Manipulation of temperature modifications could become an important control mechanism. Evidence indicates that higher temperatures favor disease in fish and in some other organisms. Relatively little is known of this phenomenon and additional information in this area would be useful. Relationship between temperature-salinity and susceptibility to disease should be examined. Are there other contributory causes (Ordal and Pacha, 1963), and if so, of what significance are they?

Estuarine organisms are known to have been pumped through power stations where they were subjected to abrupt thermal change which did not prove lethal. After-effects from such exposures have not been adequately pursued. Conseqently, nothing is known of the ultimate outcome of this stress situation other than immediate survival. For example, practically nothing is known about phytoplankton around thermal discharges.

The effects of introduced heat on the chemical and physical parameters of the envrionment have not been discussed but should be touched upon. Stability of the water column, solubility, buffering capacity, chemical reactions, rates of chemical change, current patterns, and water movements should be considered as they relate to alteration of the environment and subsequent temperature effects. The potential of chemical reactions with the sediments may be sufficiently significant to merit concern. There is a need for an exhaustive study on water-temperature patterns as influenced by man-made alteration of the natural water environment (Sylvester, 1963). Any consideration of the influences of heated effluents on the physiochemical environment should recognize these influences as contributory to the effects on the organisms. Cleansers and chlorine from thermal effluents may have an aggregate effect on the biota.

When discussing thermal additions, the impact of desalinization plants in estuarine areas should not be overlooked. Effluent brines from desalinization plants will be both heated and more saline than the receiving waters. Theoretically, they may be at nearly neutral density and, hence, subject to rapid mixing and dispersal. Actually, this may not occur, for while salinity is a conservative property of seawater, temperature is not. This means that heat may be withdrawn from the effluent plume more rapidly than the salt, leaving a mass of dense, hyper-saline water that may well sink to the bottom and form a relatively stable layer. Depending on other hydrographic features of the estuary, this could result in an altered circulation and stratification pattern, with subsequent ecological alterations. Attention should be directed toward determining whether this situation could arise and its consequences to the physio-chemical-biotic environment.

The National Technical Advisory Committee for Fish, Other Aquatic Life and Wildlife, in its "Summary of Research Needed for the Determination of Water Quality Requirements for Aquatic Life and Wildlife" treats temperature and dissolved oxygen together. The Committee recommends that laboratory and field research be carried out to determine answers to the following questions:

1. What is the significance of daily and seasonal fluctuations in temperature and dissolved oxygen?
2. How rapid may changes be without exerting an adverse effect?
3. Does it matter if the fluctuations are out of phase with the natural cycle?
4. What are the effects of wide fluctuations above or below the daily mean, and how much can temperatures be raised and dissolved-oxygen concentrations lowered from natural conditions without harming the biota or causing a change in its makeup?
5. At what temperatures and dissolved-oxygen levels will desired species be replaced by undesirable competitors?
6. What are the trade-offs relative to increased growth and total productivity versus parasites, disease, and mortality of different life stages?
7. What levels of temperature and oxygen are harmful, tolerable, and favorable for all the important species at all life stages, including egg, nymph, larvae, fry, etc.?
8. What are the effects of changes in daily and seasonal temperatures on migration, scope for activity, and the toxicity of waste and materials?
9. What are the maximum temperature and minimum oxygen levels that can be tolerated for short periods without harm?

10. What are the allowable over-all increases in temperatures at different seasons that do not produce undesirable effects on the biota?
11. What are the allowable and favorable levels of temperature and oxygen for all important species?
12. What are the requirements for low temperatures in the life cycle of important organisms?
13. What are the effects of stress due to pollutants on the temperatures which are lethal, tolerable, or favorable?
14. What are the beneficial effects of the addition of heat?

Eutrophication is an important process which may well be influenced by temperature. Other conditions being satisfactory, an increase of temperature will increase phytoplankton production, which will, in turn, disturb a habitat so that it becomes completely unsatisfactory for other aquatic life.

Royce (1967), when discussing research needs in the area of pollution and marine biology, points out that many people work separately on the fragments and have not dealt with research strategy. *How,* not *What,* is the root of the problem. Although not discussing thermal effects, he suggests that a need exists for somehow balancing the research of a dozen or more disciplines and bringing the findings regularly to bear on the decisions. The Federal Water Pollution Control Act recommends coordination of research, investigations, experiments, demonstrations, and studies relating to the causes, control, and prevention of water pollution. This, obviously, includes thermal effects, a situation which demands immediate and concentrated effort to resolve the severe problems that are already a reality.

It is essential that better methods be developed for measurement of thermal effects. Appropriate interpretation of measurements is a requisite to evaluation and any usable findings. Positive procedures for validly predicting the effects of thermal additives must be deveolped.

Research should be planned so that there are no gaps, the program is sufficiently complete, and predications can be made within a reasonable period of time.

SUMMARY

Research needs for freshwater, marine, and estuarine life have not been treated separately, for the needs are similar. Knowledge of the organisms occupying any of these environments is so inadequate that we need the same kinds of investigations. Design of such studies would be adapted to meet the unique and diverse conditions of a given situation. Immediate attention should be directed toward the determination of the effects of temperature on reproduction and growth of important fishes, shellfish, or other desired harvestable crop. This should be followed closely by, or simultaneously with, the determination of thermal effects on food-chain organisms. This approach is required for all life-history stages.

Research should reduce the lack of knowledge on thermal effects in conjunction with other parameters, particularly on long-term effects, direct and indirect. This includes their environmental requirements, relationships to and within the ecosystem. More data on thermal effects on all life-history stages and developmental processes of the organism are required. Effects on other physiological processes, migratory behavior, ecological stability, increased sensitivity to toxicity, susceptibility to disease, and predator-prey relationships must be considered. Long-term effects should be stressed.

More definitive work should be focused on temperature effects on the alteration of the environment, including salinity, temperature, current patterns, and chemical exchange. Although much data are available, especially on salinity and current patterns, more refined techniques are required to analyze the implications and interpretations of these data. Definitive surveys must be initiated in areas of temperate, subtropical, and tropical waters, where thermal loading is a distinct possibility. The influence of increased temperatures on the nutrient cycle, with subsequent eutrophication as a strong possibility, should receive due recognition.

It is essential that effects be measured appropriately so that they can be evaluated and interpreted to predict requisites for pollution control. Thermal loading should be considered as it relates to beneficial and non-beneficial water uses.

REFERENCES

American Fisheries Society. 1960. "A List of Common and Scientific Names of Fishes from the United States and Canada." Special Publication No. 2.

Committee on Public Works Staff Report. 1963. "A Study of Pollution—Water" Washington, D.C.: U.S. Government Printing Office.

Everts, Curtiss M. 1963. "Temperature as a Water Quality Parameter." In *Proceedings Twelfth Pacific Northwest Symposium on Water Pollution*, 2–5.

Mihursky, J. A. and V. S. Kennedy. 1967. "Water Temperature Criteria to Protect Aquatic Life." *A Symposium on Water Quality Criteria to Protect Aquatic Life*, American Fisheries Society Special Publication No. 4:20–32.

Naylor, E. 1965. "Effects of Heated Effluents upon Marine and Estuarine Organisms." *Advances in Marine Biology* 3:63–103.

Nicol, J. A. Colin. 1967. *The Biology of Marine Animals*, 2d ed. London:Sir Isaac Pitman and Sons, Ltd.

Ordal, Erling, Jr., and Robert E. Pacha. 1963. "The Effects of Temperature on

Disease in Fish." In *Proceedings, Twelfth Pacific Northwest Symposium on Water Pollution.* 39–55.

Patrick, Ruth. 1968. "Water Research Programs—Aquatic Communities." U.S. Department of the Interior, Office Water Resources. [Washington, D.C.: U.S. Gov't. Printing Office.]

Royce, William F. 1967. "Population and Marine Ecology: Research Needs and Strategy." *Pollution and Marine Ecology* 53–60.

Sylvester, Robert O. 1963. "Effects of Water Uses and Impoundments on Water Temperature." In *Proceedings, Twelfth Pacific Northwest Symposium on Water Pollution,* 6:27.

U.S. Executive Documents. President's Science Advisory Committee. 1965. *Environmental Pollution Panel Reports: Restoring the Quality of Our Environment.* [Washington, D.C.: U.S. Gov't. Printing Office.]

PARTICIPANTS IN INFORMAL DISCUSSIONS

JAMES R. ADAMS
Pacific Gas and Electric Company
Emeryville, California

JOHN S. ALABASTER
Principal Scientific Officer
Ministry of Agriculture, Fisheries,
and Food
London, England

GERALD R. BOUCK
Pacific Northwest Water Laboratory
Federal Water Pollution Control
Administration
Corvallis, Oregon

GEORGE E. BURDICK
New York Conservation Department
Albany, New York

JOHN CAIRNS, JR.
Research Professor, Department of
Biology
Virginia Polytechnic Institute
Blacksburg, Virginia

STEVEN CARSON
Maspeth, New York

WILLIAM A. CAWLEY
Acting Director, Pollution Control,
Technology Branch
Division of Research
Federal Water Pollution Control
Administration
Washington, D.C.

ROBERT L. CORY
USGS
Oxon Hill, Maryland

CHARLES C. COUTANT
Pacific Northwest Laboratories
Battelle-Northwest
Richland, Washington

PETER DOUDOROFF
Professor of Fisheries and Wildlife
Oregon State University
Corvallis, Oregon

PHILIP A. DOUGLAS
Sport Fishing Institute
Washington, D.C.

GEORGE J. EICHER
Portland General Electric Company
Portland, Oregon

REX A. ELDER
Director, Engineering Laboratory
Tennessee Valley Authority
Norris, Tennessee

CARLOS FETTEROLF
Chief, Water-Quality Appraisal Unit
Michigan Water Resources
Commission
Lansing, Michigan

NORRIS R. FITCH
Northern States Power Company
Minneapolis, Minnesota

JOHN W. FOERSTER
Marine Research Laboratory
University of Connecticut
Pawcatuck, Connecticut

RICHARD F. FOSTER
Pacific Northwest Laboratories
Battelle-Northwest
Richland, Washington

WALTER A. GLOOSCHENKO
Florida State University
Tallahassee, Florida

J. A. ROY HAMILTON
Pacific Power and Light Company
Beaverton, Oregon

393

JOEL W. HEDGPETH
Professor of Oceanography
Yaquina Biological Laboratory
Marine Science Center
Oregon State University
Newport, Oregon

CHARLES HODDE

C. P. IDYLL
University of Miami
Miami, Florida

MARC J. IMLAY
Federal Water Pollution Control
Administration
Duluth, Minnesota

LOREN D. JENSEN
Johns Hopkins University
Baltimore, Maryland

DONALD R. JOHNSON
Bureau of Commercial Fisheries
Seattle, Washington

MAX KATZ
University of Washington
Seattle, Washington

DANIEL F. KRAWCZYK
Pacific Northwest Laboratories
Battelle-Northwest
Corvallis, Oregon

PETER A. KRENKEL
Chairman, Department of Environ-
mental and Water Resources
Engineering
Vanderbilt University
Nashville, Tennessee

FRED A. LIMPERT
Bonneville Power Administration
Portland, Oregon

FRED MARCH

W. C. MASON
University of Washington
Seattle, Washington

DONALD I. MOUNT
Director, National Water-Quality
Laboratory
Federal Water Pollution Control
Administration
Duluth, Minnesota

KRYSTYNA MROZINSKA
World Health Organization
Katowice, Poland

R. E. NAKATANI
Director, Ecology Section
Pacific Northwest Laboratories
Battelle-Northwest
Richland, Washington

WHEELER J. North
Associate Professor of Environ-
mental Health Engineering
California Institute of Technology
Pasadena, California

RAY T. OGLESBY
Ithaca, New York

GERALD T. ORLOB
President, Water Resources Engi-
neers, Incorporated
Walnut Creek, California

RUTH PATRICK
Curator, Department of Limnology
Academy of Natural Sciences
Philadelphia, Pennsylvania

ROBERT PHILLIPS

DAVID R. SCHINK
Isotopes, A Teledyne Company
Palo Alto, California

ROBERT D. SMITH
Bendix Marine Advisers, Incor-
porated
San Diego, California

GEORGE R. SNYDER
Program Leader, Prediction of
Environment of the Columbia River
Basin
Bureau of Commercial Fisheries
Seattle, Washington

JOHN C. SPINDLER
Montana State Department of
Health
Helena, Montana

KIRK STRAWN
Texas A & M University
College Station, Texas

JOHN B. STRICKLAND
Research Oceanographer
Scripps Institute of Oceanography
La Jolla, California

LEON A. VERHOEVEN
Pacific Marine Fisheries

E. JACK WEATHERSBEE
Oregon State Sanitary Authority
Portland, Oregon

EUGENE B. WELCH
Supervisor, Biological Section
Tennessee Valley Authority
Chattanooga, Tennessee

JOHN G. WILSON
Chairman, Waters and Dams
Commission
Oregon Wildlife Federation
Portland, Oregon

CHARLES WOELKE
Washington Department of Fisheries
Olympia, Washington

CHARLES B. WURTZ
Chairman, Department of Biology
La Salle College
Philadelphia, Pennsylvania

INDEX

Academy of Natural Sciences of Philadelphia, 179
Acclimation: definition, 234; time, 235, 241; to temperatures higher than ambient, 364
Acclimation temperatures, 243, 255
Acclimatization: definition, 234; research needed, 387
Adriatic marine fish: critical line of temperature, 245
Adriatic Sea, 82
Advected energy, Q_v, 42
Aeration: provided by turbulence, 6; local, 66, 68; temperature-dependent, 369
Air pollution: and water pollution, 63
Air temperature: data available from weather agencies, 90; influence on gonad development and spawning, 90
Alevin. *See* Fry
Alewives, 217, 225
Algae: and condensers, 70; increased growth from warmer waters, 146; filamentous, 147; freshwater, 161–198; thermaphylic, 163; and thermal discharges, 172, 182. *See also* Aquatic algae
Algae community: effect of increased light and temperature, 171, 182
Algonquin Park study, 201, 207
Alimentary canal: motility in killifish and skates, 259
Allatoona Reservoir, 19, 67
American Electric Power, 128
American Fisheries Society, 385
Amertop process, 69
Amphipods, 69–70, 112
Anadromous fish: effects of heated discharges, 294–353
Anchovy: temperature effects, 246
Animal parasites: effects of temperature, 263–264

Antioch Power Plant, 107*n*
Aquaria: growth of fish, 149–150
Aquatic algae, 161–198
Aquatic environment: effects of temperature, bibliography, 295, 371
Aquatic life: in freshwater systems, 68; ecology, 85; effects of heated water, 102–107; thermal requirements, 140
Aquatic-life subcommittee, 222, 223
Aquatic vertebrates: oxygen consumption, 6
Aquiculture: of benthic organisms in heated waters, 108–112; "raceway," 109
Arago, Cape, 89
Arctic fish, 239, 242
Atomic Energy Commission: Act of 1954, 376; figures of power stations, 25
Avoidance reactions, 144
Ayu: temperature effects, 246, 249

Bachillariophyta: diatoms, 161, 181
Bacteria: seasonal cycle, estuarine, 97; incubation increased by temperature, 129; in water, 147
Baikal, Lake, 187, 250
Barnacles, 90, 94, 99, 101
Barracuda: temperature tolerance, 242
Bass: growth rates, food, 150; largemouth, metabolic activity in brain tissue, 233; striped, adult, 234; striped, deformed fins, 263; effects of salinity on, 267
Bathythermographs, 257
Bays, named: Coos, 89; Tillamook, 90; Narragansett, 97; Chesapeake, 105; San Francisco, 106, 106*n;* Morro, 112; Newport, 121; Turtle, 121, 125; Biscayne, 154

397